D1233457

Bending Science

BENDING SCIENCE

How Special Interests Corrupt Public Health Research

Thomas O. McGarity

Wendy E. Wagner

HARVARD UNIVERSITY PRESS
Cambridge, Massachusetts, and London, England 2008

Printed in the United States of America

Library of Congress Cataloging-in-Publication Data

McGarity, Thomas O.
Bending science : how special interests corrupt public health research / Thomas O. McGarity, Wendy E. Wagner.
p. ; cm.
Includes bibliographical references and index.
ISBN-13: 978-0-674-02815-9 (alk. paper)
1. Public health—Research—United States. 2. Environmental health—Research—United States. 3. Environmental policy—United States. 4. Medical policy—United States. 5. Fraud in Science—United States. 6. Pressure groups—United States. I. Wagner, Wendy (Wendy Elizabeth) II. Title.
[DNLM: 1. Public Health—United States. 2. Research—United States. 3. Environmental Health—United States. 4. Jurisprudence—United States. 5. Public Policy—United States. 6. Scientific Misconduct—United States. WA 20.5 M478b 2008]
RA440.87.U6M24 2008
362.1—dc22 2007043110

Contents

Acknowledgments

We have many people to thank for their generous contributions to this project. At a very early stage in the book's development, our former colleague, Doug Laycock, provided both insightful comments and encouragement that helped push the project into full gear. Other colleagues at the University of Texas School of Law also provided valuable comments that helped shape the book, particularly in forcing us to think more realistically about reform possibilities. We are especially grateful to David Anderson, Jeff Civins, Mark Gergen, Jack Getman, Dick and Inga Markovits, and John Robertson and to participants at the University of Texas School of Law Friday colloquium for thoughtful suggestions and gentle criticisms. Our excellent students enrolled in a "Controlling Science" seminar (spearheaded by John Robertson) also provided helpful comments on early chapters. Former Dean Bill Powers, Interim Dean Steven Goode, and Dean Larry Sager provided both the sabbatical time and the financial support to bring the book to completion.

Colleagues outside of Texas also provided comments on early versions of the book that went well above and beyond the normal expectations of academic collegiality. Most extraordinary was Tim Lytton's organization of a one-and-a-half-day invitational conference dedicated exclusively to critiquing the first draft of our book at the Albany Law School. The participants at that "Bending Science" conference in December 2006 not only read a first draft of the book, but also provided thoughtful comments, new ideas, and dozens of documents to help sharpen our argument. Several of the participants even provided detailed comments on individual pages of the draft. We are particularly grateful to participants Joe Cecil, Curtis Copeland, Mark Frankel, Bernie Goldstein, Susan Haack, Tim Lytton, Greg Mandel, David McCaffrey, and Peter Strauss for comments that greatly improved the quality of the final product.

We are also grateful to the University of Wisconsin Law School faculty, particularly Stephanie Tai and Neil Komesar, for insightful and important comments on a later stage of the manuscript. Through Wisconsin Law School's Ideas and Innovations Workshop, we received many thoughtful reactions and constructive suggestions that ultimately convinced us to write a final concluding chapter for the book. We also presented small pieces of the book at a number of venues during which participants provided helpful comments that shaped the book, including conferences at the Fordham Law School, the Center for Disease Control Annual Meeting in Atlanta, the Duke University's Integrated Toxicology and Environmental Health Program seminar series, the American Association for the Advancement of Science Forum on Science and Technology Policy, and the Society for Risk Analysis Annual Meeting.

Our tireless research assistants Samantha Cox, Howard Fomby, Lyn Reinhardt, and Cindy Yang provided rafts of research, documentation, and helpful comments and reactions to the book. Dottie Lee, our assistant, helped us through many phases of the drafting and publication process.

Elizabeth Knoll at Harvard University Press provided valuable suggestions at all stages of the drafting process and helped move the book into print. Our copyeditor, Martha Ramsey, at NK Graphics, provided a scrupulous edit. We are also grateful to three anonymous reviewers for their helpful suggestions and comments on the first draft of the book.

Finally, we want to thank our respective families. Tom's father Owen and daughter Kristi provided helpful comments and edits, and his wife Cathy generated a constant stream of citations to useful articles and books. Wendy's family, Will, Rebecca, and Mike Walker, and her parents, Rhys and Marianne Wagner, were wonderfully supportive and selfless, each providing their own, unique contributions to help move the book towards completion.

Bending Science

CHAPTER ONE

Introduction

For quite a while now judges, legal scholars, and prominent scientists have lamented the difficulties that courts and lawmakers encounter in distinguishing reliable science from cleverly manipulated, but ultimately worthless scientific junk. Inundated by experts-for-hire who flood the legal system selling their sponsors' wares, legal decision-makers have struggled to develop more rigorous tools for assessing the reliability of the scientific information that informs health policies. The solution, most have agreed, is for decision-makers to rely more heavily on the scientific community for oversight and assistance. Indicia of scientific community approval, often taking the form of peer reviewed publications and reports from consensus panels, now offer guidance to courts and regulatory agencies charged with screening the reliability of scientific research that relates to social problems.

Yet the simple solution of deferring to the scientists can be frustrating for legal decision-makers and even more precarious for the institutions of science. Accounts of "bending" science—where research is manipulated to advance economic or ideological ends—are now not only prevalent in the corridors of courts and legislatures, but also beginning to emerge from deep within the inner sanctum of science. Editors of a number of top scientific journals complain that they are encountering great difficulties policing significant conflicts of interest in published articles. Well documented allegations of bias and skewed membership have strained the scientific credibility of a number of prominent expert panels assembled by private sector entities and government agencies to advise legal decision-makers and the public. Rather than providing much needed reinforcement to the efforts of the scientific community to fend off these unwelcome intrusions, the legal system has armed outside advocates with increasingly powerful tools for infiltrating the quality control procedures of science and manipulating the ultimate outputs.

In this book we argue that the institutions of science are under attack in ways that apparently neither the legal system nor the scientific community is willing to acknowledge or prepared to repel. Dozens of sophisticated strategies are available to advocates to co-opt the science that informs public health and environmental policy. Yet as quickly as their tricks are exposed, the advocates are able to devise even more insidious ways to penetrate legitimate scientific processes to advance their ideological and economically motivated goals. Indeed, so many points of entry into the scientific process exist and so few resources are available to oversee outsider-induced scientific distortions that it is no wonder that advocates have produced a vast array of techniques for confounding research to their advantage in ways that impair both the science and distort the resulting policy.

After detailing the numerous ways in which outside advocates can corrupt legitimate scientific processes in the area of public health and environmental research, we conclude that policy-makers can no longer expect the scientific community to detect and filter out the distortions without assistance from the legal system. The pressures are too great, the stakes are too high, and the incentives to game the scientific process too intense for scientists to stay abreast of the latest tricks in the advocates' bag.

The Problem

The trouble begins with a widely held perception by a number of respected legal analysts that the mainstream scientific community can successfully inoculate itself against infiltration by determined advocates. Opinions issued by the Supreme Court of the United States and rules issued by the Executive Office of the President governing the use of scientific information in judicial proceedings and regulatory programs convey this widely held perception of a world divided into two separate and distinct territories: the realm of legitimate "science," which produces scientific research through rigorous scientific methods and procedures, and the realm of "policy," which frequently relies on this scientific research to address broader legal and policy issues.[1] Precisely because both research and critiques of that research must survive the scientific community's own internal vetting system, legal decision-makers are apparently prepared to assume that the final product has been purified by professional oversight. In its most important ruling on the use of scientific evidence in judicial proceedings, for example, the Supreme Court observed that "submission to the scrutiny of the scientific community is a component of 'good science,' in part because it increases the likelihood that substantive flaws in methodology will be detected."[2]

In what we characterize as a separatist view of the world (see Figure 1), scientific research is sufficiently reliable for use in legal proceedings and important public policy deliberations after it flows out of the realm of science through a pipeline in which those inhabiting the realm of science (often referred to as the "scientific community") have the opportunity to screen the relevant scientific studies to ensure that they are produced in accordance with the norms and procedures of science. Within the pipeline, the following scenario ordinarily plays out: scientists conduct research in accordance with a predetermined research protocol; they gather the relevant data; they analyze and interpret the data; they write up the data, analysis, and conclusions and submit a manuscript to a scientific journal for publication; the journal, prior to acceptance, subjects the paper to peer review; the scientists respond to the reviews by making any necessary changes; and the journal publishes the paper. If the paper has important scientific implications and attracts sufficient attention, it will undergo additional scientific scrutiny. Other scientists may publish letters in the journals critiquing or expanding on some aspect of the paper; it may become incorporated into a "review article" summarizing the published work in a particular area of research and drawing additional conclusions; it may play a role in a report prepared by an expert body like a "blue ribbon panel" examining a scientific topic of some social relevance; and summaries of the article or the panel report may get broadly disseminated by the media to scientists, decision-makers, and the public.

The separatist view does not seem to contemplate the possibility that the otherwise legitimate process of generating and refining scientific studies within the pipeline could become contaminated by outsiders with an economic or ideological stake in the ongoing legal, regulatory, and policy debates. Instead, it assumes that the scientific community ensures, through rigorous peer review and professional oversight, that the scientific work that exits the pipeline is reliable and generally uncontaminated by biasing influences like the economic or ideological preferences of the sponsor of the research. Separatists are quick to acknowledge that once science enters the realm of policy, it becomes the plaything of advocates. And from that point on, it is Katy-bar-the-door. As attorneys in litigation and interest groups in regulatory proceedings and policy debates seize on research that is relevant to their causes, they manipulate it however they can to achieve preferred judicial, regulatory, and legislative outcomes.[3] But separatists seem confident that these outcome-oriented strategies are readily identified and excluded within the realm of science before they have a chance to corrupt the pipeline's output.

We believe that this separatist view is the dominant weltanschauung of judges, regulatory policy-makers, scientists, and, to the extent that it ponders such issues, the general public. In this view, bias and suppression in science are filtered out of published studies, review letters, and consensus statements used to inform legal decisions by the quality control techniques traditionally employed in the scientific process. When an advocate for a special interest is caught distorting published studies or skewing scientific consensus panels, the incident is generally considered an isolated lapse in oversight and not the result of a larger pattern of distortion.

If it accomplishes nothing else, this book should demonstrate that the separatist view is an idyllic, even Polyannaish view of the interaction between science and policy in the hotly contested areas of public health

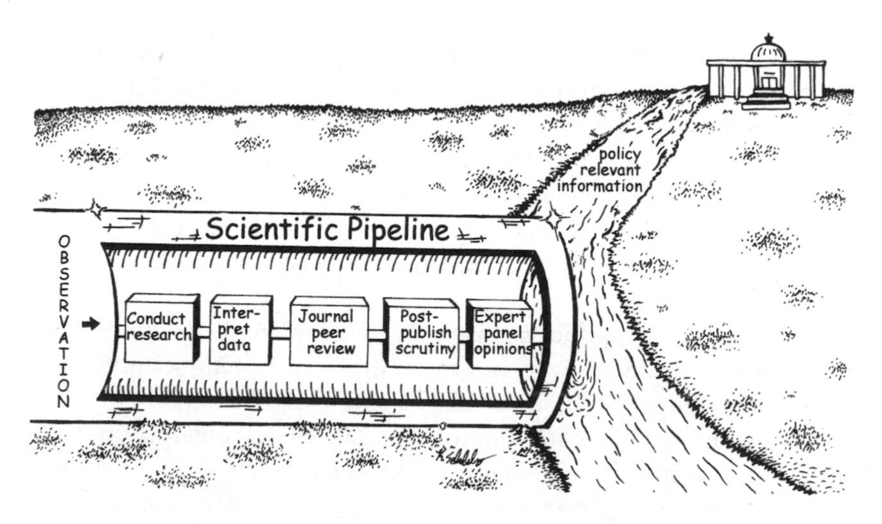

Figure 1: The separatist view of science and policy

and environmental regulation and litigation. In the real world,[4] each step in the science pipeline can become seriously contaminated by the efforts of advocates who are responding to incentives and pressures from within the realm of policy. In this more realistic view, advocates infiltrate the pipeline and secretly situate themselves at various key points to ensure that the research, the critiques, the panel reports, and the overall messages match their ends. In other cases, they undermine the work of scientists who persist in conducting research that runs contrary to their goals. After the science exits the pipeline, the advocates continue their efforts but shift their focus to manipulate public perceptions about the research as it becomes available for use in the policy realm.

In this alternate view, science no longer proceeds in a straight and insulated path through a closely supervised pipeline of scientific oversight; the pipeline is instead much more porous and vulnerable to a range of tricks developed by determined advocates to bend science to their own ends. Figure 2 illustrates how special interests can penetrate the pipeline and manipulate normal processes of science, often without detection by the scientific assembly line of internal checks and review stations. Instead of a process that has managed to insulate itself from the unruly world of policy, then, sacred scientific processes are contaminated by determined advocates who understand the power of supportive science in legal and policy proceedings. Indeed, since published

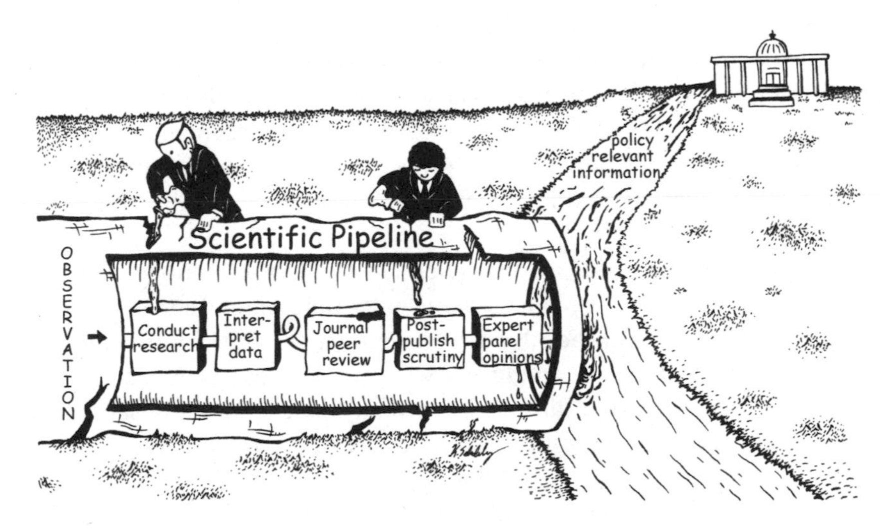

Figure 2: The bending science model of science and policy

results and peer review statements carry the highest indicia of reliability and authority, it is on these products that advocates tend to focus most intensely.

Perhaps because they assume that the pipeline produces results that are largely insulated from contamination by advocates, the institutions of government—the courts, executive branch agencies, and Congress—employ processes that not only neglect to deter efforts by advocates to *bend* science but might actually make it worse. Each of these legal institutions depends heavily on advocates to select the scientific information used in their decisions, and they employ surprisingly few mechanisms to screen out unscientific outcome-oriented research. Worse, the adversarial nature of the legal system actually encourages advocates to invade the realm of science—surreptitiously, because they are not especially welcome there—to commission research and manipulate scientific quality control procedures in outcome-oriented ways.

Ironically, this in turn creates an even greater distance between independent researchers and bent science. Since outcome-oriented research is antithetical to science, independent scientists generally avoid both the science and the scientists they know are involved in producing it. This leads to an adverse selection of sorts, with the institutions in the policy realm equipped with only limited trusted expert assistance in separating legitimate scientific research exiting the pipeline from research that has become contaminated by advocates. In many cases, scientists who leave the scientific realm to make their expertise available to institutions in the policy realm are in fact employed by advocates for the purpose of advocacy, not for the purpose of helping governmental institutions arrive at the scientific truth. Worst of all, some independent scientists whose work informs policy may find themselves on the receiving end of harassing attacks from other scientists whom advocates have hired to deconstruct their work.

The reluctance of independent scientists to involve themselves in the scientific review of and policy debates over outcome-oriented research plays into the hands of the advocates because it allows them to maintain the separatist illusion, knowing full well that it does not reflect reality. Although the advocates did not by any means originate the separatist worldview, they carefully nurture it in the realm of policy because it allows them to take the position that decisions in the policy realm should be based on "sound science" rather than on precautionary policies, while at the same time retaining a powerful influence over that "science." The final result is a train wreck of sorts with regard to a broad range of public health and environmental issues, as inaccurate separatist idealizations of science collide with the reality of what has become known as "science-based advocacy." This book documents the collision and explains the forces that have produced it. The final three chapters offer suggestions for cleaning up the resulting mess and avoiding similar collisions in the future.

How Advocates Bend Science

To understand why advocates' influence on scientific research can be both illegitimate and easily accomplished, we must begin with the scientific ideal to which researchers in the realm of science aspire. Science is knowledge that results from testing a hypothesis with an open mind using methods that scientists have accepted as valid and generally capable of replication.[5] While philosophers and sociologists of science may debate some of the precise qualities that define science, they all agree that research conducted with a predetermined outcome is not science.[6] Indeed, the productivity of the scientific enterprise depends on scientists' commitment to perform and critique research in a disinterested fashion, meaning at the very least that they do not manipulate their methods and analyses to produce predetermined outcomes.[7] Scientific studies must be replicable, and many are in fact replicated. But when scientists can build on the objective efforts of trusted peers without pausing to replicate that work, they can advance the scientific enterprise without wasting precious research resources. Indeed many individual scientists are so wedded to this norm of open-mindedness that they decline to participate in any legal or political activity, at least insofar as their research is concerned, to ensure that their findings are not tainted by even the appearance of outcome-oriented bias.[8] Disclosures of conflicts of interest have also become standard in science, with some of the most prestigious journals denying publication rights and status as peer reviewers to scientists who appear significantly compromised by sponsor bias.[9]

The scientific ideal is, of course, just that, and deviations are not uncommon, even in independently conducted basic research. Scientists face pressures to bend research in outcome-oriented directions quite apart from incentives that emanate from the realm of policy. Scientific careers in academia and industry depend on successful completion of research and, in the case of academia, frequent publication of research output. Research that yields unexpected outcomes or "breakthroughs" can generate great rewards, both reputational and financial, for the scientists who produce the research. Scientists are not above yielding to these pressures by surreptitiously departing from scientific norms.[10] Indeed, reports of overt scientific fraud have stimulated activity in the legal system to deal with such issues in the context of publicly supported research.[11] Even apart from these rare instances of overt fraud, well-intentioned scientists are human beings with their own biases, which may be reflected in their work product.[12] Many scientists have formed very strong opinions about how the universe functions in the particular microsegment of the universe to which they devote their research efforts, and they may consciously or unconsciously avoid or denigrate research outcomes that diverge from those preconceived opinions.[13] A scientist who has demonstrated a commitment to a particular view of the world is, of course, a prime target for advocates, who are always on the lookout for experts whose views line up with their economic or ideological interests and who might be willing to engage in research, sit on government advisory committees, and serve as expert witnesses.

The bias in scientific research that results when the scientist works backward from a preordained result is outside the realm of legitimate science because it lacks a fundamental characteristic of science—the open-minded pursuit of truth. When a scientist engages in such outcome-oriented research at the behest of an advocate who has intervened in an otherwise legitimate scientific process, we may properly characterize the output as bent research—rather than independent research—because the scientist has bent it in the direction of an outcome that advances the advocate's economic or ideological interests. The mere fact that research has received support from an entity that is an advocate does not, standing alone, render it bent, so long as the scientist pursues it independently in accordance with the norms and procedures of science. To be relegated to the category of bent research, it must result from a process in which an advocate has exerted direct or strong indirect control with an eye toward producing a result favorable to the advocate's interests.[14]

Bent science can also occur at later stages of the research pipeline. Scientists regularly interpret and critique the research of other scientists, write review articles and books incorporating many individual studies into a broader body of research, and sit on advisory committees that summarize research in reports designed to be comprehensible to lay decisionmakers and the public. When advocates intervene in this process by commissioning books and review articles, assembling "blue ribbon panels," or nominating scientists whose views comport with their interests to advisory committees, they are bending science in an indirect and perhaps less influential way than when they bend the research itself, but the output is nevertheless bent science.

Advocates and the scientists who work with them employ one or more of a number of strategies to bend science in outcome-oriented ways (Figure 3). At the most general level, they can bend research, and they can bend perceptions about research. When advocates attempt to bend research in the realm of science, their goal is to ensure that the research that exits the pipeline into the policy realm will be useful (or at least not harmful) to their efforts to influence common law courts, regulatory agencies, legislatures, and other governmental bodies that have the power to affect the advocates' economic or ideological interests. More specifically, an advocate can *shape* research by commissioning studies that work backward from the desired outcome through the processes of data interpretation, data analysis, data gathering, and study design. The outcome-determined results are then available for use in legal decision-making and public policy deliberations. If the commissioned research fails to produce the desired results, the advocate who commissioned it can then *hide* it, by either halting it if its unwelcome aspects become apparent early enough, or by keeping it out of the pipeline altogether.

Advocates can also influence the research that emerges from the pipeline by hiring or persuading scientists to *attack* unwelcome research during, or sometimes even after the time it is undergoing scrutiny by the scientists in the realm of science. The attacks might come in letters to journal editors, commissioned reevaluations of the underlying data, and other criticisms targeted at scientific audiences. If the attack on the research fails, advocates may harass the scientists who produced it with the aim of destroying their credibility (hence the perceived reliability of their research) or, longer term, of discouraging them and others from engaging in such research in the future. Both strategies are aimed at controlling the flow of scientific research through the pipeline in an outcome-oriented fashion.

Attempts to bend research are particularly pernicious because they may pass through the scientific community unnoticed. Predictably, the advocates who bend science are eager to obscure their roles to enhance the perceived legitimacy of the research and critiques that they generate. When one cannot discover whether a sponsor controlled a study, a letter to the editor critiquing independent research, or the membership of a "blue ribbon panel of experts," the only way to determine whether the output is badly biased may be to replicate it or, in some cases, engage independent experts to scrutinize it in detail. These are costly interventions and therefore rare.[15] The peer review process most scientific journals employ

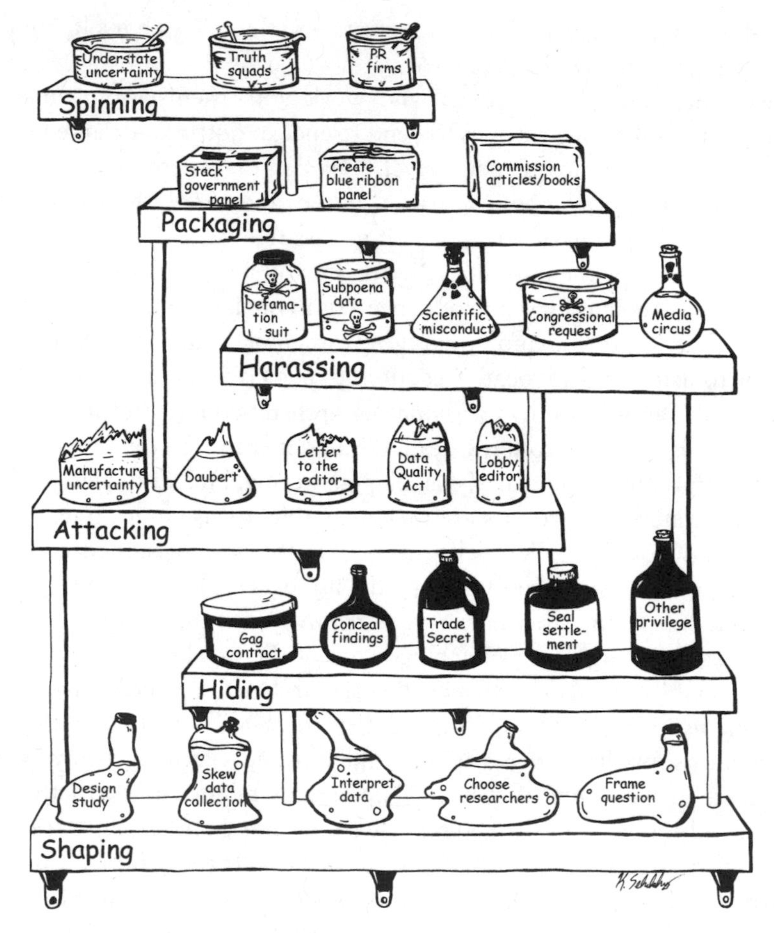

Figure 3: Tools for bending science

has also proven insufficiently rigorous to detect ends-oriented bias in research when the biasing influences are carefully hidden behind assumptions, methods, and interpretations that appear legitimate on the surface.[16] Indeed, the very fact that the journal peer review process has proven incapable of consistently identifying bent science makes journals particularly attractive targets for the advocates, because scientific publications provide an imprimatur of reliability and authority that is often quite useful in the policy realm.[17]

Making matters worse is the fact that the type of science that is most relevant to the policy realm—a subset of science we will call "policy-relevant science"[18]—is not always terribly interesting to a sufficiently large body of independent peers. As Sheila Jasanoff discusses in her seminal book *The Fifth Branch,* policy-relevant research often addresses theoretically thin questions arising directly from the policy realm and yields knowledge that is largely disconnected from both mainstream and "cutting-edge" science.[19] It therefore may attract little attention from the independent scientists whose research agendas are driven by prior research in their own fields.[20]

The foregoing characteristics of policy-relevant research are a particular boon for advocates, who focus most of their attention precisely on that region of the scientific realm that policy-relevant science occupies. A research project that a sponsor has prematurely terminated because the findings were inconsistent with a desired outcome is not likely to be noticed in this area of science, where even completed projects receive little rigorous peer oversight. Since policy-relevant research often requires expertise from multiple scientific disciplines (for example, toxicologists, epidemiologists, geneticists, and risk assessors), advocates can exploit ethical ambiguities inherent in overlapping interdisciplinary projects to justify unconventional designs and methods that have been crafted with particular research outcomes in mind. Similarly, advocates can exploit legitimate disagreements arising out of the undeveloped theoretical state of policy-relevant science to provide cover for unwarranted attacks on unwelcome research.

At the end of the day, however, bending science is only a means toward a larger end. If the goal of the scientist is to pursue truth in the realm of science, the advocates' goal is to influence decision-makers in the policy realm who have the power to affect the advocates' or their clients' interests. The fact that the resulting bent science is not really science at all, because the process that produced it was ends-oriented, is beside the point if legal decision-makers and the public perceive it to be science. And the fact that legitimate science was created and vetted in accordance with the norms and procedures of science is irrelevant if advocate-initiated attacks convince policy-makers that it is invalid.

It is quite convenient for the advocates, then, that legal decision-makers and the public are generally in no position to distinguish bent research from legitimate research. They cannot detect the bending that goes on in the scientific realm because they lack the expertise necessary to discern evidence of bending that might be obvious to the scientist. Even the scientists themselves will not be able to screen out bent science as it moves through the pipeline if the advocates are successful in concealing the incremental choices that ensure that the research supports their predetermined ends. Ultimately, advocates may succeed in bending research because the scientists, the legal decision-makers, and the public devote few resources to monitoring what goes on within the pipeline with an eye toward uncovering the advocates' ends-oriented influences. Perhaps one reason institutions in both realms have devoted so little attention to bent science is that the extent to which advocates have infiltrated the pipeline has itself been a carefully guarded secret.

While advocates pay a great deal of attention to bending research as it flows through the pipeline, they are not done with their bending once it has emerged from the pipeline. At this point, their goal shifts from bending the research to bending the perceptions of that research in the policy realm. They can, for example, *package* both commissioned and independent research by hiring or persuading scientists to write review articles or sit on "blue ribbon panels" and advisory committees that distill a body of research into a more concise package that is easily available for use in the policy realm. Although this strategy does not affect the content of the research, the aim is to affect the understanding of the policymakers and even the scientists who consume the articles and reports.

Advocates can also *spin* the science emerging from the pipeline—whether legitimate or bent—in sophisticated ways that subtly advance their own narrow ideological or economic ends. For example, advocates may present science with a precision that makes it appear to be a determinative factor undergirding a particular policy position, when in fact the science is far too uncertain to support that or any other position. For those who can afford their fees, adept public relations consultants have many powerful tools at their disposal for framing the issues and presenting policy-relevant science to legal decision-makers and the public in ways that advance their clients' interests. As we shall see in Chapter 9, government agencies and even impecunious public interest groups have also proved quite adept at spinning coarse science into golden policies.

The Role of Legal Institutions

Bending science is primarily the work of advocates, but legal institutions also bear responsibility for the phenomenon, because they provide the incentives that motivate the advocates. In large part, this unfortunate development is an unintended consequence of the commendable but sometimes naive commitment of American legal institutions to full participation and rights protection. Legal processes in the courts and agencies are premised on the conviction that interested parties deserve a seat at the table when their rights are at stake. More important, these institutions depend on the parties who are seated at the table to play a prominent, even dominant role in supplying and critiquing the research that informs their deliberations over any scientific issues that arise. The courts, for example, depend almost exclusively on scientific evidence proffered by litigants through expert testimony in making science-based determinations about such critical issues as the existence of product defects or the causal relationship between a defendant's negligence and the plaintiff's damage. The familiar battle of the "experts" (more accurately characterized as "forensic advocates") in court has been extensively explored in the legal literature, the news media, and even popular novels and movies. It should therefore come as no surprise to discover that a significant portion of the scientific information that enters the courtroom is bent science.

Since regulatory agency decision-making procedures are also adversarial in nature, a great deal of bent science winds up there as well. For several decades, American administrative law has relied on an *interest group pluralism* model of decision-making, which is built on the proposition that the best decisions arise out of the interplay of interested parties in all phases of the regulatory process.[21] A regulatory agency differs from a court in several notable ways: an agency comes equipped with its own artillery of experts; it can collect its own information and even sponsor its own research; and it enjoys considerable discretion in developing a final regulatory output. Nevertheless, the role advocates play in agency decision-making is by no means trivial. In practice, agencies often lack sufficient resources to conduct or commission independent research and are therefore forced to rely heavily, and in some programs almost exclusively, on regulated parties to produce and review the scientific research needed to support final agency action.[22] Moreover, since regulated parties are often the only entities with sufficient resources to participate actively in the regulatory process, agencies may accept the scientific research proffered without probing its scientific bona fides, simply to avoid the time- and resource-consuming battles that inevitably result when they question it.[23]

In the realm of science, independent scientists typically discount research when they suspect that a sponsor influenced the outcome, and they may even take steps to remove that research from the pipeline altogether when they discover a significant level of sponsor control. In the policy realm, however, the same compromised research may not only be used, but may be the only evidence available to information-starved decision-makers. Moreover, because deeply embedded outcome-oriented biases in scientific research are often not readily apparent to lay decision-makers and the public, the legal institutions of the policy realm may fail to uncover the extent to which advocates have controlled key research outcomes. Courts and agencies may even unwittingly facilitate advocates' efforts to undermine unwelcome research by providing the legal tools—like overbroad subpoena powers—that advocates then use to attack the research or harass the unfortunate scientists who produce legitimate, but unwelcome research.

The underlying message from the legal institutions to the advocates, then, is to bend research in the scientific realm and to do it surreptitiously. Successful advocates therefore use highly sophisticated trickery to conceal the provenance of commissioned research, to suppress commissioned research that produces unfavorable results, to recruit scientists from the best universities to critique unwelcome independent research, and to use available legal tools to disrupt the ongoing work of independent scientists whose research points in unfavorable directions. In an environment in which both the legal system and independent scientists are unable to muster the resources or resolve to engage in meaningful oversight of bent science, hollow calls for "sound science" in regulation and litigation can only make the situation worse.

A Prefatory Warning

The many examples of bending science related in the following pages do not establish that all policy-relevant science, or even a large proportion of it, is tainted. Honest scientists employing sound scientific practices and adhering to high moral standards produce a great deal of legitimate scientific research on which legal decision-makers and the public can comfortably rely. Furthermore, on numerous occasions both courts and agencies have successfully distinguished legitimate from bent science and have based their final decisions on reliable and even cutting-edge research.

So why focus on the bad actors, the illegitimate science, and the resulting misinformed policy decisions? Because they matter. They matter to the victims who suffer adverse health consequences when the agencies do not protect them. They matter to innocent companies when courts force them to compensate people for harms they did not cause because plaintiffs' experts have bent the relevant science. They matter to the independent scientists who do not bend science and who, in these resource-lean years, can ill afford the loss of public trust in science that results from periodic media exposés. And they should matter to the rest of us who believe in the ideal of good government and a responsible civil justice system. Current legal arrangements are not capable of policing the scientific pipeline for abusive practices, and they offer few incentives to others to take on that task. If we do not take steps soon to address the legal arrangements that encourage advocates to bend science, the abusive practices will only increase over time.

What remains unclear from these accounts is just how prevalent the problem of bending science really is. Few quantitative measures of the pervasiveness of this phenomenon exist, and the measures that do exist are far from definitive.[24] Understandably, special interest advocates are disinclined to share case studies of their efforts to bend science with outsiders and we suspect that some and perhaps many of these efforts still remain hidden from view and are thus absent from our accounts. As a result, the evidence presented here is necessarily qualitative and limited to the reports that have surfaced in the literature over the years. Many may conclude from this evidence that the problem is already quite serious. Those who would conclude otherwise should bear in mind that the evidence presented here is likely to be underinclusive in at least two respects. First, we focus only on those cases that involve either intentional or reckless ends-oriented strategies to manipulate science. We do not consider research and criticism sponsored by advocates to be bent science if the advocates lack a means of controlling the outcome to ensure that it is favorable to their interests. Thus, while we relate statistical analyses conducted by others of the biasing effects of sponsorship on research, we do not attempt to isolate real-world examples of subtle psychological biasing effects on individual researchers induced by outside sponsorship. Recognizing that the job of an advocate is to present scientific evidence to decision-makers in a way that most favorably reflects the client's position, we have likewise not included examples of overly aggressive advocacy that do not involve clear efforts to bend science in outcome-oriented ways. Second, even within this worst-case set, we have focused primarily on the best-documented accounts and have excluded many frequently cited examples that have weak factual support.

It is also worth noting that the problems we highlight do not necessarily have equivalent implications for either law or science. The consequences of bending science vary dramatically. At one end of the spectrum, bending activities can lead to significant harms, approaching what some would characterize as homicide, while at the other end of the spectrum, the consequences involve only minor impediments to regulatory protections or judicial relief. Our goal has not been to rank techniques for bending science according to their relative social importance. Indeed, we doubt that we would have succeeded had we tried, because we suspect that the consequences of the same technique may differ in different regulatory and judicial contexts. Our goal has been to isolate and identify the techniques in a number of illustrative settings, both as a matter of purely academic interest and in the hope that the effort will caution future participants in regulatory and judicial proceedings and public policy-making to be on the lookout for bent science.

Finally, although we explore the impact of bent science on policy-making generally, our primary institutional focus is on regulatory agencies and courts. The advocates who bend science are frequently interested in influencing the general public as well as judicial and administrative decision-makers, and those efforts are relevant to this book for the simple reason that agencies are deeply influenced by public perceptions and courts rely on juries to find facts and assess blame. We do not, however, include Congress and other legislative bodies within the scope of our institutional analysis, primarily because of the absence of formal procedural vehicles for presenting scientific information to those bodies. It is difficult enough to ascertain the influence of bent science on institutions in which advocates employ relatively transparent legal tools to make scientific information available to decision-makers. We leave to others the considerably more challenging task of ascertaining the influence of bent science conveyed by lobbyists to elected officials through covert channels. We expect, however, that at least some of the recommendations in Chapters 10 and 11, such as greater disclosure and greater use of expert advisory bodies, are as applicable to legislative bodies as to courts and agencies.

Lessons

Our research has led us to a number of conclusions, several of which are worth sharing at the outset. Most important, the evidence we have gathered reveals that the problem of bent science is not simply limited to a few bad actors or to unique institutional settings. Rather, the evidence suggests that advocates for nearly every category of participant in health and environmental regulation and litigation have attempted to bend science or use bent science to their advantage at some point, and the legal system encourages and often rewards this undesirable behavior.

Prior research on the interface between science and law has identified and elaborated on many specific aspects of the problem that we address in this book, and our research necessarily builds on this important work of others. It does, however, differ from previous work in several respects. First, we have endeavored to take a broad view of all aspects of the problem, which encompasses both litigation and regulation and which includes the efforts of advocates to produce unpublished forensic studies to support expert testimony and, beyond that, to influence core scientific processes, such as journal peer review and blue ribbon panels. We also consider a wide range of actors, including an uncomfortable mix of industry, plaintiffs' attorneys, nonprofit public interest groups, and government officials. Thus, while Peter Huber's book *Galileo's Revenge* is perhaps the most influential work to underscore the frailties of the legal system in tolerating what he characterizes as junk science, Huber focuses on only one portion of the legal system and primarily on plaintiffs' attorneys and their experts. As the research in this book makes clear, however, one cannot justifiably assign blame for bent science exclusively to plaintiffs' attorneys, Republicans, "rational" corporate executives, or member-hungry public interest groups.[25] These groups all participate in bending science to some degree, and their modus operandi and toolkits are supplied, perhaps unwittingly, by our legal system. In many cases, the individuals who must operate within its twisted incentive structure are simply acting rationally. In some situations, the advocates may even discover that they have an ethical obligation to bend science to advance the legitimate interests of their clients. Hence, bent science will continue to plague the legal system until reforms are implemented to modify the incentive structure.

Second, this broader perspective on the complex interactions between bent science and independent science in a variety of legal contexts reveals that scientists also have a role to play in addressing the problem of bent science. Policy-relevant science is already unstable, due to limited oversight and engagement by independent scientists. Bent science finds ample room to hide within the larger body of policy-relevant research, and it can successfully mimic legitimate policy-relevant research by obscuring the influence of sponsor control. Making matters worse, scientists may actively avoid policy-relevant science altogether because it is messy and often uninteresting from a theoretical standpoint. In some cases, then, the absence of serious peer scrutiny may not so much reflect the efforts of clever advocates as it manifests a professional vacuum surrounding much policy-relevant science.

Finally, a broad perspective on the many techniques that advocates have employed to bend science strongly suggests that all "science" is not equal and that more care should be taken to distinguish scientific information that results from independent research from unscientific information that results from outcome-oriented attempts to bend science.

Bent science is in reality the antithesis of science, even though on the surface it may be indistinguishable from independent research. Indeed, the very fact that many scientists avoid areas of policy-relevant research that may attract the attention of advocates provides a compelling reason for distinguishing it from the rest of science and subjecting it to more careful scrutiny.

The Broader Message

Our findings also provide broader lessons for legal institutions that extend beyond the limited arenas of environmental and health regulation and common law litigation. Most obviously, our account reinforces the significance of legal rules and the legal culture from which they derive in explaining both the existence and the persistence of bent science in the American legal system.[26] In particular, any adversarial decision-making setting in which highly complex technical information is required to resolve high-stakes issues can create a climate that encourages advocates to bend science. In such a setting, advocates have strong incentives to control the course of understanding and discussion by manipulating the underlying information or "facts." And the more they manipulate, the more respected independent experts will avoid becoming involved. While the legal system is generally prepared for run-of-the-mill deception and has, in fact, instituted layers of procedures and evidentiary rules to prevent advocates from influencing outcomes, these screens function poorly when they encounter esoteric information that, in the end, only independent experts can properly evaluate.

The account of bending science provided here also contains a larger message about the influential role the legal system plays in spurring both legitimate and bent research. Although most studies of science funding examine the competence and motives of Congress and executive branch funding agencies, like the National Institutes of Health (NIH) and the National Science Foundation (NSF), our research suggests that common law courts and regulators may be at least as important institutionally in channeling research monies, albeit in an ad hoc way, toward specific types of projects. In fact, the role of courts and regulators in encouraging entities in the private sector to generate policy-relevant research likely dwarfs the more meager but deliberate allocation of public monies to finance that branch of research. Yet, in contrast to Congress and the funding agencies, the research priorities these institutions induce are incoherent and often counterproductive. Solely on the basis of examples we have identified, it is possible to conclude that the current legal institutions are inducing some private sector actors to funnel considerable research money toward questionable scientific practices that appear designed to obfuscate rather than advance scientific understanding. At the same time, these efforts squander the precious time and resources of peer reviewers, journal editors, independent scientists, and other equally important players in the regulatory and judicial arenas.

This discouraging assessment, however, also offers a ray of hope. If institutional rules and processes currently induce advocates to bend science, then they should be responsive to adjustments designed to counteract that trend. Over the last decade, legal academics have devoted a great deal of attention to the problem of asymmetric information in the legal system, and they have thought hard about how to restructure the rules to adjust to that problem.[27] Similar restructuring may be required in the area of health and environmental litigation and regulation. Our findings suggest that when legal rules and procedures approach technical information in an adversarial fashion, the consequences can be quite destructive given the wide disparities among affected parties in available resources and ability to engage in the legal process. The parties presenting information to the relevant legal institutions will sometimes succeed in bending the relevant science without detection, and the independent scientists, who are the most reliable adjudicators of disputes about the reliability of such information, will flee the scene. The best response to this unfortunate state of affairs may thus demand more direct and focused legal intervention, such as targeted sanctions against bending practices and legal reinforcement for independent scientists to identify and condemn bent science.

CHAPTER TWO

Why Bend Science?

The Players, the Setting, and the Consequences

A classic nineteenth-century play depicts the plight of a scientist who discovers bacterial contamination in the city's lucrative hot springs.[1] The scientist expects to be honored with ceremonies and voted a raise in salary for preventing an epidemic of life-threatening illness;[2] instead, he is effectively run out of town by the mayor (his brother), a newspaper reporter, and his neighbors for threatening the city's economy. The closing scene depicts the scientist and his family vowing to remain strong and "stand alone," as they collect rocks thrown through the windows of their home by the townspeople.[3]

A twenty-first-century remake of Henrik Ibsen's *Enemy of the People* would read like any one of the following. One version opens with scientists doing path-breaking research showing that human activities are changing the global climate in ways that may have disastrous consequences. They arrive at their laboratory one morning to find a six-page letter from several congresspersons demanding to see every laboratory record generated during the scientists' long careers of research as part of the congresspersons' investigation into purported "methodological flaws and data errors" in the research.[4] In another version, a toxicologist is making path-breaking discoveries in endocrine disruption by environmental contaminants, only to find his research ambushed by scientists paid by the company that manufactures the implicated products. To ensure that no one uses the research to protect the environment at the expense of product sales, the company files formal complaints with the government demanding that the results of the research be expunged from all of its databases and excluded from all governmental decision-making, because the research depends on "invalidated tests that have no proven value."[5] In a third version, a scientist discovers that cigarette smoke causes cancer in nonsmokers and ends up working patiently to follow up on his research in the face of a multimillion-dollar "topline, unified and synergistic" public relations campaign developed by a multinational public relations firm to attack his research and any other research that yields such unwelcome discoveries.[6]

The timeless human reality captured in Ibsen's play is that when society asks scientists to provide answers to pressing policy questions, those who will be affected by the answers to those questions will exert whatever influence they can to control how that research is created, interpreted, and used. We need the guidance of science, but we also seek to control it, to ensure that what it reveals is not too frightening or unfavorable to the status quo. This book addresses that reality in the important context of public health and environmental policy in two institutional settings—government regulation and private tort litigation.

In today's legal climate, science has become the most respected and therefore the most powerful influence on domestic health and environmental policy-making. This exalted status does not come without significant risks to the broader scientific enterprise. The parties who will be affected by public policy decisions are anxious to bend the relevant science to advance their own economic or ideological ends, and they are not constrained by scientific norms or professional principles that bind the scientific community. Even when policy-relevant research is conducted by independent scientists and emerges from the scientific pipeline uncontaminated, its reception in the chaotic realm of public policy can be anything but hospitable—not an orderly assembly of congregants expectantly gathered to use the insights offered, but a frenzied mob of interest groups seeking to control what the research reveals and manipulate how it is perceived and used in the policy-making process.

Some of the more familiar examples of orchestrated efforts to bend science come from the tobacco industry's recently revealed program of the 1960s through the 1980s to control and cover up research on the health effects of smoking and nicotine addiction, and the pesticide industry's highly publicized attempts in the 1960s to intimidate Rachel Carson and belittle the science underlying her influential book.[7] Sadly, efforts to bend science are far more widespread and insidious in environmental and health policy today. In 2001, for example, 50 percent of environmental epidemiologists responding to a survey administered by the International Society for Environmental Epidemiologists reported that they had been harassed by someone with an economic or ideological interest in their research.[8] Numerous empirical studies of industry-sponsored research reveal strong associations between the results of the research and sponsor control.[9] Incidents of pharmaceutical manufacturers' suppression of information about the health hazards of prescription drugs regularly make the headlines.[10] And the demand for consulting groups that distort science to promote the interests of their clients has been on the rise, creating profitable business niches for contract research organizations, ghostwriting services, product defense firms, and public relations consultants.

From a theoretical perspective, the contemporary popularity of bending science is not surprising. Basic rational choice theory and common sense converge on a simple truth: if the scientific research can be manipulated in ways that significantly benefit the interests of a party in a legal or policy dispute, that party will invest in manipulating the research up to the point at which the last dollar expended just equals the expected benefits of the manipulation.[11] The inherent value of scientific truth may not figure into the accounting. The primary drivers of attempts to bend science—legal and market pressures—not only feed these practices but also help explain their development over time.

Bending Science in Historical Perspective

The first meaningful evidence of bending science in the legal system comes from the early 1900s when several large industries, most notably asbestos, tobacco, pharmaceutical, and pesticide manufacturers, tried to control the scientific bad news about their products in response to potentially devastating liability and market pressures.[12] Suppression of industry-sponsored research and occasional harassment of independent scientists were the primary methods employed during this low-tech era of sponsored distortions of science. In most other industrial settings, manufacturers were able to follow a much less expensive path to avoid liability—they simply avoided scientific research altogether, at least with regard to the latent hazards their products posed.[13] Since victims bear the burden of proof in tort cases and since individual victims typically lack sufficient resources and information to document health threats from consumer products or industrial activities, the largely unregulated industrial sector quickly learned that ignorance is bliss.[14]

The federal government played a very modest oversight role during these early years. In 1906, Congress enacted a pure food and drug statute aimed primarily at ensuring that food and drugs were not "adulterated" with poisons and disease-spreading microorganisms and that their labels were not "misleading."[15] In 1938, in response to a report of dozens of deaths caused by the drug sulfanilamide, Congress amended the law to require premarket safety testing, but not premarket approval, by the Food and Drug Administration (FDA). Drugs could still go on the market unless the FDA assumed the burden of proving that they were unsafe.[16] In 1947, Congress enacted the Federal Insecticide, Fungicide and Rodenticide Act, which required companies to obtain a registration for pesticides, and in 1954 it amended the statute to allow the FDA to require premarket testing of pesticides that left residues on food.[17] Since the agencies administering the new laws were more concerned with getting new pesticides into the hands of farmers than protecting consumers and the environment from the adverse effects of pesticides, they produced only limited incentives for bending science.

This all changed in the mid-1960s, with the Johnson administration's strict implementation of the Kefauver-Harris Amendments to the Food, Drug and Cosmetics Act and the tidal wave of health and environmental regulation that hit the United States in the 1970s when Congress weighed in with a series of laws creating several new regulatory agencies to promulgate regulations providing anticipatory protection for potential victims of harmful products and wastes. The Kefauver-Harris Amendments, enacted in 1962, finally required premarket FDA approval of drugs, and the manufacturers had the burden of demonstrating with "adequate and well-controlled investigations" that they were safe and effective.[18] In addition to establishing some upfront regulatory controls, this new legislation placed more responsibility for conducting testing on the pharmaceutical and pesticide industries. A new Toxic Substances Control Act empowered (but did not require) the newly created Environmental Protection Agency (EPA) to order toxicological testing for all new and existing chemicals that met certain toxicity and exposure thresholds.[19] Many of these laws also required manufactures and polluters to self-report particularly bad adverse effects, spills, or other activities that threatened public health and the environment.[20]

Public support for scientific research was strong during this period, and the major federal research agencies shifted funding priorities to focus greater attention on the adverse health and environmental impacts of a growing population, an aging industrial infrastructure, and a huge outpouring of new products and technologies designed to meet the demands of a consumer-oriented society. University scientists desiring to conduct research in the new and rediscovered fields of environmental health sciences and ecology for a time found federal funding much easier to come by. This independent government-funded research began to shine a fresh light on the hazards posed by some products and industrial activities, much to the chagrin of the companies that were responsible for them. At the same time, advocates of stronger government controls and more powerful tort liability regimes sometimes pressed the existing science too far in their enthusiasm to hold companies accountable for harms they did not cause.[21]

The regulated industries reacted to the new onslaught of federal regulation and the simultaneous outpouring of federally sponsored research with a second generation of much more sophisticated science-bending strategies. To limit the damage that could result from the new government-imposed testing requirements, affected manufacturers developed intricate strategies for retaining significant control over the design, conduct, interpretation, and publication of the studies they financed while maintaining the outward appearance of scientific objectivity. Entrepreneurs rushed to fill the demand for sponsor influence over research with massive animal testing laboratories and sophisticated clinical research arrangements. Contract research organizations (CROs) employed physicians and business executives, rather than more independent-minded academic scientists, to conduct pharmaceutical testing under contracts that gave sponsors more control over the methods, data collection, analysis, and publication of the study results than the more common, university-based arrangements of the past had allowed.[22] Boutique firms baldly advertised "ghostwriting" services: anonymous science writers who draft scientific articles to be signed by prominent scientists who are paid handsomely for lending their reputations and a modest amount of their time to the effort.

As an increased flow of independent research generated in universities from around the world unveiled new and unexpected environmental and health hazards in the 1980s and 1990s, public pressure mounted on regulatory agencies to take action, and lawyers for alleged victims began to file hundred-million-dollar lawsuits against the companies allegedly responsible for those hazards. A mini-industry of "product-defense" firms sprung up with—to quote one such firm's pitch letter—"capabilities in assessing the scientific facts, developing appropriate responses or sound scientific messages, building a team of world class experts to deliver those messages, and implementing a strategy to limit the effect of litigation and regulation."[23] As a more enlightened public began to clamor for greater regulatory controls to meet newly emerging threats, advocates and even government agencies, with the help of multimillion-dollar public relations firms, developed sophisticated techniques for spinning favorable research in misleading ways to advance economic interests or ideological policy agendas. Consistent with broader trends toward mergers and acquisitions, large public relations firms began to buy up ghostwriting companies and CROs so that they could offer a full menu of science-bending services to their clients.

The multiplicity of differing legal requirements flowing out of the 1970s wave of regulatory programs also led advocates to diversify their portfolios of strategies for bending science. While the premarket testing requirements created incentives in the pharmaceutical and pesticide industries to commission controlled research and suppress studies containing bad news, companies in the chemical, tobacco, food, and other industries that faced few premarket testing requirements developed techniques for attacking independent research with bad implications for their business and for harassing the independent researchers who engaged in that research. Rarely needed prior to the 1970s, these strategies came into their own at the fringes of the contentious regulatory battles of the 1980s and 1990s.

The litigation arena did not remain static, either. The greater availability of experts who were prepared to testify that exposures to hazardous substances caused individual plaintiffs to contract particular diseases and the willingness of some courts to manage large class action lawsuits combined to produce a tsunami of mass tort litigation that threatened to overwhelm the judicial system and bankrupt whole industries. The trickle of asbestos and lead lawsuits filed by workers in the 1920s and 1930s became a flood by the 1980s, and they were joined in the 1990s by similar suits against companies in the tobacco, pesticide, chemical, pharmaceutical, and food and other products industries.

Some of the targets of mass tort claims were heavily regulated, and some were not, but all of them, along with the plaintiffs' attorneys who brought the litigation, had a strong incentive to adapt emerging techniques for bending science to the modern litigation. Mobile diagnostic screening operations could examine hundreds of potential plaintiffs in a short period of time, under arrangements in which plaintiffs' attorneys paid the entrepreneurs who ran them only for "positive" diagnoses.[24] Advocates began to employ tools that are only available to litigants, like third-party subpoenas, to harass scientists with burdensome demands for underlying data, laboratory notes, and even early drafts of articles. Attempting to erect some limits on the use of so-called junk science in litigation, the Supreme Court in 1993, in *Daubert v. Merrell Dow Pharmaceuticals, Inc.*, instructed lower court judges to play the role of "gatekeeper," ensuring that all scientific information presented in court through expert testimony passed threshold legal tests for relevance and reliability.[25] The case had the salutary effect of greatly reducing the number of specious claims, but it also generated a powerful incentive on the part of potential defendants to "manufacture uncertainty" about the implications of well-conducted scientific studies by launching vigorous "corpuscular" attacks on every minute aspect of every relevant study cited by the experts, rather than on the scientific reliability of the experts' overall conclusions.[26] These attacks undermined the perceived reliability of an important subset of scientific information.

A third wave of science bending emerged at the turn of the twenty-first century out of frustration on the part of regulated entities with their general failure to constrain what they believed to be "regulatory excess," in contrast to the relative success they had achieved in courts, where the judges were inclined to play their "gatekeeper" role aggressively to exclude scientific evidence from jury consideration. Drawing on their successful experience in civil litigation in using subpoenas to obtain the data underlying studies and in using the *Daubert* precedent to exclude expert testimony, advocates sought legislation that would give them similar advantages in their dealings with the regulatory agencies. Within the span of four years—from 2000 to 2004—Congress had enacted two important statutes, and the White House Office of Management and Budget (OMB) had implemented two new programs aimed at agency use of policy-relevant science. Each of these developments expanded the menu of tools available for bending science, with no apparent correlative benefit for improving health or environmental protections.[27]

Who Does the Bending?

A wide range of advocates participate in this vigorous new world of bending science, but the primary players are heavily regulated industries and the scientists and consultants they employ, trial attorneys and their scientific experts, public interest groups, and government officials located both inside and outside the regulatory agencies. The supporting cast includes a variety of public relations firms, industry-supported think tanks, and boutique companies like the CROs, ghostwriters, and product-defense firms briefly described earlier.

Although an advocate will attempt to bend science under a wide range of circumstances, two factors help to predict when bending is most likely to occur. First, the advocate must believe that the costs to it of any adverse scientific findings will be higher than the costs spent undermining the research. When the stakes are very high—for example, when the scientific information might increase regulatory burdens, lead to costly litigation, or threaten a product's marketability—an advocate may conclude that some sort of manipulation or intervention is worth the cost.[28] Second, in addition to the motivation, the advocate must have sufficient resources to mount the desired attacks. As we shall see in Chapters 4 and 8, some bending techniques, like commissioning biased research or assembling a skewed group of experts, are very expensive and therefore available only to advocates with access to a lot of money.

The fact that advocates may be well positioned, both in terms of resources and motivation, to bend science does not guarantee that they will in fact do so. Ethical principles, the lack of scientific know-how, or the perception of serious reputational damage arising from public exposure of overly aggressive tactics may caution against self-interested distortions of policy-relevant science. Moreover, perfectly legitimate challenges may be sufficient to refute potentially damaging scientific findings that are not in fact scientifically reliable. When an advocate determines it is necessary and cost-effective to work backward from a predetermined end to design studies, critiques, panels, or otherwise bend science, the stage is set for the science bending drama.

Industry

Regulated companies generally have the most to lose from scientific discoveries that suggest that their products, wastes, or activities are riskier or less effective than originally supposed. Many corporations also have considerable resources to invest in ensuring that the available science is as favorable as possible, especially when this investment is compared to the lost profits that would result from the broad dissemination of uncomfortable scientific truths. The tobacco industry is the poster child for bending science, and its often path-breaking strategies will be featured throughout this book. Unfortunately, that industry seems to have served as a role model for other industries that manufacture products and engage in other activities that pose risks to public health and the environment.

"IGNORANCE IS BLISS"

During the late 1960s, the R. J. Reynolds Corporation (RJR) initiated a multimillion-dollar four-year study in which laboratory animals were forced to inhale high doses of tobacco smoke. Soon after the American Cancer Society held a press conference to announce the results of a study it had funded—Dr. Oscar Auerbach's now-famous "smoking beagle" study, which revealed that direct inhalation of tobacco smoke caused serious adverse health effects in the animals—RJR reacted to the news by abruptly canceling its own study, seizing the researchers' notebooks, and firing the twenty-six employees who were operating the laboratory. One of RJR's earlier studies had already indicated that smoking could cause emphysema in rabbits, and Auerbach's study only increased the risk of more forthcoming bad news. Many years later, one of the scientists working on the project observed that "it wasn't about bad science or a business decision. The decision . . . was made because Reynolds did not at that time want to be collecting information that might be detrimental to itself. . . . Ignorance is bliss."[29] ❒

As RJR understood, the bad news that emerges from research can be exceedingly costly for a corporation. These costs can result from three separate sources of institutional pressure that often occur simultaneously. First, the marketplace may react to news that a product is either more dangerous or less effective than expected. Preliminary research indicating that a painkiller might increase the risk of heart failure,[30] or that an herbicide transforms male frogs into ovary-bearing hermaphrodites,[31] is not likely to increase the attractiveness of those products to consumers. Indeed, many observers believe that consumers are so vulnerable to risk-related scares that preliminary adverse reports can cause large reductions in market shares, even when the public's reaction is neither well-informed nor rational.[32] Manufacturers who agree with this characterization of consumers will have a strong incentive to respond quickly and effectively to scientific discoveries that denigrate their product in the marketplace.

Second, companies may understandably worry that research indicating that a product or activity is more dangerous than expected will lead opportunistic lawyers to generate expensive lawsuits. Like the marketplace, the common law tort system can be finicky. It may take only a single preliminary, but apparently incriminating study to precipitate mass litigation against a manufacturer. Once the litigation is initiated, however, the effects can cascade, leading to lawsuits that sometimes include far more plaintiffs than the number of possible victims and seek damages at levels that are so substantial that they threaten the company's economic future. In a number of familiar cases, including asbestos, tobacco, and drugs and medical products like the Dalkon Shield, diethylstilbestrol (DES), and thalidomide, adverse scientific research resulted in liability that drove the product from the market and threatened the future economic viability of its manufacturer.[33] Companies cannot realistically expect to be repaid for the expense of defending themselves in court or the marketplace if it later turns out that they have been unjustly accused. Indeed, in some litigation, companies have paid not only compensation but additional "punitive" damages, even though scientists later concluded that preliminary research was wrong and their products were not so harmful after all.[34]

Third, scientific research may spawn heightened oversight and attention from regulators. The regulatory system is less reactive than the tort system and the marketplace, and it employs a cadre of expert staff and scientific panels to vet preliminary research for flaws. Regulatory agencies are therefore less inclined than the other institutions to take precipitous action based on unreliable scientific information. The regulatory system also moves more slowly. It took the EPA nearly ten years in the 1980s to implement a ban on asbestos products, despite the fact that tort liability and consumer disinterest had already eliminated most of the market for asbestos products, and even then, the agency's partial ban was ultimately remanded by a court of appeals for further factual findings.[35] Nevertheless, concerns that regulators may take action or, perhaps more likely, publicize unflattering scientific findings can motivate companies to bend science. The OMB estimates that the cumulative costs of current regulatory requirements exceed $4 billion per year,[36] and at least some of these regulatory burdens are based on scientific research that demonstrates that particular products or wastes can cause harm. Moreover, U.S. regulatory agencies have on rare occasions banned offending products in whole or in part after scientific research revealed significant hazards. Research on DDT, PCBs, and asbestos was critical in supporting the EPA's attempts to ban those products.[37]

Plaintiffs' Attorneys, Defense Counsel, and Their Experts

Since the companies that produce potentially harmful products and engage in potentially risky activities generally bear the economic burden of health and environmental regulation and are subject to large compensation awards in litigation, they have a stronger incentive to bend science than many of the other players in the legal arena. They also typically have the greatest economic resources to devote to the enterprise and face fewer practical barriers to combining their resources through trade associations, research consortia, and think tanks. Hence, the regulated industries tend to be the dominant players in the science bending drama.[38] They are, however, occasionally upstaged by other actors who are reacting to equally powerful incentives and have considerable resources of their own to devote to bending science on a more ad hoc basis. In particular, the civil justice system provides powerful incentives for plaintiffs' attorneys to bend the relevant science to their advantage.[39]

SUBJECTIVE SILICOSIS SCIENCE

When Judge Janis Graham Jack learned that doctors working for plaintiffs' attorneys had identified over 20,479 silicosis cases in Mississippi over a three-year period when only 4,000 would have been expected on the basis of national averages, she suspected something was amiss.[40] As the plaintiffs' experts testified in her consolidated federal court proceedings in Corpus Christi, Texas, the scientifically shaky foundation of the plaintiffs' lawsuit crumbled. Rather than relying on a range of reputable, neutral doctors for screening and diagnosing silicosis in thousands of potentially exposed individuals, attorneys for the plaintiffs had financed an assembly-line operation conducted mostly by medical entrepreneurs, often operating under the direction of the attorneys themselves. Silicosis, which generally results from exposure to silica dust, can cause lung fibrosis and emphysema. Since genuine silicosis victims are entitled to what can amount to substantial damages, particularly when many claims are aggregated into a single lawsuit, and since attorneys in such mass tort litigation may expect a hefty slice of those awards (usually between 20 and 30 percent), the attorneys in Judge Jack's case had devised a silicosis diagnosis regime that was biased toward identifying all possible victims, even at the risk of including a few (or, as it turned out, a great many) who did not in fact have the disease at all.[41]

The attorneys had hired a screening firm that received remuneration only for positive diagnoses and collected occupational histories through a questionnaire administered by a nonphysician. The firm then commissioned forensic doctors to review x-ray scans and produce assembly-line readings. The doctors knew that the attorneys who were paying the bills were eager to receive positive diagnoses, and that is what they got. One of the leading physicians in the case produced positive diagnoses at an alarmingly higher rate than that of the other physicians. Most of the experts, many of whom lacked formal training in diagnosing silicosis, later confessed that they had not examined a single patient (or, in their revealing terminology, "client"), and had based their diagnoses instead exclusively on their readings of the x-rays. In some cases, the physician diagnoses were even cut and pasted from a menu that was provided to their assistants. While x-ray reading is inherently subjective and therefore requires scientific judgment, the dramatic bias demonstrated by the plaintiffs' experts, coupled with their sloppy procedures for screening and diagnosis, virtually guaranteed that the output would consist of bent science. In one of the most straightforward condemnations of science bending rendered by a decision-maker in the legal system, the presiding judge concluded that "these diagnoses were driven by neither health nor justice: they were manufactured for money."[42]

The broader ramifications of this sort of assembly-line medical practice are perhaps even more disconcerting. If Judge Jack's candid assessment casts all silicosis diagnoses into doubt, those who actually do suffer from the disease and the reputable doctors who perform their diagnoses may be tarred with the same brush.[43] This is particularly unfortunate because it appears that in the general population, silicosis is generally underdiagnosed and undertreated.[44] Equally worrisome is the possibility that the victim-mill approach to screening individuals as possible plaintiffs has become a staple practice in modern mass tort litigation. Some observers suspect that a fair proportion of asbestosis plaintiffs have been questionably identified by similar assembly-line diagnoses, and there is evidence that these tactics have also been employed in the fen-phen diet drug litigation.[45] As long as it pays plaintiffs' attorneys and their scientific consultants to be unscientifically optimistic about the incidence of mass diseases—usually because the opposing party finds it more cost-effective to settle thousands of cases at once rather than to challenge thousands of medical diagnoses individually—this practice is likely to continue. ❐

Although some attorneys, like the silicosis lawyers just described, have hired entrepreneurial diagnosticians to produce a greater number of medical claims, a great deal of manipulation is possible without resorting to such extreme tactics. The adversarial process thrives when attorneys strive to make the best case for their clients. This, in turn, demands that the attorneys identify experts who will make the strongest scientific and factual presentations in support of their clients' positions, while still retaining sufficient credibility to survive predictable judicial challenges to their scientific reliability. The fact that the outcome might precede the scientific analysis is, quite simply, not their problem. In our adversarial legal system, it is the system's problem.

Dubbed "litigation science" by Judge Alex Kozinski, scientific research and analysis commissioned solely for the purpose of advancing a party's interest in private litigation has become both an annoyance to the judicial system and a threat to the legitimacy and respectability of its judgments.[46] Although no quantitative measures are available, a fair portion of litigation science may well be bent science, because lawyers usually hire experts they expect will advance their clients' positions on the scientific issues. As will become apparent in Chapter 4, some of these experts are willing to work backward from a predetermined result to provide testimony that advances the interest of the party who hired them. Perhaps more frequently, attorneys on both sides of the courtroom commission ends-oriented critiques of research to undermine their opponents' litigation positions. In some cases, the hired experts not only exaggerate the flaws in mainstream research in their trial testimony but also publish those critiques in scientific journals to enhance their legitimacy.[47]

Attorneys and their clients also engage in other strategies for bending science. They can harass scientists whose research is producing evidence that weakens their cases, a problem we will take up in Chapter 7. Plaintiffs' attorneys may even be complicit in suppressing relevant research when it suits their purposes. For example, Judge Jack upbraided the plaintiffs' attorneys and their experts for suppressing scientific information, noting that "not a single doctor even bothered to lift a telephone and notify any governmental agency, union, employer, hospital or even media outlet, all of whom conceivably could have taken steps to ensure recognition of currently-undiagnosed silicosis cases and to prevent future cases from developing."[48] Finally, victims, with their attorneys' assistance, occasionally settle litigation with agreements to seal the records and evidence, thereby reinforcing precisely the kind of scientific suppression that they condemn in their complaints.[49]

The Public and Its Advocates

Members of the general public participate far less frequently than regulated industries and litigation-oriented attorneys in bending science, because individuals usually lack sufficient resources to mount sophisticated attacks and are generally not as highly motivated to bend scientific research toward predetermined ends. Consequently, their distortions of science are both fewer and much easier to identify: they consist primarily of either exaggerating (often unintentionally) the immediate implications of preliminary research for public health or rejecting credible research on unscientific (often ideological or religious) grounds.[50]

REJECTING SACCHARIN SCIENCE

In the 1970s, a series of laboratory tests revealed that high doses of saccharin caused a surprisingly large number of tumors in laboratory mice. Standard public health protocol suggested that given this reaction, humans were also at risk for cancer, and the relevant regulatory statute required the FDA to ban use of the substance as a food additive.[51] A storm of protest followed the FDA's announcement that it would ban saccharin for most uses, and it came not just from saccharine manufacturers but also from the consumers of saccharin and foods containing saccharin. "All day long we've been taking calls from people, some of them in tears, demanding that we leave saccharin alone," an FDA official complained the day the agency announced its proposed ban.[52] Given the public's affinity for the artificial sweetener—because it seemed to help people limit weight gain and provided diabetics with a taste of sweets—this scientific research was precisely the kind of news the public did not want to hear. And so the public and its elected representatives dismissed, distorted, and chided the research for being unreliable.[53]

The revolt was not so much against the FDA's legally required proposal to ban the product as against the science itself. People refused to believe that saccharin was hazardous, and they focused their frustration on the science—the bearer of the bad news—rather than coming to terms with the policy implications of the discovery. The president of the Atlanta-based Calorie Control Council, for example, condemned the proposal as "an example of colossal government overregulation in disregard of science and the needs and wants of consumers."[54] Twenty years and millions of dollars of research later, scientists ultimately agreed that the mechanisms for saccharin's carcinogenicity in rodents were in fact specific to rodents.[55] Humans ingesting modest levels of saccharin were therefore not likely to contract cancer. This scientific finding, unlike the earlier one, was widely accepted by the public and quickly incorporated into the policy-making process. ❒

When scientific research drives a wedge between the public and its artificial sweeteners, a mass revolt is likely to erupt. Even more dramatic clashes result when science has implications for deeply held values; the conflict between the science of evolution and the beliefs of some religious groups is just one example. When the source of the risk is an environmental contaminant, rather than a favorite dietary supplement, the public's general aversion to involuntary exposure to cancer risk can create a groundswell of demands for more stringent regulation following even preliminary discoveries suggesting such a risk.[56] In the case of conflict with strongly held beliefs, the public may demand that elected officials take steps to suppress offending science. In both situations, the pressures can be quite strong for a period of time, and they can produce bad decisions on the part of government agencies.

Beneath the surface of these sporadic outcries for public action, one may also find self-appointed public interest advocates eager to advance their own public policy or religious agendas. While the resources available to public interest groups are typically far less abundant than to other classes of participants, the scientific staffs of such organizations can be highly sophisticated; they are certainly capable of disseminating intricate distortions, despite their smaller numbers. And since their primary goal is to catalyze public action, once they succeed in sparking the public's attention, their subtle bending of the relevant science can be insidious and difficult to reverse. Critics have accused environmental public interest groups, for example, of implying that their risk-averse positions are supported by a greater body of scientific evidence than in fact exists.[57] Public interest advocates have also been charged with presenting unbalanced reviews of the existing literature that emphasize evidence that is more favorable to their positions than other equally robust evidence.[58] Finally, some critics complain that public interest advocates are too quick to attack the results of sponsored research without providing any evidence of actual bias in the reported results.[59] Thus, while public interest groups may not be able to keep pace with their industrial opponents in the race to bend science, the evidence detailed in the following chapters suggests that these groups are not entirely above the fray.

Political Actors

Government agencies and departments often find themselves in conflict over scientific questions concerning the health and environmental risks posed by government actions. The National Environmental Policy Act requires agencies to prepare environmental impact statements detailing the health and environmental impacts of major federal actions, and other federal statutes like the Endangered Species Act require federal departments charged with developing public resources to limit those activities when they conflict with particular health and environmental goals.[60] The conflicting statutory obligations inevitably lead to conflicts over the scientific information that informs on how to proceed in particular cases. These conflicts can be just as adversarial as the conflicts over science in the private sector. For example, internal disputes can arise over the health risks associated with military and energy development operations, because the federal government operates as both the entrepreneur and the regulator. Contractual arrangements at all Department of Energy (DOE) weapons and radioactive waste projects put the government in just such an internal conflict with itself. The DOE is responsible for ensuring a continuing supply of materials for military operations and disposal capacity for the wastes resulting from the manufacturing processes at the same time that other agencies like the Occupational Safety and Health Administration (OSHA) and the EPA are responsible for protecting workers, neighbors, and the environment from the health and environmental risks those activities pose. The resulting incentive on the part of the DOE to bend science is no less reprehensible because it does not stem from the profit motive.

PROTECTING WORKERS OR BUILDING MISSILES

During World War II, the federal government entered into contracts with the Brush Wellman Company and another company to provide beryllium to several government-run laboratories that were secretly building atomic bombs as a part of the massive Manhattan Project.[61] It soon became clear that a small proportion of the workers who were exposed to beryllium dust were suffering from a lung-scarring disease called berylliosis. Government health officials recommended that measures be taken to reduce workplace exposures to beryllium, and the federal facilities began to supply respirators to workers. But soon after the war ended, a secret report was circulated within the newly created Atomic Energy Commission (AEC) cautioning that if the incidence of berylliosis in workers became known outside the defense establishment, the outbreak "might be headlined, particularly in non-friendly papers, for weeks and months," and this might in turn "seriously embarrass the AEC and reduce public confidence in the organization."[62] Rather than risk embarrassment and potential interruptions in beryllium supplies, the AEC and its contractors decided to keep berylliosis under wraps.

After a 1943 outbreak of berylliosis among workers and neighbors of a beryllium plant in Lorain, Ohio, caused a local furor that threatened exactly the public relations fiasco the government feared, the AEC took steps to reduce exposures to beryllium in the vicinity of beryllium processing and weapons manufacturing plants. Scientists at the AEC determined that neighbors should be exposed to no more than 0.01 microgram per cubic meter ($\mu g/m^3$) in the atmosphere. The workplace exposure standard was more difficult, because it would have been prohibitively expensive to limit exposures in many workplaces to 0.01 $\mu g/m^3$. As a result, a standard two hundred times higher (2.0 $\mu g/m^3$), which originated in a taxicab conversation between an AEC scientist and a medical consultant, became the federal workplace limit.[63] That standard remains in place to this day in private sector workplaces.

For the next twenty-five years, the original standards were enforced not through regulations backed up by civil and criminal penalties but through clauses in the contracts between the AEC and its successor agency, the DOE, and private contractors. Indeed, the very same official was in charge of purchasing beryllium for the AEC on the one hand and enforcing the safety provisions in the purchase contracts on the other. When this official later threatened to cancel a contract because of safety violations during the 1960s, a general called an agency official to ask: "What are you, out of your goddamn-picking mind? I've got submarines out there. We need missiles." The official soon left the agency to become a top executive at Brush Wellman. We will follow the beryllium story as it progressed through the next three decades in subsequent chapters, but for now, the lesson of beryllium's early history is that government agencies appear to be just as susceptible as private sector actors to pressures to bend science in outcome-oriented ways. ❐

The DOE is by no means the only federal agency that has an incentive to bend science to advance its own institutional and political agendas. A similar conflict exists between the EPA and the United States Department of Agriculture (USDA) in its role as promoter of the country's agricultural interests.[64] It is thus not surprising that, after the USDA microbiologist Dr. James Zahn wrote up a paper detailing air emissions of potentially dangerous antibiotic-resistant bacteria near hog confinement operations in Iowa and Missouri, his USDA superiors repeatedly prohibited him from publishing his findings or presenting them at scientific conferences. They relied on a February 2002 USDA directive that required staff scientists to seek approval from superiors prior to publishing any research or speaking publicly on "sensitive issues," a term that included "agricultural practices with negative health and environmental consequences."[65] Similarly, officials in the National Oceanic and Atmospheric Administration severely regulated and sometimes flatly prohibited communications between agency scientists and the media about global warming, and more than 120 scientists from across seven federal agencies reported that upper-level officials had pressured them to remove references to "climate change" and "global warming" from a range of documents.[66]

Sometimes the government's bending activities grow out of its support of certain private sector actors, rather than its own internal policies. During the late 1990s and early 2000s, the FDA was under a great deal of political pressure to approve promising new drugs and new uses of existing drugs. Yielding to this pressure, agency officials occasionally cooperated with drug companies and portrayed underlying scientific research in ways favorable to the companies. For example, after the manufacturer of the drug Rezulin complained to FDA management that the scientist assigned to review its "fast track" application was interpreting the underlying safety research too skeptically, the agency replaced him with scientists who reached conclusions that were more to the company's liking.[67] The scientist's superior also insisted that he cancel a previous commitment to discuss his concerns about Rezulin's liver toxicity at a public conference, an intervention for which the manufacturer privately expressed its appreciation.[68] The agency was forced to withdraw its fast-track approval three years later, however, after concluding that the drug could have caused at least sixty-three deaths due to liver failure, just as the original reviewer had feared.[69] Likewise, the Deputy Assistant Secretary for Fish, Wildlife and Parks during the mid-2000s became an aggressive advocate for the views of developers and the energy industry on scientific issues involving endangered species, and he rejected the recommendations of the agency's senior scientists. She was later caught altering scientific field reports and emailing confidential scientific documents to an oil company and a property rights litigation group that frequently challenged the Interior Department's endangered species decisions in court.[70]

Faceless bureaucrats are not the only public officials who bend science on behalf of the government. As we shall see in Chapter 7, prominent congresspersons can also be aggressive participants in efforts to distort science, using Congress's broad investigatory powers to intimidate climate change researchers. Highly placed political appointees like the chief of staff of the President's Council on Environmental Quality have revised scientific information about climate change in government reports at the behest of powerful political actors.[71] These abuses by government officials are especially disturbing because they suggest that in some circumstances, even public servants cannot resist the temptation to manipulate science when it serves their interests or those of their favored constituencies.

While this book focuses exclusively on the research used to inform public health and environmental policy, it appears that similar types of bending—motivated by the same types of incentives—occur in other important areas of public policy, the most visible of which is the forensic science that supports criminal convictions. The reliability of scientific methods underlying such investigatory techniques as handwriting evidence, hair comparisons, fingerprint examinations, firearms identifications, and intoxication testing has come into question in recent years.[72] Yet the originators of this body of research—crime laboratories of the Federal Bureau of Investigation, for example—have apparently developed and even promoted their work with little meaningful oversight from the larger scientific community.[73] Since defendants in most criminal cases lack the resources to mount effective challenges, much less undertake their own counter-research, it is certainly possible that some of this science is also badly bent.

Strategies for Bending Science

Once convinced that it is in their interest to bend science, advocates can employ a wide array of techniques to manipulate, undermine, suppress, or downplay unwelcome scientific research. Chapters 4–9 provide in-depth descriptions of how these tools work, and all the chapters give examples of how advocates have creatively deployed these strategies in the national policy-making process. For now, here is a bird's-eye view.

Shaping Science

One of the most elaborate and expensive ways to distort science is to commission research designed to produce a particular outcome. By hiring scientists willing to collaborate closely (for example, under contracts that provide for sponsor control of the research), sponsors have historically been able to exert considerable control over research outcomes. Sponsoring companies are often heavily involved in framing the research question, designing the study, even editing and occasionally ghostwriting the resulting published article so as to produce a desired research result. If the distortions are carefully crafted, the advocate's hand is difficult to discern. Such ends-oriented outcomes generally come to light only when independent scientists take the time to scrutinize the research and discover flaws.

Hiding Science

When efforts to shape science fail, an obvious second-best option is to suppress the unwelcome findings. Suppressing bad news is not only relatively inexpensive for the party with the relevant information but also quite difficult for outsiders to discover. Thus, the benefits of hiding damaging scientific findings can often be quite high and the costs relatively low. Since the strategy is, of course, only available to someone with scientific information to hide, it tends to be employed primarily by governmental entities and companies that can afford to support scientific research.

Attacking Science

Another common approach to bending science consists of launching illegitimate attacks on damaging research. Groundless challenges to methodology, data interpretation, or review processes can be used to discredit high-quality technical research in the minds of regulatory officials, the general public, and even some scientists. Since the practice of science is not always well disciplined, the normal critiques and vetting of science are often brutally frank and therefore hard to separate from illegitimate attacks that challenge well-established methodological choices. In some settings, the simple fact that a study has been subjected to a lengthy and credible-sounding attack may be enough to impair its perceived reliability in policy-makers' minds. In others, carefully orchestrated attacks from multiple sources can lend the appearance of scientific legitimacy to baseless criticisms.

Harassing Scientists

An attack on the quality of research can easily escalate into a full-scale assault on the integrity of the researcher. Unsupported allegations of scientific misconduct, harassing subpoenas or depositions, and burdensome data sharing requests (often facilitated by public records statutes) aimed at scientists whose research has adverse implications for advocates can provide sufficient distraction to prevent those scientists from conducting follow-up studies or even intimidate them into leaving the field to other scientists who have the stomach for controversy.

Packaging Science

Advocates can package science by commissioning review articles that purport to summarize existing research on a topic but are in fact intended, even contractually guaranteed, to portray existing research in the light most favorable to the sponsor. The practice of convening handpicked or stacked expert panels to gather and reach a "scientific consensus" on important scientific topics is even more popular. Far from representing a balanced assembly of scientific viewpoints, these panels are consciously structured to yield conclusions that support the convener's interests. These practices can both create and reinforce a slanted view of research that is very difficult for other, more objective observers to dislodge.

Spinning Science

Like any other aspect of reality, the results of scientific studies are subject to sophisticated efforts by advocates or their surrogates to interpret or spin them in ways that are not accurate or faithful to the research but instead advance economic or ideological goals. In the hands of public relations firms with expertise in manipulating public opinion, spinning science also involves portraying damaging research as "fatally flawed" or "tentative" in an effort to convince the public to ignore it and to generate pressure on decision-makers to discount it. To the extent that the press fails to dig beneath the surface of these campaigns to mischaracterize acceptable science as junk science, their complacent coverage serves only to reinforce the perceived legitimacy of the distortions.

The Consequences of Bending Science

If the consequences of attempts by advocates to bend science were limited to the realm of science, the phenomenon might be of critical interest to the scientific community, but of little relevance to the rest of us. Since the reason that advocates go to the trouble of bending science is to affect decisions in the policy realm, however, their efforts are likely to have consequences for the broader public. If the consequences were limited to making their clients wealthier, the phenomenon would be disturbing, but perhaps not so much so that action would be warranted. After all, information has been manipulated in the marketplace for as long as there have been markets. There are strong reasons to believe, however, that the efforts to bend science documented in this book can have serious adverse consequences for human health and the environment, as well as for the economic well-being of legitimate businesses. Although it is impossible to establish a cause-and-effect connection between every attempt to bend science and some adverse effect on human health or a company's business, the connection is clear enough in many cases to warrant serious concern and to suggest the need for action to discourage such practices.

A PREVENTABLE TRAGEDY

According to the Rand Institute for Civil Justice, at least 225,000 people in the United States will have died prematurely by 2009 as a result of their exposure to asbestos.[74] Nonfatal injuries resulting from asbestos exposure afflict many more. As of 2005, tens of millions of people in the United States had been exposed to significant levels of asbestos, and over six hundred thousand people had filed claims for damages.[75] A significant portion of these fatalities and injuries would have been prevented if the asbestos industry had been forthright about the risks of asbestos when it first learned of them. As early as 1930, the industry understood from its own suppressed research that animals exposed to asbestos contracted a serious lung disease called asbestosis. One company scientist, for example, persuaded the editor of a trade magazine in the mid-1930s that the growing body of scientific studies on asbestosis should "receive the minimum of publicity." Even late into the 1950s, the Johns-Manville Company adopted a policy of refraining from informing sick employees of the precise nature of their health problems.[76] When Dr. Irving Selikoff published compelling epidemiological research in 1965 concluding that asbestos was a potent human carcinogen,[77] the industry mounted a sustained multiyear attack on Dr. Selikoff and his research.[78]

The truth ultimately won out, however, and by 1990, most of the prominent asbestos companies of the 1960s were in bankruptcy. The claims filed against them far exceeded the value of their remaining assets, leaving the victims and society at large to hold the bag. In the United States alone, the costs of asbestos injuries are estimated to reach at least $200 billion, including the expense of litigating and deciding the claims.[79] ❐

The adverse health consequences of the asbestos manufacturers' successful attempts to bend science are by no means unique. As we shall observe, suppression and distortion of scientific research on the addictive properties of cigarettes, the adverse side effects of a number of pharmaceuticals, and the presence of toxins in water supplies and other media have led to countless cases of unnecessary deaths and injuries. While it is impossible to calculate the cumulative costs to public health that are attributable to the imaginative efforts of advocates to bend science documented in this book, there can be little doubt that the health and environmental consequences have been both immense and preventable.

Workers and the general public are not the only victims of conscious attempts to bend science. A number of large and small businesses have found themselves losing sales and investing in damage control because of publicity campaigns that bent science to exaggerate the risks of their products.[80] Companies have settled cases for large sums in which, it appears in hindsight, attorneys manipulated the science to assemble classes of plaintiffs that included alleged victims who were probably not in fact harmed. In some cases, people have even stopped using life-saving drugs or other useful products because of deliberate or reckless manipulation of scientific information.[81]

Although the costs to science and the scientific community are not easily monetized, they are yet a third source of significant social loss. Scientists have lost funding and have been threatened with lawsuits when their research produced results that threatened powerful economic interests. These cumulative pressures take a toll on researchers who engage in policy-relevant science. Dr. Donald Kennedy, the editor of *Science,* observes:

> Many [scientists] are wary of work that may find use in some regulatory proceeding. They wonder whether the data underlying their findings may be subject to examination and reinterpretation, perhaps with some "spin" supplied by the revisionists. They know that charges of research misconduct could arise from hostile access to their scientific work. They know they are vulnerable to personal attack from those whose interests may be adversely affected by the product of their research.[82]

These pressures also take a toll on the scientific journals, which are key targets in several strategies for bending science. Some journal editors are finding it difficult to police compliance with their conflict of interest disclosure requirements or to craft them in a way that does not leave room for creative evasion.[83]

Finally, advocates' successful attempts to bend science can lead to bad policy decisions, poor public understanding and participation, and erosion of the public's overall confidence in science. News reports that portray a constant stream of scientific manipulations may leave the lay public with the uneasy feeling that one cannot trust any claims science makes, at least in the short term. The public might also question whether elected officials have implemented the right set of laws to deter these types of scientific abuses. Finding themselves similarly adrift in their attempts to navigate the battle of the experts, agency decision-makers and courts may simply ignore or exclude valuable scientific information as they deliberate over important legal and policy disputes.

In the end, the susceptibility of policy-relevant science to outside pressures reflects the simple but ugly reality that even scientific knowledge can become contaminated by ideology and money. The greater the private interest in thwarting unwelcome research, the more likely it is that properly conducted science will be eclipsed by bent science. Ultimately, the information regulatory agencies rely on to protect society and that courts rely on to administer corrective justice may not reflect the objective work of talented scientists, but something much less. Nevertheless, we argue in the final part of this book that there is room for hope that the merits of high-quality science can still win out in the long term. Shedding even a little light on how advocates bend policy-relevant science could go a long way toward remedying these problems. Indeed, precisely because the advocates have overtaken the law in this area, heightened attention to the social costs of bending science could itself precipitate significant change.

CHAPTER THREE

Where Are the Scientists?

Distorting Science without Oversight or Penalty

Scientists would seem to be in the best position to prevent distortions of science and stave off illegitimate attacks. And from all outward appearances, scientists are vigorously engaged in this collective oversight. Common references to the "scientific community" and "peer review" give the impression that an invisible jury of qualified experts sits in judgment on every scientific study to ensure its credibility and reliability. Media further accounts reinforce this impression. Scientists charged with falsifying data or otherwise violating tenets of the profession are publicly flogged, even if they previously contributed to science in ways deserving of a Nobel Prize. Denials of tenure to budding young academic scientists seem almost routine and appear to reflect the judgment of the scientific community that their research does not cut the mustard. Competitive grant processes overseen by scientists provide only the most capable researchers with sufficient funding to engage in meaningful projects and leave the rest scrambling to find support. Sociologists speak fondly of the "norms" that invisibly govern scientific conduct. Therefore even as the specific mechanisms for this professional oversight remain illusive, the scientists—as a collective—appear to control what research ultimately passes for good science and who are deemed worthy scientists.

In the realm of policy-relevant science, however, reality does not match these idealized perceptions. Policy-relevant research—perhaps more than any other body of science—is just not that interesting to the scientific community. Policy-relevant research often lacks theoretical grounding, is frequently unoriginal, and is usually contested by advocates with nonscientific agendas.[1] For scientists who dislike courtrooms, bureaucrats, and politicians or who believe that becoming involved in real-world legal and policy battles will compromise their objectivity, policy-relevant science is a particularly risky area of research.[2] As a result, some prominant scientists are worried that independent scientists may begin to avoid this bread-and-butter research that informs environmental and health regulation and litigation. Yet in the absence of active involvement by numerous independent scientists, much policy-relevant science is effectively relegated to orphan status and must make its own way in the world, even if that entails violating the rules that play an essential role in keeping the rest of its prestigious scientific family together.

The resulting dearth of rigorous professional oversight for some policy-relevant research presents a serious problem not only because it allows poor-quality research to escape undetected but also because it provides ample opportunities for advocates who are so inclined to bend science with abandon. Bent science consists of ends-oriented research and critiques of research that work backward from a result, and for that reason it is not true science. Yet because bent science is produced for the policy realm, it remains an important subcategory of this larger category of policy-relevant research.

Bent science is in fact even more likely to fall through gaps in peer oversight than is independent policy-relevant science. When scientists embed ends-oriented choices deeply within the design of a study or critique, the most effective ways to detect them are for other scientists to replicate the study or for independent experts who are well versed in the relevant scientific discipline to scrutinize the work very carefully. Since policy-relevant science often lacks this vigorous independent review, however, the ends-oriented biases built into bent science are likely to escape unnoticed. Even when independent scientists question the quality of bent science, their concerns may look to outsiders more like one side of a legitimate point-counter-point debate in the scientific literature than the consensus view of independent scientists that the research may be unreliable.[3] Still more perversely, when independent scientists appreciate that science has been badly bent, they may simply ignore the research rather than challenge it, because it has no relevance to their research agendas. As a result, bent science seems to encounter a perfect storm of perverse forces—strong incentives for advocates to distort science, limited internal peer involvement to catch and publicize abuses, and few cultural or institutional mechanisms to punish such abuses.

This chapter challenges the prevailing assumption that the existing system of voluntary scientific oversight of policy-relevant science is adequate to expose and expunge the bent science. We hasten to add this should not be taken as a criticism of scientists or the scientific community. To the contrary, the lack of interest on the part of independent scientists in bent science is a testament to their single-minded pursuit of the truth. Yet if we are correct in concluding that immediate action is required to reduce the role that bent science plays in regulatory and judicial decision-making, then the scientific community will have to assist in designing and implementing the necessary reforms. We will relate in Chapter 4 how the scientific community has already taken some steps toward exposing potentially bent science by requiring disclosures of sponsor control and assurances of authorship as a condition to publication in many prominent journals. We will suggest additional reforms in Chapters 10 and 11 that also require the active support and participation of the scientific community.

Self-Governance in Science

Although many people associate science with the quaint image of a lonely researcher hunched over a microscope or Bunsen burner, science is in reality a sophisticated social undertaking.[4] Modern technology allows a single scientist to produce a great deal of scientific research, but the reliability and legitimacy of that research depends on how the scientist's colleagues receive it and how they perceive it to fit with their own work.[5] The scientist's peers determine through debate and research the most promising methods and theories, and they provide a critical degree of quality control over the research that tests these theories.[6] While philosophers and sociologists of science understand that the collective scientific judgment about the reliability of research can sometimes be wrong or badly biased, nearly all agree that peer oversight is essential to ensuring scientific quality.[7] In addition to weeding out questionable research, vigorous peer oversight encourages scientists to be rigorous and objective.[8] Peer reviewers will not always catch and condemn bias and related errors, but the realistic possibility that they will provides a strong incentive to avoid them. When research presents particularly novel findings, scientists frequently attempt to replicate the results voluntarily. Self-governance in science thus consists of both peer scrutiny and the possibility of having a study replicated.

Despite its theoretical importance to the scientific enterprise, self-governance in science tends to be informal and even permissive in its actual implementation.[9] There are no generally accepted written standards against which all research may be judged, and only a few research specialties have written codes of professional ethics.[10] Resource limitations account for a great deal of slippage that occurs in scientific self-governance in the real world. Replication can be extraordinarily resource-intensive, so it is typically reserved for the most important studies and for cases in which research findings are particularly suspect. Both scrutinizing and replicating the work of fellow scientists are also uncompensated services. Consequently, a good deal of research may pass through the system with only cursory review.[11]

Fortunately, a series of informal norms of good scientific practice supplements active professional oversight of research quality. Although compliance with these norms does not necessarily ensure that the resulting research will be of high quality, violation of the norms casts serious doubt on a study's reliability. The first norm expects that scientists engaged in creating or reviewing research will conduct their activities in a disinterested way, free from interference by sponsors and other interested entities.[12] A second consensual principle insists that researchers be willing to share data and methods with other scientists so that their work can be evaluated fully and replicated if necessary.[13] Third, the scrutiny applied to ongoing research must be disinterested, be collaborative, and occur from a diverse and representative group of peers.[14] While deviations from these norms are legend, they remain the established objectives that define good research practices.[15]

Policy-Relevant Research and Limited Peer Oversight

Although scientists have an unwritten professional obligation to participate in the informal network of peer oversight that governs most science, voluntary engagement by independent scientists in policy-relevant research can be quite limited. Policy-relevant research is often designed to answer questions raised by legal institutions, not scientists, and the studies as a general rule do not test cutting-edge theories or even incrementally advance an existing body of scientific work. As a result, it takes either financial incentives or some other inducement, like the prospect of an expenses-paid trip to an interesting locale, to prod most scientists into reviewing some kinds of policy-relevant research. Not surprisingly, the review and oversight that do occur are usually routine and formal, not spontaneous and collegial.

Not all policy-relevant science, however, is equally unattractive to research scientists. One form of policy-relevant research, which we will call "standardized science," consists of prescribed tests that are routinely conducted on individual products and byproducts. Because it has little prospect for advancing the theoretical underpinnings of science, standardized science is the least interesting to independent scientists. The second type of policy-relevant research, which we will refer to as "general applied research," can involve more open-ended studies that do have some prospect of advancing scientific theories. This research tends to be more interesting for the scientists who undertake it, but it can also have features that can discourage active scientific oversight.

Standardized Science

Standardized research is designed to answer very specific questions about localized problems that occur with some regularity in regulation or litigation. In environmental and health regulation, standardized science includes a number of routine toxicological tests for pesticides and toxic substances, some clinical trials for drugs, common fate and transport studies that follow a pollutant through the air, water, or soil to determine its physical and chemical characteristics in those media, and routine environmental science research conducted in specific settings to characterize risks or patterns of movement peculiar to those settings.[16] The tests are generally either required or strongly recommended by regulatory agencies, and they tend to come up in litigation only after they have already spent time tucked away in an agency file drawer. Agency-promulgated protocols typically specify most of the study details, including the number of test animals, the food, the container size, and so forth. Even when they leave some discretion to the researchers—for example, the dosing levels—the prescribed procedures typically offer specific guidance to limit the extent to which an investigator's exercise of judgment results in deviations from prescribed patterns.

Standardized science leaves the researcher with little discretion to design and conduct studies in creative ways. Its defining feature is therefore its unoriginality. This feature helps regulators ensure that the results remain unbiased, even though the researchers or their sponsors have a strong interest in their practical implications. But it also means that standardized research is the antithesis of the innovative, cutting-edge kind of science that attracts interest and invites scrutiny from the broader scientific community.

PESTICIDE TESTING FRAUD

Under a 1972 U.S. pesticide law, manufacturers were required to provide data for about 650 existing pesticides sufficient to meet modern toxicological protocols within a three-year period. The EPA in 1975 wrote detailed testing protocols to ensure that testing results would be useful to agency scientists who would rely on the research to evaluate the health and environmental risks of the tested products. The new requirements generated a huge demand for toxicological studies, and a company called Industrial Bio-Test Laboratories began taking orders from many pesticide manufacturers to conduct hundreds of animal studies. Before long, that company was conducting almost 40 percent of all pending animal studies in the United States. The company became overwhelmed and, in order to meet customer demands, began to forge its reports, often providing the same data tables for more than one pesticide. A special EPA investigation conducted during the late 1970s revealed that despite the care that the agency had taken in crafting testing guidelines, the company had perpetrated fraud on a massive scale. Without the unusual EPA investigation, the fraud might never have come to light.[17] ❐

In hindsight, it is not surprising that the larger scientific community did not become engaged in the boring work of pesticides toxicity testing, despite the critical social need for such testing in the early 1970s, nor should it come as any surprise that independent scientists did not review the work of Industrial Bio-Test Laboratories as it came off the assembly line. The sponsors of standardized research need a scientific product that they can use for the practical purpose of getting their commercially useful products approved by regulatory agencies and accepted by consumers. They are not especially interested in advancing the theoretical underpinnings of environmental toxicology. Likewise, the purveyors of standardized research, like Industrial Bio-Test Laboratories, rarely submit their studies for publication, and the scientific journals are in any event not eager to publish unoriginal work of this nature.[18] In fact, sponsors typically classify much of the critical information underlying their standardized testing as trade secret protected.[19] As a result, most standardized studies reside quietly in government files, where they may be accessed only through a formal Freedom of Information Act (FOIA) request.

This is hardly a situation that is conducive to independent peer review and collective scientific oversight. Even if independent scientists had easy access to this body of work, they would probably shun most of it because it is done pursuant to relatively simple, cookbook-like protocols and cannot be used to test theories at the outer edges of the discipline. The localized nature of most standardized research, which usually focuses on individual products or geographical areas, also makes it less interesting to the broader expert community.[20] No scientist is likely to gain academic credit, much less tenure, at a major university for conducting research like that performed by Industrial Bio-Test or for reviewing its output.

Even the somewhat more intellectually stimulating process of designing the protocols for standardized research may involve limited input and review from independent scientists. The process of standardization rarely involves novel questions or cutting-edge methods. Regulators prefer protocols that are well established, easily repeated, and as free from controversy as possible. None of this is especially interesting to scientists engaged in path-breaking work. The fact that agencies typically solicit the views of the broader scientific community through tedious *Federal Register* listings, rather than in scientific publications, makes it unlikely that scientists will learn about the work at all. The decision-making process therefore tends to be dominated by scientists representing regulated companies like product manufacturers, polluters, and public interest groups. For example, when the EPA developed neurotoxicity testing procedures for several toxic substances, it received input almost exclusively from more than a dozen scientists working for the industrial sector.[21] The primary source of independent scientific input, beyond the agency staff, is likely to be an agency-appointed scientific advisory committee, which may also be dominated by scientists who work either directly or indirectly for the regulated companies or nonprofit groups.

In the final analysis, the prescriptive rigidity of standardized research ensures that the larger scientific community will tend to play only a limited role in scrutinizing the development of standardized tests and the quality of the research those tests yield. Although standardized research is unquestionably science, its unoriginal nature and poor dissemination keep it out of the sightlines of most members of the scientific community.

General Applied Science

The second general category of policy-relevant research consists of applied research that responds to pressing policy questions as they arise in nonroutine ways. Because the need for residual discretion is unavoidable in this sort of research and because the agencies do not have the leisure of waiting until they have developed standardized protocols, these settings provide researchers with greater leeway in determining the methods and carrying out the experiments. When a pesticide appears to be capable of causing harm to particular native waterfowl, for example, more open-ended testing may be required to investigate its adverse effects. In this follow-up study, researchers will need to be alert for a variety of symptoms and signs of disturbance, some of which may be identified in advance and some of which will emerge through trial and error.[22] A similar degree of discretion characterizes most clinical research on drugs, which for a number of practical and ethical reasons cannot be forced into predetermined, cookbook-like protocols. While the dividing line between standardized and general applied research is not a sharp one, the two are roughly separated by the degree to which the regulatory authority has constrained research discretion with standardized protocols.

Like standardized research, general applied research is quite often sponsored or even conducted by an entity with an interest in its outcome. Since the greater discretion to design and conduct studies with particular outcomes in mind may prove quite tempting for the parties that sponsor this kind of research, much bent science tends to be of this variety. The lack of prescribed protocols, combined with the inability of outsiders to police such research for compliance with scientific norms and procedures, thus makes peer oversight centrally important for ensuring the reliability of this type of research. But active peer engagement can again be quite limited, even when research has interesting implications for cutting-edge scientific research—as did Dr. Tyrone Hayes's frog studies in the following case.

HERMAPHRODITIC FROGS: SCIENCE FACT OR FICTION?

When he made the surprising discovery in 2000 that male frogs exposed to low levels of the herbicide Atrazine developed ovaries, Dr. Tyrone Hayes found himself feeling somewhat alone. His source of funding for the research was Syngenta Crop Protection, the manufacturer of Atrazine, and he alleged that the company strongly discouraged him from communicating the results to the larger scientific community.[23] After skillfully maneuvering the company's lawyers into presenting the results of his research in public (thus lifting the gag order in his contract with the company), Hayes published his path-breaking findings in several venues, including the prestigious and widely read journal *Nature*.[24]

Other scientists found Hayes's results interesting and important, but the NSF and other granting agencies do not routinely disburse funds simply for the replication of studies. Consequently, Syngenta's scientists were able to step into the void. Within two years after the publication of Hayes's research, company-funded researchers reported they were not able to replicate the laboratory results, thereby throwing Hayes's study into question.[25] Prestigious journals like *Science* reported a great controversy among researchers as to whether Hayes's research was valid.[26] To parse through the competing claims in this research battle, an independent scientist would have to invest considerable energy and resources into carefully examining the methods of all of the studies and, if necessary, reviewing and analyzing the underlying data. It did not appear that any scientist was prepared voluntarily to make such a commitment, or at least not as swiftly as Syngenta's scientists.

Because of the important implications of the research for its pesticides program, however, the EPA intervened to sort out the battle between experts. The agency tasked its Science Advisory Panel, a permanent advisory body required by the federal pesticide statute, with scrutinizing the literature on Atrazine, including both the Hayes and the later Syngenta-funded studies.[27] After nine months of deliberations, the panel determined that most of the negative Syngenta-funded studies contained flaws that rendered the results unreliable.[28] In one, for example, the test animals uniformly suffered high mortality, and the water quality in the laboratory aquariums was poor.[29] Subsequent research by scientists, including one team funded by Syngenta, largely validated Hayes's original findings, but not until after considerable conflict had ensued.[30] ❐

As Dr. Hayes discovered, reinforcement from peers can be slow in coming, particularly in comparison to research funded by an advocate. We examine below three overlapping explanations for this lack of peer oversight, all of which are endemic to science in general but are aggravated by the peculiar features of policy-relevant research.

Limited Oversight of Technical Details

Voluntary scrutiny of most scientific research—whether policy-relevant or not—generally occurs at a rather coarse level of review. Scientific peers typically consider only the researcher's own written description of the methods and data and attempt to evaluate the study without duplicating the research or reviewing the underlying data.[31] As a result, the accuracy and reliability of the minutiae of the research, such as the comprehensiveness of data collection efforts and even more technical methodological choices, may be taken for granted.[32] Errors in the details of methodology or data collection are instead generally revealed only in a few, sporadic efforts by other scientists to replicate studies. Arnold Relman, a past editor of the *New England Journal of Medicine* (NEJM), concludes that "[t]here is no practical alternative to this presumption of honesty in research, because it would be impossible to verify every primary datum and every descriptive statement in a research report."[33] In the case of private sector–sponsored studies, however, this leaves journal editors and their peer reviewers "in the untenable position of having to trust that . . . sponsors have accurately and completely reported their findings."[34] Reports of periodic scientific misconduct seem to confirm suspicions that important sources of bias and error remain undetected by peer scrutiny and that the prospect of peer review is not always an adequate deterrent to the scientists committing these errors.[35]

While this superficial level of peer oversight is commonplace throughout science, its inadequacy takes on particular social significance in the unique context of policy-relevant research. The high stakes riding on the results of much policy-relevant research provide strong incentives for sponsors to tweak methods and tease data to obtain more favorable results. Advocates know that if the distortions are carefully hidden in the details of the study, superficial review can be the equivalent of no review at all and pernicious sources of bias in research design and execution can pass through the peer review process without correction or penalty.[36] They also know that many second- and third-tier journals are hungry for submissions and will publish articles that receive minimal peer scrutiny both before and after publication.[37] Finally, they know that research that is vetted and published can give lay judges and regulatory officials the erroneous impression that a study has been adequately scrutinized and is therefore reliable.[38]

At the same time, this cursory form of review may allow ends-oriented critiques of reliable research to go largely unexamined. Commissioned experts may therefore succeed in discrediting research that is actually of high quality. Because most scientists have neither the time nor the inclination to engage in detailed review, such disputes are generally waged at the ground level between a limited number of interested scientists. Some of the counter-studies that challenged Hayes's Atrazine results, for example, were published in peer-reviewed journals and characterized in prominent scientific publications, like the widely read weekly magazine *Science,* as raising legitimate questions about Hayes's research.[39] The studies were only rejected after the EPA's staff scientists remained unconvinced and the agency specifically asked its Science Advisory Panel to review the controversy.[40] Perversely, the courts and regulatory agencies thus find themselves saddled with the primary responsibility for assessing the quality of much of this policy-relevant science, at least in the short term.

Every rule has its exceptions, and independent scientists will sometimes vigorously scrutinize the details of their colleagues' policy-relevant research and even provide robust oversight of potentially bent science in a very short time frame. A good example of such voluntary oversight was the discovery that a postdoctoral student at Tulane University, who reported that synergistic interactions between chemicals led to endocrine effects far worse than expected, had fabricated the data. The fraud was uncovered when John McLachlan, the professor in charge of the laboratory, and other scientists not affiliated with Tulane were unable to replicate the results and became suspicious.[41] Because the study had significant implications for both science and policy, it precipitated vigorous and almost immediate scrutiny from colleagues.[42] McLachlan swiftly withdrew the paper from *Science,* and the student was duly sanctioned for committing scientific fraud. The Tulane story demonstrates that policy-relevant research can sometimes become important enough to the scientific community to warrant immediate interest from peers, particularly when replication does not entail a major investment in resources.

Isolating the Scientists' Collective Judgment

A different and less easily surmounted impediment to tapping into scientists' collective views about the quality of policy-relevant research stems not so much from the limited nature of scientific scrutiny as from the difficulties legal decision-makers face in isolating the scientific community's actual judgment on a specific research project or body of research. For the scientific community to provide meaningful assistance as gatekeepers monitoring the quality of policy-relevant research, its collective judgments must be explicit and must reflect collaborative deliberations among unbiased scientists. However, with the exception of specially created panels, such as committees appointed by the National Academies of Sciences (NAS) and the EPA's science advisory panels, these judgments often remain elusive. Rarely does the collective vetting that typically attends peer review of scientific research result in anything resembling a consensus statement, and even if it did, the conclusions would ordinarily be heavily qualified and annotated with sources of internal disagreements about features of the research, rather than the bilateral up-or-down statements of approval or disapproval that judges and policy-makers would rather see.

Complicating matters still further is a counterintuitive feature of collective scientific scrutiny that can be intensely frustrating to judicial and regulatory decision-makers. As already noted, the scientific community actively oversees the quality of important and topical research because it tends to be both more interesting and more relevant to the research interests of scientists at the cutting edge. By contrast, studies that sponsors control in obvious ways may attract little attention from independent scientists because they are reluctant to spend their valuable time refuting studies that most of their peers know to be unreliable. Simply observing the debates as they are waged within science, then, will tell an outsider little about the underlying reliability of most policy-relevant research. Yet policy-makers, the media, and the public may perceive silence from the scientific community as a form of tacit approval and erroneously presume that the research is of high quality because peers have not challenged it.

The difficulty of tapping into scientists' collective views on the quality of research has stymied the courts, in particular, for nearly a century. To ensure that lay jurors are exposed only to reliable scientific evidence and expert testimony, the courts have struggled to develop a suitable test for extracting from scientists what they think about specific research in their field—is it reliable, consistent with their methods, and otherwise scientific? Until ten years ago, federal judges determined whether scientific testimony was reliable in large part on the basis of whether the testimony and underlying studies were "generally accepted" by the scientific community or some relevant subset thereof.[43] Yet the courts had difficulty defining the relevant "scientific community" and determining what was "generally accepted" in that community, and they were especially perplexed by novel research that might be perfectly sound, but had not percolated in the scientific community for a sufficient time to become "generally accepted."[44] In 1992, the Supreme Court reframed the test in *Daubert v. Merrill Dow Pharmaceutical Co.* to ask, from an external vantage point, if the science was "reliable" based in part on whether the expert's conclusions were capable of being empirically tested.[45] It is fair to say that the courts continue to struggle with this vaguely defined task, which is not one for which the judges' legal training naturally prepares them.[46]

Regulators have managed to circumvent this difficulty to some extent by assembling ad hoc scientific panels to draft "consensus statements" and by relying heavily on internal staff scientists and permanent science advisory boards to provide a proxy for this larger process of scientific scrutiny and review. These efforts yield a second-best accounting of scientific views, but they are nevertheless critical to regulatory decision-making because of their capacity to provide a trustworthy expert anchor in hotly contested debates over the quality of policy-relevant science.[47] While agencies can manipulate expert panels to provide political cover, these science advisory boards provide important counsel when properly employed. For higher profile controversies, the NAS also serves this function of collecting and presenting the views of the larger scientific community. Since its reputation rests on its ability to provide neutral advice, it has generally been less amenable to overt politicization. Still, even the NAS has received criticism for overreaching and for failing to provide a sufficiently diverse scientific perspective on some issues.[48] More recently, NAS staffers have complained about challenges they face in recruiting qualified neutral experts to participate on panels addressing controversial policy-relevant scientific topics.[49]

The Greater Contestability of Policy-Relevant Research

Perhaps the most challenging impediment of all to decision-makers seeking to discern the collective judgment of a community of scientists concerning the reliability of policy-relevant studies is the prevalence of situations in which a clear-cut scientific judgment is unachievable. Policy-relevant science is characterized by tentative and evolving methods that may not have been used before, much less subjected to widespread scrutiny by a larger body of scientists. At the same time, applied policy-relevant research often proceeds in the absence of a motivating scientific theory because the questions the research is designed to answer do not arise naturally out of a discipline's pursuit of deeper scientific understanding but are instead posed by outsiders who are more interested in materializing and resolving disputes.[50] As a result, a good portion of applied policy-relevant research is fairly characterized as little more than sophisticated trial and error.[51] Yet without theoretical connections, the experimental methods of policy-relevant science become tentative and subject to intense disagreements within the scientific community.

Moreover, in some areas of public health and environmental research, scientists are in no position to validate the accuracy of a particular study's methodological approach at the time its findings are announced and pressed on agencies or courts. This is particularly true in the case of research that attempts to be predictive—as, for example, by estimating the rate of cancer in a population twenty years in the future or the extent of ecological damage that will occur over the next decade. Adding to the indeterminacy and potential for error is the fact that the objects of concern are often not subject to direct scrutiny and manipulation by the scientists. Because environmental settings have far too many natural variables to allow ecologists to isolate most adverse effects in the field, for example, the scientists are often relegated to experiments in artificial laboratory environments that may or may not reflect real-world conditions. Humans are even more difficult to study directly, because prevailing ethical norms prohibit human experimentation unless the effects are benign or the benefits are likely to outweigh the risk of harm and the individual test subjects consent to encountering that risk.[52] Since science is not capable of validating many of the critical features of public health and environmental experimentation as either the "correct" or even the "best available" approaches, the scientific community's willingness to accept the results of these tests is usually little more than a fragile agreement that is always open to renegotiation.

The artificially short timeline for policy-making likewise renders policy-relevant research tentative in nature. Once a study, however contestable, suggests the need for immediate governmental action, regulators usually need to make critical decisions before the larger community of experts has time to reach consensus or, in some cases, even deliberate on the relevant scientific issues. This lack of closure on the most reliable experimental methods renders policy-relevant science even more vulnerable to both legitimate and illegitimate criticism.[53]

The Inscrutable Case of Bent Science

Bent science may be especially immune to peer oversight for two different reasons. First, when the researchers and their sponsors carefully hide the ends-oriented research in ways that require either replication or extraordinary scrutiny to detect, the underlying flaws in the results are likewise likely to escape detection by the normal processes of peer review, even when journal editors are especially vigilant. As a clearly exasperated Dr. Catherine DeAngelis, the editor-in-chief of the *Journal of the American Medical Association (JAMA),* lamented: "The degree I hold is an MD, not an MDeity; I have no ability to know what is in the minds, hearts, or souls of authors. Furthermore, I do not have, nor desire to have, the resources of law enforcement agencies, but I do know that the accuracy of lie detector tests is questionable."[54] To the extent that independent scientists fail to scrutinize or replicate this research after publication, some of the more cleverly bent science will remain in the scientific literature and enjoy the appearance of scientific legitimacy.

Second, when researchers and sponsors are not so clever in their efforts to bend science, their research ironically tends to escape peer scrutiny and rebuttal precisely because its ends-oriented qualities are obvious. In these situations, independent scientists are inclined to ignore the research because it is uninformative and unreliable. Rather than writing critical letters to the editor or otherwise bringing its scientific faults to the attention of other scientists, they may conclude that such conscientious acts are only likely to draw them into the underworld of advocates and taint their own reputations for disinterested work. Worse, they could be personally attacked by advocates for having the audacity to engage in vigorous oversight of badly biased research. More than one scientist has learned the hard way not to question the reliability of commissioned research.

In the final analysis, though, individual scientists may avoid policing the quality of potentially bent science for the simple reason that the benefits of engaging in such oversight do not begin to outweigh the financial and institutional costs. A similar cost-benefit calculation may explain a related tendency of scientists to avoid reporting incidents of scientific misconduct such as data fabrication by their colleagues. Surveys and other accounts reveal that many scientists have witnessed incidents of misconduct, but very few have been willing to report the incidents because they have concluded that the professional costs would be too high.[55] To the extent that exposing bent science may entail similarly high costs, scientists may tend to resist engaging in this vigorous oversight. One commentator has even suggested that scientists may look askance at whistle-blowing peers because publicizing episodic scientific transgressions jeopardizes the scientific establishment's status in the eyes of elected officials, private sector funders, and the general public.[56]

Fortunately, while many scientists may have little to do with visibly bent research, journal editors like Dr. DeAngelis appear quite eager to exclude this tainted work from their publications. The editors of many of the top scientific journals have attempted in recent years to enforce conflict of interest disclosure and peer review policies more vigorously, scrutinize letters to the editor more carefully, and insist on retractions and administer other sanctions more frequently to ensure that their scarce journal pages are dedicated to research that is reliable and advances scientific discovery. Nevertheless, this flawed science may still end up in some impressively named publications. Dr. DeAngelis diplomatically describes one instance in which a paper that *JAMA* declined to publish because it did not provide an independent statistical analysis was "published elsewhere shortly thereafter and . . . received much media coverage." She could only "hope that the decision by the sponsor was based on something other than not wanting an outside analysis of data that might have uncovered flaws in the original analysis."[57] That may have been a vain hope. As we shall see in Chapter 8, some less prominent scientific journals are heavily sponsored by regulated industries, are edited by industry consultants, and publish large numbers of industry-sponsored papers.

Consequences for the Legal System

For a number of perfectly legitimate reasons, independent scientists may not find it professionally rewarding to invest their scarce resources in overseeing policy-relevant research, absent some tangible inducement like a prestigious appointment to an NAS committee or a government scientific advisory board. Researchers typically derive little personal or professional benefit from probing the technical features of a study unless they are engaged in research exploring the same or similar questions, and even then, any resulting interactions may take years to play out. The limited frequency with which scientists disagree about the parameters and appropriate methods for policy-relevant research does not imply that every approach is valid or that every criticism is fair. It does, however, provide easy opportunities for self-serving generators of dubious studies and bogus critiques to maneuver below the scientific community's radar.

The most obvious implication of the lax system of scientific oversight for policy-relevant science is the opportunity it presents for advocates to design and execute research to achieve predetermined ends with little risk of detection. Standardized research designed exclusively for consumption by regulatory agencies and courts may also escape the attention of the larger scientific community because it is never published or, worse, is actively suppressed. The general absence of serious oversight of policy-relevant research combined with the scientific tradition of vigorous peer scrutiny in other areas of research allows advocates to finance blistering critiques of unwelcome studies with little fear of detection or rebuttal from within the scientific community. The prospect of unfair criticism and even active harassment may in turn even discourage scientists from engaging in the kind of policy-relevant research that provokes such attacks, thereby leaving the field to be occupied by the advocates and their consultants. The scientific community's general lack of engagement in policy-relevant science also provides dangerous opportunities for advocates to misrepresent scientific consensus or even manufacture their own artificial consensus for public consumption. Since silence from the scientific community may be viewed as passive assent, the limited involvement of independent scientists in overseeing policy-relevant research leaves a critical void that can be filled by the opportunistic and strategic behavior we will explore in more detail in the following chapters.

CHAPTER FOUR

Shaping Science

The Art of Creating Research to Fit One's Needs

When science will not cooperate with an advocate's economic or ideological needs, the most expeditious course of action may be to shape future scientific research to fit those needs. The strategy begins with a desired outcome, rather than a hypothesis to be tested, the starting point for most independently conducted research. With the assistance of cooperative scientists, the advocate designs study protocols, specifies a process for carrying them out, carefully oversees that process, and devises a publication strategy that is likely to yield the desired result. When the study produces the expected findings, the advocate ensures that it is widely publicized and used in regulation or litigation. When it does not, the advocate may have to resort to contract terms, financial threats, and a variety of other techniques to keep the unwanted results under wraps and to render the scientist financially unable to pursue the matter further—the strategy of hiding science, to which we return in Chapter 5. Having shaped the contours and outcomes of individual studies, the advocate may then hire additional scientists to "package" the science in review articles and books, a topic we will take up in Chapter 8. Among the suite of tools available for bending science, shaping science is one of the more expensive, but it is also among the most effective when things go according to plan.

The law tends perversely to reward attempts to shape policy-relevant science, and it does little to deter them. Research showing that a pesticide will not harm wildlife allows that pesticide to move more rapidly through the EPA approval process than research revealing serious ecological risks. A case relying on impressive-looking research showing that the defendant's product injured the client is more likely to survive preliminary motions to dismiss than a case built on circumstantial evidence of causation. Absent clear evidence of fraud, moreover, the law does little to penalize affected parties who produce unreliable science for use in regulation and litigation. The federal courts insist that scientific evidence be "testable" in accordance with judicially crafted "scientific" criteria, but they do very little, beyond allowing the give and take of cross-examination and competing expert testimony, to screen for significant indications of ends-oriented bias in the research itself. Regulatory agencies have the expertise to exert a greater degree of oversight over similar biases in regulatory science, but their willingness and ability to delve deeply into the bona fides of the scientific studies that they routinely review and use is surprisingly limited. Thus, both courts and agencies leave ample room for parties to shape science.

This chapter provides multiple examples of how sponsors can shape research, either directly by collaborating with the scientists undertaking the research or more discreetly by selecting friendly scientists or favorable research projects. After considering when shaping will be most successful, we consider the individual stages of research during which shaping can take place. Frequently employed techniques include adjusting study design, manipulating the data or statistical methods, "creatively" interpreting the data, and misrepresenting the principal investigators' conclusions. We then consider several indirect strategies advocates use to control or shape the trajectory of available scientific knowledge about a regulatory or litigative issue. We conclude the chapter with some tentative explanations of why scientists might allow their research to be abused in this fashion.

The Recipe for Shaping Science

The recipe for shaping science includes three main ingredients: one or more savvy scientists, a desired outcome, and sufficient finances to support the enterprise. When combined, these ingredients can change the course of scientific understanding and the content of regulations and jury verdicts, at least for a while.

FIGHTING FUNGUS WITH SCIENCE

For the last half of the twentieth century, the primary treatment for potentially deadly fungal infections in immune-compromised patients was amphotericin B, a drug derived from spores found in the Orinoco River region of Venezuela. The drug is not, however, kind to patients, causing them to tremble violently and on rare occasions to suffer fatal kidney damage. Because the generic version of the drug is inexpensive, pharmaceutical companies in the 1990s came up with "new drugs" that were, in fact, patentable alternative ways to deliver amphotericin B that they hoped would be more effective and patient-friendly. An advisory committee of the FDA held two public meetings in 1994 and 1995 to hear proposals by scientists from the manufacturers and the National Institutes of Health (NIH) to cosponsor human clinical trials probing the safety and efficacy of the new drugs.[1]

At these meetings, Dr. Thomas J. Walsh, an NIH scientist and one of the country's foremost experts on treating fungal diseases, presented the case for testing the drugs in cancer patients whose low white blood cell counts and persistent (four-day) fevers made them likely to have fungal infections. One test for the efficacy of the drugs would be whether the fevers abated. When members of the committee objected that it would be preferable to use patients who were confirmed to suffer from fungal infections, Walsh explained that "financial and logistical limitations" precluded that approach because the most deadly kind of fungal infections, those caused by the fungus *aspergillus,* were notoriously difficult to diagnose. Another point of contention was the level of amphotericin that would be given to the patients in the control group, since using a placebo was not an option when the patients in both the test and control groups were suffering from a life-threatening disease. The study would have to compare the new drug with amphotericin. Dr. Walsh assured the committee that the dose of 0.6 milligram per kilogram (mg/kg) of body weight that the study protocol specified was the standard dose of amphotericin in these situations. Another expert told the advisory committee that the standard amphotericin dosage in such situations was probably between 0.6 and 1.0 mg/kg. He acknowledged that increasing the dosage of amphotericin did increase the risk of kidney failure, but he argued that the prospect of knocking out an *aspergillus* infection could easily outweigh that risk. Four years earlier, Dr. Walsh himself had coauthored a recommendation that dosages of amphotericin ranging from 1.0 to 1.5 mg/kg should be employed to treat suspected *aspergillus* infections, and he had published a paper stating that 0.6 mg/kg for patients with suspected but unproven *aspergillus* infections was "obviously . . . not sufficient." The FDA ultimately acquiesced in the proposed study design.

The first multisite clinical trial, sponsored by the Fujisawa Corporation, compared amphotericin administered at 0.6 mg/kg to Fujisawa's

AmBisome administered at 3 mg/kg in 687 patients. It was published in *NEJM,* with Dr. Walsh listed as the lead author. Reporting that 10.5 percent of the amphotericin patients died, compared to 7.3 percent of the AmBisome patients, the paper concluded that AmBisome was "as effective as" amphotericin and was "associated with fewer breakthrough fungal infections, less infusion-related toxicity, and less nephrotoxicity."[2] In a July 1997 presentation to the FDA advisory committee, Dr. Walsh reported that AmBisome "was more effective in preventing proven invasive fungal infections and fungal-infection-related deaths" than amphotericin. Hours later, the committee voted to recommend approval. Although an FDA statistician concluded that the study presented "inadequate scientific grounds" to conclude that either drug was superior, the FDA approved AmBisome in August 1997. A month after that, Dr. Walsh announced at a conference of physicians and research scientists that AmBisome was "the first agent shown to be superior to amphotericin B in reducing proven, invasive fungal infections in cancer patients." Two years later, he recommended that doctors consider amphotericin levels of 1 to 1.5 mg/kg when *aspergillus* infections were suspected.

The publication of the study results, however, attracted criticism from other scientists. In a letter to the editor of *NEJM,* three German scientists argued that "most institutions in Europe and the United States would agree that treatment of this patient population requires a dose of at least" 1.0 mg/kg of amphotericin. They concluded that it was "very likely" that if the higher dose had been used, there would have been fewer fatalities in the control group.[3] A 2001 reference book published by the British Society for Haematology likewise warned that the use of the lower dose in the study may have "bias[ed] the results in the favour of AmBisome" and could "entirely explain the difference observed."[4] In a reply letter to *NEJM,* Walsh and his coauthors argued that the 0.6 mg/kg complied with "the standards of care" at the participating institutions and noted that liver toxicity considerations had prevented some of the testing physicians from administering "appropriate doses" of amphotericin to some of the patients.[5]

Within two years after FDA approved AmBisome, Dr. Walsh had signed up with the Merck Corporation to conduct similar clinical trials on its new antifungal drug, Candida, in cancer patients. As with AmBisome, Walsh helped Merck persuade the FDA advisory committee to approve a study that was not limited to patients with confirmed fungal infections. Collaborating with Merck scientists, Dr. Walsh drafted protocols in which patients receiving 3.0 mg/kg of AmBisome orally were compared with patients receiving 50 mg/kg of Candida through a much different intravenous route. In September 2004, Dr. Walsh and his coauthors reported in *NEJM* that the mortality rate for patients receiving Candida was 10.8 percent, compared to 13.7 percent in the AmBisome-treated patients, and they concluded that Candida was "as effective as and generally better tolerated than" AmBisome.[6] The FDA approved Candida in September 2004.

Once again, outside scientists wrote letters to the editor of *NEJM* criticizing the Merck study. One letter cautioned that the patients receiving AmBisome "may have received suboptimal doses."[7] Another letter noted with some surprise that the response rate of the AmBisome-treated group was the "lowest ever reported in the treatment of *aspergillus*" with a drug related to amphotericin.[8] Commenting on both the AmBisome and Candida studies, Dr. Curt Furberg, a professor at Wake Forest University, remarked: "When you set up studies with controversial comparisons, you risk misleading everybody—regulatory agencies, physicians and patients."[9]

A full-time NIH employee, Dr. Walsh maintained throughout that he had not served as a consultant to Fujisawa or Merck during the time that he was engaged in testing their drugs and helping them secure FDA approval, but a Fujisawa representative recalled differently.[10] Ultimately, a confidential NIH review of Dr. Walsh's outside activities conducted in 2006 concluded that he had engaged in "serious misconduct" by failing to obtain agency authorization for dozens of private arrangements with drug companies and failing to report over $100,000 in outside income from those companies. Worse, he failed to comply with NIH regulations because he knew that if he sought approval, he would not receive it. In circumventing NIH regulations, Walsh "fail[ed] to acknowledge that the reason for" the NIH conflict of interest rules was "to prevent the integrity of the agency and its science from being called into question."[11] Despite these powerful accusations, NIH was powerless to do anything about Dr. Walsh's transgressions because he was a commissioned officer in the Public Health Service and could only be punished by the Commissioned Corps of that entity. The NIH therefore referred the matter to the Commissioned Corps for further action.[12] ❒

The AmBisome and Candida tests should make it clear that scientific research is much more malleable than elementary science textbooks imply. The critical judgments embedded in designing the research question, establishing the methods, collecting the data, interpreting the data, and communicating the results are an inescapable aspect of any scientific research project.[13] Consequently, multiple layers of microbiases can become so intimately woven into a study that they are difficult to identify, even with vigorous scrutiny by sophisticated peers. Researchers who are so inclined can carefully calculate methodological decisions so that they lead to a preferred research result. They can collect data and analyze them under circumstances that diverge substantially from the open-minded inquiry that typifies rigorous research practice. Scientists and their sponsors can obscure significant financial ties or related conflicts of interest so that readers are not put on notice of potential sources of bias.[14]

The research that yields policy-relevant science is particularly susceptible to ends-oriented manipulation. Applied research often begins with questions that outsiders find important, making it more difficult to excise outside influences from the research process.[15] The absence of strong theoretical grounding or accepted methods for much of the research in environmental and health sciences also leaves considerable elbow room for devious manipulations that can pass as credible research.[16] Because research financed by an entity with an economic stake in a regulatory decision or lawsuit is unlikely to be replicated by independent scientists, the misleading implications of the original work may persist for a long time.

The extensive documentation of scientific manipulations in the following pages should not, however, be read to imply that all policy-relevant science is biased or that high-quality policy-oriented research is impossible. On the contrary, a great deal of excellent research currently informs regulatory policy and tort litigation. Independent scientists can and regularly do conduct policy-relevant research in a disinterested way, without input from affected parties or financial inducements that cause them to tilt or skew the research toward a particular end.[17] Furthermore, scientists can and do carry out sponsored research in accordance with the norms and procedures of science without being influenced by the sponsors' economic interests. Only when the stakes are very high and parties become desperate does the temptation to shape science sometimes become irresistible.

The Techniques for Shaping Research

Once a sponsor decides to shape science, it is apparently not that difficult to locate scientists who are willing to violate the scientific norm of disinterestedness by knowingly participating in such outcome-determined research. In a 2005 survey of thirty-two hundred U.S. scientists, 15 percent of the respondents admitted that they had changed the design, methodology, or results of a study under pressure from a funding source. Moreover, an increasing share of the clinical trials that support FDA drug approvals are now conducted by corporate CROs that in turn hire doctors and scientists to do the work.[18]

Sponsor control over these scientists is sometimes informal and tacitly asserted through periodic opportunities to "review" the progress of the research, but it can also be formally established in a research grant or contract if the researcher and his or her organization are willing to sign on the dotted line. In either case, the sponsor may enjoy a great deal of influence over how the research proceeds. And when the private sponsor of a multisite clinical drug trial designs the research plan, specifies the protocols, has the final say over the reported interpretations of the results, and gets to decide whether to submit the study to a peer-reviewed scientific journal or to send it to the FDA in a plain brown envelope, the potential for bias is enormous.[19] A simple disclosure that an affected party provided support for a study does not begin to alert peers to the extent to which the sponsor may have influenced the conduct and outcome of that research.

In most research settings that involve testing for health hazards, virtually every stage of a study is vulnerable to ends-oriented shaping by a determined sponsor. Our tour of shaping techniques therefore follows the natural progression of research, beginning with study design and then moving to data collection and analysis, interpretation, and publication.

Manipulating Study Design

Framing a research question and structuring a study to resolve that question are complex tasks that require both scientific expertise and considerable discretion. As a check on this discretion, the scientific community commits itself to the norm of "disinterestedness."[20] Conducting research in an open and honest way helps prevent embedded biases from skewing the numerous discretionary judgments that are an inevitable part of research design. Since limited scientific resources are put to better use when they support scientists' efforts to build on the objective work of their peers, instead of scrutinizing and replicating previously conducted studies, the disinterestedness norm also promotes the overall efficiency of the scientific enterprise.[21] When the potential for bias in research cannot be completely purged from a study, as is often the case for policy-relevant research, scientists at the very least expect the researchers to identify sources of bias by fully disclosing any potentially significant conflicts of interest.[22] Presumptively disclosable conflicts include external influences such as employment, financial gain, and direct constraints on experimentation imposed by contracts for research.[23] Sponsor attempts to shape research always involve such conflicts of interest, and they are often not disclosed. Confirming the scientific community's worst fears, the resulting badly conflicted research usually yields results that are either misleading or flatly unreliable. The FDA's attempt to digest olestra research provides a good introduction to the problem.

APPROVING OLESTRA

Manufacturers of food additives must demonstrate to the FDA with scientific studies that their products are "safe" for use in food.[24] In 1971, the Procter and Gamble Corporation filed a food additive petition with the FDA for the fat substitute olestra. At the time, scientists knew that while olestra could reduce fat intake, it also inhibited absorption of some vitamins and other nutrients. The company's petition contained 150 human and animal studies and 150,000 pages of data from studies related to the effects of olestra on absorption and excretion of drugs and related topics, virtually all of which had been conducted by the company and its contractors. The studies invariably found olestra to be free of any health risks. Many of them, however, were designed to last only a few weeks or months, not nearly long enough to reveal any long-term effects on digestive functions or depletion of essential vitamins.

In claiming that the research designs for many of the olestra studies were "years behind" those employed in other laboratories, critics contended that this olestra research was plagued by small sample size, short duration, and several other methodological weaknesses. A subsequent review of published articles on olestra found "a strong association between authors' published opinions about the safety and efficacy of olestra and their financial relationships with the food and beverage industry."[25] Fully 83 percent of the articles acknowledging the support of olestra's manufacturer were supportive of olestra, and none were critical. Despite the conclusion of a study undertaken by a Procter and Gamble competitor (and likewise somewhat suspect) that high doses of olestra caused gastrointestinal problems and six thousand anecdotal complaints of similar problems, the FDA decided to allow its sale with an accompanying warning. ❐

When a preferred outcome drives a study, the researcher can tweak microfeatures of the design—either consciously or subconsciously—to point the study toward that outcome. Unless a balanced group of scientist-peers carefully scrutinizes the research—a fate that at least some policy-relevant research escapes—these biases can penetrate the study design so pervasively, yet so invisibly, that they may escape the attention of agency regulators, untrained judges, and the other gatekeepers who are supposed to be responsible for ensuring the reliability of the research.

Scientists conducting sponsored epidemiological research have employed a veritable toolkit of bad practices in designing studies to guarantee favorable outcomes for their sponsors.[26] An epidemiologist can hedge against the likelihood of finding an association between human exposure to a potential toxin and disease by simply doing a poor job of separating the categories of "exposed" and "unexposed" subjects.[27] For example, the researcher can count all "employees" at a manufacturing plant (including, for example, lawyers who work in offices not on site) as "exposed" and "nonemployees" (including contractors who work in high-exposure areas on a daily basis) as "unexposed."[28]

Biased designs are also evident in some research prepared for litigation. For example, a scientist hired by plaintiffs' attorneys hoping to prove that a fire at a manufacturing facility had caused its neighbors emotional harm employed methods for enlisting participants in his study that badly biased the results.[29] The content of a flyer designed to enroll a random sample of subjects living near the fire read: "If you were exposed to the Smoke and Fumes from the Fire and Toxic Release****Have Concerns about your Health and Safety****We would like to interview you."[30] Participants in the study were also notified in a variety of subtle ways that a plaintiffs' firm was sponsoring the study.[31] In reality, the study examined not a random cross-section of residents living near the fire, as its conclusions implied, but a very biased subsample of the most affected who were also potentially the most eager to receive compensation.

Shaping science through study design is by no means limited to after-the-fact epidemiological studies and surveys. Pharmaceutical manufacturers develop the protocols and fund the data collection for the prospective clinical trials they use in seeking advance FDA approval of their products.[32] Even in this heavily regulated area, biased study protocols are not at all uncommon.[33] For example, drug manufacturers have commissioned clinical studies that enroll a disproportionate number of young subjects in trials of drugs intended primarily for older people, thereby reducing the likely incidence of detected adverse side effects and enhancing the product's marketability.[34] Likewise, a sponsor can design efficacy studies to compare a high dose of its drug with a less than efficacious dose of its competitor to enhance the apparent efficacy of its drug, a technique Dr. Walsh was accused of utilizing in his fungus research.[35] Or a sponsor can design the study to test its drug in people who are healthier than the subjects taking the competing drug.[36] The short duration of the trials can also enable manufacturers to avoid undertaking a meaningful assessment of a drug's long-term toxicity.[37]

New business arrangements in the lucrative field of pharmaceutical testing have made it much easier for sponsors to exert a heavy, self-serving hand in research design. As some universities during the 1990s began to restrict the degree of control that they permitted outside sponsors to exert over research, the pharmaceutical industry quietly shifted many clinical research studies away from university-affiliated medical schools and into a much less visible private sector system in which individual physicians conduct the vast majority of clinical trials under contract with drug companies or CROs.[38] A CRO typically establishes and supervises a network of individual physicians in multiple locations who administer the drug to their patients pursuant to protocols designed by the sponsoring drug company.[39] Since the physicians are not technically trained researchers, they "simply do what they are told—or risk losing their lucrative deals."[40] In 2001 doctors received an average $7,000 per patient, with occasional bonuses for rapid enrollment.[41] For enthusiastic physicians, this can amount to as much as a million dollars a year.[42]

The final analysis of the combined data in these "multisite" clinical trials is performed by statisticians who are separately hired by the manufacturer or the CRO. Individual researchers are generally not entitled to receive information from their counterparts at other locations, and they are not involved in data analysis.[43] Confidentiality clauses in individual contracts with physician researchers further ensure that they do not attempt to analyze and publish even the limited data they produce. The sponsoring drug companies believe that this degree of control is necessary to prevent unqualified individual researchers from publishing information based on improper statistical analysis, but it also gives the companies exclusive control over how the data are analyzed and presented to the scientific community and the public.[44]

According to a former editor of *NEJM,* these new arrangements have allowed pharmaceutical companies to "exercise a level of control over the way research is done that was unheard of" in prior decades, "and the aim was clearly to load the dice to make sure their drugs looked good."[45] This alone may account for the surprisingly rapid movement of clinical trials out of academic settings and into the much less visible world of private CROs. From 1990 to 2001, the percentage of industry-sponsored research conducted in academic institutions dropped from about 80 percent to less than 40 percent, and the trend continued through 2004, when it was a mere 26 percent.[46] Pharmaceutical testing has become so lucrative that several major CROs are now competing heavily for the business, creating an atmosphere that is not necessarily conducive to the disinterestedness norm.[47] At the same time, at least some of the university-affiliated medical centers that continue to participate in clinical trials have become dependent on the income the studies yield, and they have offered less resistance to drug companies' insistence on controlling the details of research that is conducted in the medical schools.[48]

Modern biotechnology and medical research have also given rise to the phenomenon of the small "start-up" company financed by entrepreneurs and managed by prominent researchers and set up to bring laboratory discoveries to the marketplace.[49] The companies are often located near the universities that employ the researchers in their "day jobs," and those universities, like the researchers themselves, may even have an equity interest in the companies. Researchers who are also major shareholders in the companies sponsoring their research may have an especially strong incentive to design research in biased ways. In the late 1980s, a researcher at the Harvard University–affiliated Massachusetts Eye and Ear Infirmary conducted clinical trials on a vitamin A dry-eye treatment manufactured by a small start-up company in which he held a substantial equity interest. He enrolled more than the approved number of patients, modified the doses the patients received in the middle of the ongoing investigation, and minimized the negative aspects of ongoing reports in order to maintain the appearance that the treatment was effective in reducing the patients' symptoms.[50] The final data, which were never published, failed to demonstrate that the treatment was any more effective than a placebo, but not before the researcher and his relatives earned at least $1 million from sales of the company's stock.[51]

Although replication of suspect studies can offer a highly effective check against bias, as we learned in Chapter 3, the reality is that replication rarely happens spontaneously in mundane areas of applied research. Even worse, in the low-visibility world of policy-relevant science, where research is sometimes carried out at the margins of scientific respectability, even the salutary process of replication can be abused by sponsors who would prefer to see a different result. For example, when a formaldehyde industry trade association learned that the Chemical Industry Institute of Technology (CIIT) was about to publish a study concluding that formaldehyde caused nasal cancer in rats, it arranged for a New Jersey laboratory to conduct a "confirmatory" study on the same strain of rats. The "confirmatory" study, however, exposed only one-third the number of rats and for only half as long as the CIIT study, differences that were obviously not designed to maximize the probability of detecting a cancerous effect.[52] The trade association then announced to the Consumer Product Safety Commission at a public hearing that "[a] new study indicates there should be no chronic health effect from exposure to the level of formaldehyde normally encountered in the home."[53]

Manipulating Research Data and Statistical Methods

The second stage of a research study—data collection and analysis—also entails a significant degree of scientific judgment[54] and provides still more hidden opportunities for strategic manipulation of research results, ranging from outright fraudulent reporting to more subtle fudging of accepted analytical methods to achieve outcome-oriented results.

Since data collection is merely the process of reporting observations to facilitate subsequent analysis, any shaping that occurs at that early stage typically equates to outright fraud. In the mid-1970s, the FDA appointed a special task force to investigate the integrity of the data the Searle Corporation had submitted to support new drug applications for two heavily used drugs. After studying the data extensively and performing site visits of the company's laboratories, the task force concluded that the company had "made a number of deliberate decisions which seemingly were calculated to minimize the chances of discovering toxicity and/or to allay FDA concern."[55] In another 1970s investigation, the FDA fortuitously learned from a former employee of the manufacturer of the drug MER-29 that upper-level company officials had ordered the employee to change data in animal studies that were submitted to the FDA in support of the new drug application.[56] Three of the company's employees pleaded *nolo cotendere* to charges that they had forged and manipulated the animal studies to minimize the apparent risks posed by MER-29.[57] As we learned in Chapter 3, the EPA during the same time period discovered that a testing contractor, Industrial Bio-Test Laboratories, had perpetrated systematic fraud in its routine pesticide testing operations.[58]

As the markets for promising pharmaceuticals became superheated in the 1990s, history repeated itself. The FDA's investigators discovered that the prodigious output of small-scale clinical studies that the Southern California Research Institute (SCRI) turned out for pharmaceutical companies came from a single clinic that was owned and operated by a former family-practice physician who also owned and operated the SCRI. The *New York Times* reported that the SCRI was "conducting research fraud of audacious proportions, cutting corners and inventing data to keep the money flowing from the drug industry."[59] The doctor and several accomplices eventually pled guilty to criminal fraud, but by that time the underlying scientific basis for several important drug approvals had been compromised.[60] A professor of psychiatry at the Medical College of Georgia also spent time in jail for running a similar business in the 1990s in which he served as "key investigator" in clinical trials of many of the antipsychotic and antidepressant drugs that the FDA had approved for use in part on the basis of his research.[61]

Although reported instances of outright data fraud are, fortunately, fairly rare,[62] scientists need not stoop to fraud to achieve sponsor-friendly results. In many cases, they can subtly tailor the analysis of honestly reported results in various ways, including limiting the analysis to only a supportive subset of the relevant data or simply tossing out unwanted observations as "outliers." The manufacturer of the panic disorder drug Xanax secured rapid FDA approval and huge financial success on the basis of results taken at the midpoint of an eight-week cross-national clinical trial. The abstract of the published study, however, failed to mention that the midpoint efficacy had diminished by the end of the trial and that the rate of panic attacks in the Xanax-treated group returned to pretest levels within two weeks after that.[63] Similarly, after the initial results of clinical tests on the hepatitis B drug lamivudine proved quite promising, the principal investigator discovered that some patients taking the drug for more than one year became infected by a deadly mutant strain of the hepatitis B virus that appeared to cause liver failure. The company, however, continued to publish under her name an "upbeat" scientific abstract that did not mention her concerns, and the FDA approved the drug without considering the risk posed by the mutant virus.[64]

For many kinds of modern scientific research, the raw data must undergo further statistical analysis before scientists can draw supportable conclusions. Because statistics is itself a specialty, research teams often include a statistician to perform the necessary statistical analyses on the raw data, and the study protocols usually specify the type of statistical analysis that will be applied to the data in advance. By now it should come as no surprise to learn that these statistical analyses may ultimately be manipulated to achieve predetermined results.

THE STANDARD INDUSTRY PRACTICE IS TO LIMIT ACCESS

In the summer of 2002, Dr. Aubrey Blumsohn, a senior lecturer at Sheffield University, entered into an agreement with the Procter and Gamble Corporation to head up a major research project to study the relationship between bone fractures in older women and reduction in bone turnover caused by the corporation's best-selling osteoporosis drug, Actonel.[65] The company was worried by recent studies showing that Actonel-medicated suppression of bone turnover peaked at about 30–40 percent, while its primary competitor, Fosamax, exerted a stronger effect. The corporation very much hoped that a study on the relationship between fractures and turnover would show that any additional decreases in bone turnover beyond 30–40 percent would not significantly reduce bone fractures. Under the £180,000 contract, Dr. Blumsohn's laboratory gathered data on bone samples collected over a three-year period from thousands of women who took Actonel, noting the degree of bone turnover in each of them and whether they had suffered bone fractures. The raw data were then shipped to Procter and Gamble in the United States, where company statisticians broke the code that had allowed Dr. Blumsohn's researchers to be "blind" with respect to the data and performed statistical analyses on the data.

When Dr. Blumsohn, with the assistance of a colleague who was on the corporation's scientific advisory board, asked to see the statistical analyses, the company declined on the ground that it did not want to take the time to train Dr. Blumsohn or one of his underlings in how to perform the relevant statistics. Several months later, Dr. Blumsohn learned that two company employees had submitted a paper containing the company's statistical analysis of the data for presentation at a major medical research conference in Minneapolis. At about the same time he received a communication from the company informing him that it had hired an "external medical writer" from New York to write up the results for publication in a scientific journal under Dr. Blumsohn's name. The writer was, in the company's words, "very familiar with both the [Actonel] data and our key message, in addition to being well clued up on competitor and general osteoporosis publications."[66]

After the company finally allowed him to see the relevant graphs on a company employee's computer, Dr. Blumsohn immediately recognized two things. First, the analysis supported just the sort of plateau effect in bone fractures that the company had wanted to see. Second, about 40 percent of the data had apparently been left off the bottom of the graph, and Dr. Blumsohn suspected that if they had been included, the plateau effect would have disappeared. The company ultimately decided to include the data it had excluded from the graph, but only after Dr. Blumsohn's nine month struggle to access the data.[67] In a statement for the press, the corporation strongly denied that it had in any way skewed or manipulated the data. It explained that "[f]or post hoc exploratory analyses, it is standard industry practice to limit access to the raw data by external researchers," even though they may appear as lead authors on the resulting scientific publications.[68] ❐

Whether or not the Procter and Gamble statisticians manipulated the statistical analysis (or the presentation thereof) to demonstrate the desired plateau effect, it is clear that statistical manipulation is a real possibility in retrospective analyses of existing data, particularly when the primary investigator is excluded from participating in the company's statistical analysis and from accessing the underlying data set. Thus, the International Committee of Medical Journal Editors in 2001 revised its Uniform Requirements for Manuscripts Submitted to Biomedical Journals to require that "researchers should not enter into agreements that interfere with their access to the data and their ability to analyze it independently, to prepare manuscripts, and to publish them," and to empower journal editors to request that authors of sponsored studies sign a statement such as: "I had full access to all of the data in this study and I take complete responsibility for the integrity of the data and the accuracy of the data analysis."[69] Sheffield University is by no means the only academic institution that does not insist on independent access to all data in its contracts with research sponsors. A survey of officials at 108 medical schools reported in 2002 that although all of the institutions insisted on allowing their scientists to access the data that they collected, only 1 percent of the institutions required access to all of the data taken in multisite clinical trials.[70]

The rich regulatory history of the dietary supplement ephedra offers a particularly blatant example of how a determined sponsor can statistically manipulate data. A study conducted by the University of Michigan researcher W. Jeffrey Armstrong initially concluded that there was no statistically significant difference in weight loss between subjects taking ephedra and those taking a placebo. A consultant statistician for Cytodyne Corporation, the manufacturer of ephedra, however, agreed to "try to carefully nudge [Armstrong's] interpretation/writing in Cytodyne's favor" by presenting Armstrong's data on a percentage, rather than absolute, basis.[71] The consultant explained that "using percentages can be misleading (especially when the absolute changes are small) but check this out. Damn I am good."[72] Although the absolute differences were still too small to be statistically significant, the report as amended by the consultant suggested that ephedra caused a huge percentage loss in weight. When Armstrong objected, Cytodyne representatives issued a thinly veiled threat to sue him if he refused to withdraw his version of the results and allow it to be replaced with the consultant's version.[73]

Epidemiologists also have a great deal of discretion to exercise scientific judgment at the data analysis stage of retrospective studies, and that discretion can be abused.[74] For example, in an effort to head off an OSHA proposal to regulate workplace exposure to chromium, a human carcinogen, a chrome industry coalition commissioned a study of two U.S. plants and two German plants that used a manufacturing process that resulted in lower worker exposure to that element. The original protocols for the study called for combining the data from all four plants, but when the combined data showed a statistically significant elevation in lung cancer, the coalition declined to publish the study. Instead, the scientists who prepared the original study separated the data into U.S. and German components and published two papers, one finding no statistically significant increase in lung cancer and the other finding an increase only at the highest exposure level.[75]

Interpreting Data

The critical stage at which scientists interpret the statistical analysis of collected data, place it in proper perspective, and give it meaning also offers them considerable leeway to shape science to predetermined ends. The kinds of shaping that can occur at the interpretation stage range from tried-and-true ruses like reporting study outcomes in ways that emphasize the good news and omit the bad to much more sophisticated interpretative exercises employing modern computer programs that churn data until they produce a welcome result.

Much of the documentation of early attempts to shape science through data interpretation has come to light in recent litigation over products like asbestos, lead, and vinyl chloride to which workers were exposed in the early to mid–twentieth century. In a 1929 industrial hygiene survey commissioned by several asbestos manufacturers and users, for example, the Metropolitan Life Insurance Company found that the lung disease asbestosis was prevalent throughout the workforce but did not predispose workers to tuberculosis. When Metropolitan's medical director decided to publish the results three years later, the Johns Manville

Company forwarded changes to the manuscript suggested by a New York law firm, along with a cover letter urging him to ensure that "all of the favorable aspects of the survey be included and that none of the unfavorable be unintentionally pictured in darker tones than the circumstances justify." The published study contained all of the suggested changes.[76]

A modern example of data interpretation to downplay risks is the interpretation Merck Corporation scientists provided for the results of a company-sponsored postmarketing clinical trial in which the statistical analysis showed that subjects taking Vioxx for nine months had a five times greater risk of heart attack than did those taking naproxen, another commonly used painkiller.[77] Since the tests were not run against a control population of persons not taking any painkiller, scientists working for the manufacturer could interpret these results in one of two ways. Either taking Vioxx increased the risk of heart attacks by around 400 percent, or taking naproxen reduced the risk of heart attack by about 80 percent. Although the latter interpretation seemed highly implausible, the company's scientists adopted it, and Vioxx remained on the market for another five years until another postmarketing trial comparing Vioxx with a placebo showed that Vioxx caused a twofold increase in the risk of heart attack. According to one FDA scientist, Vioxx caused between 80,000 and 139,000 heart attacks during those five years.

A final technique for skewing interpretations of published studies, frequently employed in reporting the results of clinical trials for drugs and medical devices to improve their marketability, is called outcome reporting bias. Scientists employing this strategy publish only some of the study outcomes specified in the study's protocols. The net effect of this form of reporting bias is to allow "data dredging" and selective reporting of favorable results.[78] Several empirical studies that have compared reported outcomes in published studies with the actual protocols for such clinical drug trials have found strong evidence for outcome reporting bias in several databases of published studies.[79] Since medical journals typically do not demand that authors submit protocols along with their manuscripts, there is no way for peer reviewers to police against this very subtle form of interpretative bias.

Obscuring Provenance in Publications

Most scientists would agree that when a sponsor contractually controls research or otherwise acts as a significant collaborator, the norms of science require that any published version of the study disclose at least the fact of sponsorship, irrespective of any journal conflict of interest disclosure requirements.[80] Nevertheless, scientists serving as paid consultants for companies or plaintiffs' lawyers have periodically published articles in peer-reviewed journals without disclosing this critical fact.[81] Over the last decade, in response to this trend, many, but not all, journals have required such disclosures as a condition to publication.[82] These requirements are sometimes ignored, however, and violations are very difficult for journal editors to detect and punish (matters we explore in more detail in Chapter 10).[83] Moreover, private sector sponsors of policy-relevant research have employed a number of sophisticated techniques to circumvent these requirements or otherwise obscure the true origins of shaped studies.

The most insidious technique for obscuring provenance is the practice commonly employed in the pharmaceutical industry of hiring ghostwriting companies that in turn hire bright young science writers to turn data and analyses from company-sponsored clinical trials into articles suitable for publication in medical journals under the signatures of prominent researchers. The listed authors frequently have only marginal involvement in study design and execution, and they do not collect or analyze the data. Instead, they simply provide their prestigious imprimatur, often without even acknowledging the assistance of the unnamed ghostwriter.[84] Targeting the leading journals in the field, the company typically pays the supposed "author" a fee ranging from $3,000 to $5,000 for reviewing the ghostwritten manuscript, suggesting revisions, and signing the submission letter.[85] The companies offer the same ghostwriting service for letters to the editor (typically critiquing other articles), symposium speeches, opinion pieces, and review articles. The ghostwriters are even careful to adopt the writing style of the "author."[86] Yet the "author" may not even have access to the underlying data on which the article is based.[87]

One prominent ghostwriting company advertises that it will prepare "studies, review articles, abstracts, journal supplements, product monographs, expert commentaries, and textbook chapters."[88] It also conducts meta-analyses and organizes journal supplements, satellite symposia, consensus conferences, and advisory boards.[89] As the advertisement suggests, the companies compete for business. According to a prominent scientist who is a frequent recipient of such overtures, the big pharmaceutical companies "can, for example, put out a contract for 25 articles with big-name authors to be placed in good journals, and wait for offers" from the ghostwriting companies.[90] Another prominent scientist estimated in 2006 that 90 percent of industry-sponsored studies that list a prominent academic as lead author are in fact ghostwritten.[91]

Ghostwriting is not a new phenomenon. One of the clinical studies of the infamous drug thalidomide that was published in a prestigious journal and used to support the manufacturer's new drug application to the FDA was ghostwritten by the company's medical director.[92] The incidence of ghostwritten articles in the scientific literature, however, increased dramatically during the last decade of the twentieth century. A review of the articles published on the antidepressant Zoloft in the scientific literature in 1998 revealed that fifty-five had been coordinated by a single ghostwriting company and only forty-one were written by authors it had not coordinated. Moreover, the "authors" publishing the ghostwritten articles had published on average three times more articles than the other authors, and the citation rate for the coordinated papers was three times higher than the citation rate for the uncoordinated papers.[93] Ghostwriting thus offers an efficient vehicle for flooding the scientific literature with a river of articles reflecting a single sponsor's assessment of the virtues and risks of its own product.

Because the ghostwritten articles are typically submitted without any indication of their true origin and authorship, the propensity of this practice to mislead readers of medical journals would render it ethically questionable. It is, after all, "entirely inconsistent with the norms of science (and indeed the general norms of academia) for one person to pretend to be the author of an article written by someone else."[94] Sheldon Krimsky concludes that ghostwriting of the sort pharmaceutical companies routinely sponsor "would violate the minimal standards of plagiarism we demand of our students."[95] Journals that detect ghostwriting during the editorial or peer review process can refuse to publish the articles, but it is very difficult to detect a ghostwritten article when it is done well.[96]

The practice is especially objectionable from the perspective of sound public policy-making because of its propensity to obscure serious sources of bias. In effect, the entity that collects and analyzes the data and makes all of the critical interpretational calls is not always the well-respected author but a company that has a great deal to gain from favorable results. The "author" is, of course, free to decline to sign a ghostwritten article,[97] but it will probably just appear in the same journal under someone else's name.[98] "Authors" are also free to edit ghostwritten pieces extensively, but alterations are not well received if they do not adhere to the company line.[99] If, at the end of the day, the paper no longer meets the company's needs, it can simply decline to submit the article for publication and bury it in company files.[100]

Several examples of ghostwritten articles downplaying the risks of a sponsor's products are available. The manufacturer of the weight loss drug fen-phen commissioned ten articles exploring the drug's safety and efficacy, two of which were published in peer-reviewed medical journals, and edited drafts of the articles to belittle its side effects.[101] After the "author" of a ghostwritten article on the antidepressant drug Efexor inserted a reference to the possibility that its class of drugs might cause suicide, the article went through further drafts and was submitted for a symposium issue of the *Journal of Psychiatry and Neuroscience* under his name before he was afforded another opportunity to review the manuscript. One of the changes eliminated the reference to suicidality and included instead a message at the end stating that evidence suggested that this particular selective serotonin reuptake inhibitor (SSRI) drug was most effective at alleviating depression. Although this was "at odds with everything" the "author" knew about the drug, the symposium editor had signed off on the article without even consulting him.[102]

An equally questionable technique for obscuring provenance is the practice of "redundant publication" of "several different articles of limited data from a single study to give the impression of a growing body of evidence."[103] Redundant publication of the same data may be unavoidable in the case of multiyear studies, but that is not the case when sponsors publish the same study in different journals under different author's names with no crossreferences, making it appear that the scientific support for the sponsor's product is based on several independent studies, rather than repeated publication of the same results.[104] Redundant publication can further multiply the impact of a single study in review articles that do not carefully examine the source of the data in individual published studies, and it can result in the double counting of the data in large meta-analyses of multiple studies.[105] Most journals regard multiple publication of the same data as wholly unacceptable, except in extraordinary circumstances in which the fact of multiple publication is clearly disclosed.[106] None of these objections, however, has succeeded in putting an end to the practice.

Indirect Inducements to Shape Science

Even when they do not attempt to control research outcomes directly through contractual relationships with researchers, well-endowed sponsors can indirectly shape science by carefully selecting the projects and the scientists they support to maximize the likelihood of favorable research results. In the typical situation in which only limited funds are available from government research agencies and other independent sources to support policy-relevant research, the ability of private sector sponsors to "cherry-pick" the projects and the scientists they support gives them great influence over the resulting body of available scientific knowledge.

Cherry-Picking Research Projects

In some cases, sponsors shape science indirectly through their willingness to provide generous support to a narrow set of favored projects. Scientists, whose careers necessarily depend on their ability to attract continuing research funding, can sometimes be influenced by the ready availability of private sector money to allow their interests to migrate in the direction of a particular sponsor's pet projects.[107] Private sector support may not carry the same weight at tenure time as a grant from the NSF, but it keeps the lights burning during fallow times and can keep tenured scientists well supplied with graduate assistants in the event that federal money does not come through.

Taking full advantage of the opportunity that sponsored research offers to call the piper's tune, manufacturers occasionally pool their resources in "research centers" that are structured to fund a number of individual research projects. The centers typically resemble government research agencies or private foundations, and they often employ the traditional model of soliciting and approving research proposals from interested researchers in academia or the private sector. While their structure lends the patina of neutrality, closer investigation into the actual operations of some of these centers exposes significant self-serving biases in their funding decisions. A week after the American Cancer Society and the American College of Surgeons came out strongly against smoking in late 1953, for example, six tobacco companies followed the advice of their public relations consultant and created the Tobacco Industry Research Committee (TIRC), later given the less revealing name the Council for Tobacco Research, to finance research that would undermine and challenge claims that smoking posed human health risks.[108] According to a 1976 internal company memorandum, the millions of dollars that the Council poured into health-related tobacco research "allowed the industry to take a respectable stand along the following lines: 'after millions of dollars and over twenty years of research, the question about smoking and health is still open.'"[109]

Following this model, the tobacco industry established the Center for Indoor Air Research (CIAR) in 1988 to address a rapidly growing scientific literature on the adverse health effects of environmental tobacco smoke (ETS) on nonsmokers. A comprehensive study of projects funded by CIAR found that most of the projects funded through the peer review process CIAR established were devoted to topics other than ETS.[110] Another study concluded that by "[p]ositioning ETS as just one component of the much more complex problem of indoor air quality," the tobacco industry could flood the scientific literature with studies suggesting that any health problems were attributable to other chemicals in indoor air.[111] At the same time that it allowed funded scientists to avoid the stigma of working for the tobacco industry,[112] the CIAR "allowed industry-funded scientists to produce seemingly independent results aimed at contradicting ETS findings . . . while keeping such research under industry control."[113] In the unvarnished assessment of a tobacco industry lawyer, the CIAR was "a credible and effective vehicle for conducting research that is needed to buttress the industry's position" that ETS is not a cause of lung cancer.[114]

The Cellular Telecommunications Industry Association (CTIA) followed suit in 1993, announcing that the industry would sponsor a $25 million program called Wireless Technology Research (WTR) to follow up on claims that cell phones caused brain cancer in frequent users.[115] The express goal of WTR was to "re-validate" the studies showing that cell phones were safe, but its covert purpose was to find flaws with studies that were beginning to indicate that cell phones did pose health risks. After a WTR-funded researcher made a discovery that cast doubt on much of the research the industry had previously relied on, the CTIA stopped funding WTR. When asked how much money his company was willing to devote to follow-up epidemiological studies, one industry executive replied: "Zero. Zero surveillance. We're going to do enough research so that we can prove safety—and then we can stop doing research."[116]

Industries have also participated in collaborative research ventures with government agencies that sometimes, but not always, allow industry participants to influence funding decisions. A good example is the Mickey Leland National Urban Air Toxics Research Center at the University of Texas School of Public Health in Houston, Texas, which was created by Congress in 1990 to "sponsor and gather scientific information on the human health effects caused by exposure to air toxics."[117] The federal government provides the bulk of the Leland Center's funding, and the remainder comes "primarily from corporations in the petroleum and chemical industries."[118] Unlike the more independent-minded board of directors of the highly regarded jointly funded Health Effects Institute, which consists of distinguished academics most of whom have no ties to the auto industry,[119] the eight-member board of the Leland Center, which is appointed by Congress and the president, includes only three academics, with the remainder coming from industry or law firms representing industry.[120] The Center's scientific advisory panel includes representatives from academia, industry, and government, but not public interest groups.[121] Although the Center plays an important role in determining which scientific studies on toxic air pollutants receive the very limited federal funding available for that research, it decided early on not to fund health effects research, focusing instead on analytical devices to measure concentrations of toxins in ambient air.[122] According to a Houston environmental attorney, the Center "has never done what . . . it was supposed to do," because it "seems to have always been inordinately concerned with industrial points of view."[123]

Cherry-Picking Scientists

A more hit-and-miss but still effective technique for sponsors bent on shaping research is to use generous financial incentives or inducements to endear an individual researcher to its position. Just as regulated industries were beginning to feel the full effects of the modern environmental statutes of the 1970s, two highly regarded students of the regulatory process offered the following advice:

> Regulatory policy is increasingly made with the participation of experts, especially academics. A regulated firm or industry should be prepared whenever possible to co-opt these experts. This is most effectively done by identifying the leading experts in each relevant field and hiring them as consultants or advisors, or giving them research grants and the like. This activity requires a modicum of finesse; it must not be too blatant, for the experts themselves must not recognize that they have lost their objectivity and freedom of action.[124]

The industries apparently took the lesson to heart. A longtime observer of the drug research industry concluded in 2004 that "[i]t is hard to believe that close and remunerative personal ties with drug companies do not add to the strong pro-industry bias in medical research and education."[125]

The monetary support used to befriend a scientist need not be limited to a scientific support grant. A survey of researchers at the University of California at San Francisco undertaken in the late 1990s revealed that 7.1 percent of university researchers had personal financial ties to commercial sponsors of their research. Of these researchers, a large percentage: served as paid consultants to companies whose products or activities might be affected by their research (33 percent); served in paid positions on industry advisory boards (32 percent); received fees for speaking engagements (34 percent); or received research-related gifts, such as biomaterials and discretionary funds (43 percent).[126] In some cases, these inducements may be great enough to lead less scrupulous researchers to alter or fudge their data. In the fen-phen and silicosis litigations, for example, financial incentives were sufficient for some physicians to produce badly skewed diagnostic readings. An article published in *Academic Radiology* compared plaintiff radiology experts—who were financially retained—with disinterested readers and found a diagnosis rate of silicosis as high as 95 percent for the paid consultants but only 4 percent for the disinterested ones.[127]

Perhaps the most notorious attempt to befriend scientists was the fund named Special Projects that the tobacco industry established after the 1964 Surgeon General's report on smoking and health to support a network of consultants who could be depended on to take the industry position in scientific publications, congressional and administrative presentations, and expert testimony in court.[128] The Special Projects fund was administered over the years by a "Committee of Counsel," which was composed of the general counsels of all of the major tobacco companies and attorneys from four major law firms so as to preserve the ability of both the funder and the recipients to claim the attorney-client privilege if pressed about the funding.[129] Its mission was "to seed the universities with research money and grow an unassailable crop of advocates."[130]

In addition to attracting scientists to their favored projects (and outcomes), sponsors can also use their privileged financial position to increase the output of scientists who are already doing work that advances their interests. In the same way they cherry-pick research projects, sponsors can cherry-pick consultants they believe are already on board with positions they advocate. This sometimes means that an honest scientist who takes an outlier position finds himself a popular grant recipient from sponsors whose business or ideological needs serendipitously dovetail with his research. Dr. Patrick Michaels, a researcher at the University of Virginia and the state's climatologist, is a longtime skeptic of climate change. His scientific views became lucrative when industries resistant to added regulatory controls on greenhouse gasses discovered his work and found him to be a capable and sincere spokesperson.[131]

By generously supporting a small cadre of scientists whose work supports their economic interests, sponsors succeed in shaping science by increasing their productivity, potentially well beyond what it would have been without that support. The fortunate recipients of such unconditional support no longer have to invest time writing detailed grant proposals to federal funding agencies or worrying about whether the next grant will come through. Indeed, the very different selection criteria sponsors use may cause these regularly commissioned scientists to become much less rigorous in their research and in developing their research agendas precisely because they no longer have to please a skeptical group of their peers to gain funding for each new project. The Lead Industry Association's support of Dr. Joseph Aub, a prominent Harvard University scientist, in the late 1920s to conduct medical research on lead poisoning provided a model for future sponsors of policy-relevant research. Thanks to the industry's support, "Aub's research at Harvard . . . determine[d] the nation's agenda for all toxicological research on lead" for three decades.[132] While there is no evidence of any distortions or deception with respect to Aub's research, his research agenda regrettably focused only on "a narrow range of questions" that tended to be more favorable to the industry and overlooked the effects of lead on children.[133] As a result, some of the most important discoveries in lead poisoning arose only after federal agencies began to fund lead toxicity research and the output of Aub's laboratories no longer dominated the field.

To maximize the return on their scientific investment, sponsors sensibly try to hire scientists who possess great prestige and high visibility. Dr. Aub at Harvard was an excellent choice for the lead industry because of his prominence as a researcher and the prestige associated with his institution. Many scientists were dismayed to learn in 2006 that one of the academic pillars of modern epidemiology, Dr. Richard Doll, whose path-breaking work had played a prominent role in demonstrating that smoking caused lung cancer, had been for many years a paid consultant to the vinyl chloride industry, had received £15,000 plus expenses from chemical companies to write a review article on vinyl chloride, had agreed to allow two chemical company scientists to review the article before submitting it for publication, and had not disclosed any of these facts to the scientific journal that published it.[134] Doll's defenders pointed out that he contributed much of his consulting income to Green College, Oxford, which he helped to establish while he was on the Oxford faculty, but his failure to disclose these conflicts troubled many others.[135]

The best arrangement of all is one with a prominent government scientist who is in a position not only to direct government resources toward the research but also to help private sector entities obtain valuable product approvals from sister agencies. Consider the case of the relationship between a pharmaceutical company and Dr. Richard C. Eastman of the NIH.

MOONLIGHTING AT NIH

In 1994, during the heyday of cooperative research agreements between the government and industry, the NIH launched a major $150 million nationwide study of the safety and efficacy of diabetes treatments.[136] In June 1996, the Warner-Lambert Company announced that in return for the company's pledge to contribute $20.3 million to the project, NIH had agreed to include its not-yet-approved drug Rezulin in the study.[137] Unbeknownst to the public at the time, Dr. Richard C. Eastman, the NIH scientist with "overall responsibility" for the study, had become a $150-per-hour consultant to Warner-Lambert in November 1995.[138] While the nationwide study was being carried out, Dr. Eastman also served as a paid member of the Rezulin National Speakers Bureau, a group of experts who toured the country instructing doctors on the proper use of Rezulin for treating diabetes, and as a prominent participant in the National Diabetes Education Initiative, another organization partially sponsored by Warner-Lambert.[139] Against the advice of an NIH attorney but with the approval of his NIH superiors, Dr. Eastman continued to play these multiple roles after Rezulin was included in the study, but he did abstain from at least one final vote that resulted in the selection of Rezulin for the NIH study.[140] To make matters worse, his immediate superior, Dr. Jerrold M. Olefsky, was also a paid Warner-Lambert consultant during the pendency of the study and, more disturbingly, at the time that he spoke on behalf of Rezulin at a meeting of the FDA advisory committee that unanimously recommended the drug's fast-track approval.[141] He was also listed on three patents for Rezulin and was cofounder and president of a private company that received substantial funding from Warner-Lambert.[142]

The NIH nationwide study had been in progress for little more than a year when a fifty-five-year-old patient died of liver failure caused by Rezulin.[143] Although Dr. Olefsky strongly advocated leaving Rezulin in the study,[144] NIH removed it on June 4, 1998, after it had been administered to more than one million patients. The drug remained on the market for doctors to prescribe, however, and Dr. Eastman (whose role as a consultant to the industry had not yet been revealed in the media) told the press that he was "glad to see that the drug continues to be on the market and available for people to use it."[145] By the time FDA finally withdrew the drug's approval more than two years later, it had been linked to more than ninety liver failures and sixty-three deaths and/or liver transplants.[146] ❐

The public did not become aware of the full extent to which NIH scientists were moonlighting for the drug industry until 2005, when it became a major scandal for the otherwise highly regarded institution. More than 530 of the agency's 1,200 senior scientists had taken consulting fees or stock options as compensation from biomedical companies during the years 1999–2003, and an internal investigation revealed that forty-four had apparently violated NIH rules on outside consulting.[147] Most of the violations consisted of failure to request approval for outside consulting, work on government time, or failure to report consulting income; in nine cases, however, the violations were serious enough to warrant possible criminal investigation.

Under pressure from two congressional committees, the director of the NIH, Elias A. Zerhouni, in February 2005 proposed a complete ban on outside consulting and on NIH employees and their spouses holding equity interests in related biomedical companies.[148] Soon thereafter, three of the NIH scientists who had violated the rules announced that they were leaving the government.[149] When Zerhouni announced the new policy at a "town hall meeting" of several hundred NIH employees, he encountered almost universal opposition and frequent derision from the participants.[150] The final rules require the top two hundred officials in the agency to limit their holdings in drug companies to $15,000 apiece, and six thousand more employees were required to submit their holdings for review and sell them if they posed a potential for conflict of interest.[151] A dissident group, called the Assembly of Scientists, hired at reduced rates a law firm that represented the drug industry to challenge the regulations in the D.C. Circuit Court of Appeals.[152] A survey of NIH scientists conducted in late 2006 found that 80 percent of those surveyed found the restrictions to be too restrictive, but 79 percent allowed that they were still happy with their jobs.[153]

Why Scientists Help Shape Science

When scientists participate in misleading or even dishonest attempts to shape science, they not only violate the norms and procedures of science but also betray a dark side of their professional ethic that is rarely exposed to public view. They risk public embarrassment, loss of peer respect, and even professional censure if their complicity is uncovered. Why, then, do they do it?

The most important influence may also be the most obvious—the money. The remuneration from sponsors may be so generous that it provides adequate incentive for some scientists to ignore their professional obligations to the scientific community and to the general public. Scientists working as consultants to or expert witnesses for attorneys can easily earn half a million dollars annually, and some make considerably more than that.[154] Dr. Gary Huber received millions of dollars in grants from the tobacco industry's Council for Tobacco Research to shape science at the Harvard Center for Tobacco Research, and he billed millions more in private consulting fees to the Special Projects fund administered by the tobacco industry's Committee of Counsel.[155] The consulting and industry salaries of environmental scientists and toxicologists are reportedly more generous—sometimes by as much as 30 percent—than their academic and government counterparts.[156] While this salary differential by no means implies that private sector scientists are all busily engaged in shaping science, it may explain why companies who want to shape science rarely have difficulty locating scientists who are willing to help.

The easier availability of sponsored research support may also persuade many scientists to discount seemingly abstract and distant restrictions on their scientific freedom. Research funding is vital to the continued operation of university laboratories, particularly at a time when government funds for environmental and public health research are scarce. While researchers might conclude that the hidden costs of private sponsorship are excessive after sponsors exercise their right to control how their research money gets spent, hindsight is always 20/20. Just ask Dr. Betty Dong.

REINTERPRETING DR. DONG'S DATA

In 1987, the predecessor of Boots Pharmaceutical Company faced the imminent loss of market share to generic equivalents of its highly successful hypothyroidism drug, Synthroid. The company entered into a twenty-one-page, $256,000 contract with Assistant Professor Betty J. Dong of the University of California at San Francisco in which she agreed to conduct a six-month human bioequivalency test of Synthroid and its major competitors.[157] The contract provided that all information contained in the specified research protocol and all data obtained by the researchers were to be "considered confidential and [were] not to be published or otherwise released without written consent from" the company.[158] Dr. Dong was nervous about signing a contract containing this clause, but company representatives told her not to worry about the possibility that it would restrict her ability to publish a paper in a scientific journal. This oral assurance was, of course, legally meaningless.

When the study was completed, the results were not what Boots expected—Synthroid and its generic competitors were in fact bioequivalent. This meant that Boots could not cite the study to prescribing doctors in support of its claim that Synthroid was more effective than its generic counterparts. Although its predecessor had specified the protocol for the study, Boots decided that the results were flawed, and it complained to high-level university officials.[159] The university commissioned two reviews of the study; one concluded that the study contained minor and easily correctable flaws, and the other concluded that Dr. Dong's study was fine and it was the Boots criticisms that were "deceptive and self-serving."[160]

Although this extraordinary effort on the university's part should have been the end of the matter, Boots intervened once again after *JAMA* accepted Dr. Dong's article for publication following peer review by five outside scientists. Knoll Pharmaceuticals, which had by now purchased Boots, warned her and her colleagues that publication of the results without its approval would violate the terms of the contract, and it threatened to sue the university and her if publication resulted in economic loss to the company. At the same time, Knoll's scientists wrote *JAMA* a letter detailing their criticisms of Dr. Dong's paper and warning the journal not to publish it.

After university attorneys warned Dr. Dong that they would not defend her in court, she withdrew the paper. Knoll's scientists then reinterpreted her data and published their reinterpretation, which (not surprisingly) found that Synthroid outperformed its competitors, in a less prestigious journal of which one of Knoll's scientists was an editor.[161] This publication did not even acknowledge the fact that Dr. Dong and her colleagues had collected the underlying data. Consumers and their doctors knew nothing of Dr. Dong's conclusions until an intrepid investigative reporter for the *Wall Street Journal* broke the story a year later. After receiving a stern warning from the FDA that it had violated the misbranding provisions of the Food, Drug and Cosmetics Act by suppressing Dr. Dong's paper, the company became much more cooperative and ultimately agreed to allow her to publish her own interpretation of the results in the April 1997 issue of *JAMA*. Meanwhile, the company flourished financially as consumers paid about $336 million more per year for Synthroid than for the equally efficacious generic drugs.[162] ❒

Dr. Dong is not the first, or likely the last, academic to find her research restricted by a private sponsor. The pharmaceutical industry now funds approximately 25 percent of all published academic biomedical research.[163] Two national surveys reveal that roughly 25 percent of all faculty members in the life sciences at major U.S. universities had received research support from industry by the mid-1990s, and 90 percent of firms conducting research in the life sciences had some type of relationship with a university.[164] In the same studies, 50 percent of the faculty members in life sciences had consulted for the drug industry. It is quite possible, however, that the general movement of pharmaceutical industry funding out of university research during the early 2000s has caused these numbers to shift downward.[165] Regulated industries are also an important source of funding for environmental and public health research.[166]

Rather than attempting to limit these arrangements, Congress has actually hastened this growing partnership between the private sector and academia in two mutually reinforcing ways. First, the Bayh-Dole Patent Amendments of 1980 allow universities to patent inventions resulting from federally funded research.[167] This provides an important catalyst and tacit endorsement for the university–private sector collaborations that inevitably result.[168] Most universities and many prominent research hospitals now have technology licensing offices to manage their patent portfolios and other commercial interests,[169] and university scientists and their parent institutions often form start-up companies to test and market new products.[170] In 2003, 165 universities responding to a survey reported $968.1 million in royalties from product sales as a result of their patents and related entrepreneurship.[171]

Second, as state-supported university budgets have quietly dwindled during the last twenty years of tax relief, Congress has likewise allowed direct federal support of basic research to decline. Despite the growing importance of environmental science, for example, the EPA's research budget has actually dropped by 20 percent over the past thirty years.[172] Federal funding of basic and academic research in this area has remained relatively flat, although funding for applied research has increased by almost 20 percent.[173] The agency's permanent Science Advisory Board in a March 2007 report expressed its concern that "continuing intentions to decrease EPA's support of research will erode staff morale and ultimately, if it has not already done so, harm EPA's ability to maintain national leadership in environmental science and engineering."[174] Even funding for basic cancer research, once the darling of congressional appropriations committees, is losing the battle with inflation.[175] The pharmaceutical industry now spends more on research than the entire operating budget of the NIH.[176] As a result, universities have had to rely on the private sector to help fund basic health and environmental research.[177]

In his disturbing book *Science in the Private Interest,* Dr. Sheldon Krimsky concludes that this heavy intrusion of corporate money into the pursuit and dissemination of university research has vastly increased the potential for conflict of interest among university scientists that spills over into research areas beyond traditional technology development programs. The resulting "loss of a socially valuable ethical norm—disinterestedness—among academic researchers" in turn increases the potential for biased research outcomes.[178] Once the university or its scientists become corporatized, the values guiding the research are no longer the values of science but the values of the marketplace, "where cutting corners for the bottom line, taking legal shortcuts, and standing in opposition to regulation are commonplace."[179] Krimsky's concerns are echoed in the wider literature on scientific ethics and integrity.[180]

In the wake of several highly publicized ethical lapses by university-affiliated scientists, a comprehensive 2002 NAS panel report entitled *Integrity in Scientific Research* recommended a range of curricular and related reforms to call attention to the vitally important subject of research ethics.[181] Following that advice, many universities have begun to offer (even require) courses on this topic. Several sources of research ethics training materials are now available for such courses from institutions like Indiana University's Poynter Center for the Study of Ethics and American Institutions and the Online Ethics Center for Engineering and Science at Case Western Reserve University.[182] Duke University and

Columbia University maintain similar online resource centers for teaching "chemical ethics" and "responsible conduct of research."[183] Some professional organizations, like the American Society for Microbiology and the Association of American Medical Colleges, have promulgated codes of ethics that speak directly to conflicts of interest.[184] A 2004 survey of 126 medical schools in the United States (82 percent of all U.S. medical schools) reported that the vast majority of them had established conflict of interest policies and procedures for research employing human subjects. The authors found "clear evidence of substantial responsiveness on the part of the academic medical institutions, and for many, their parent universities, in strengthening their conflict of interest policies well beyond the minimum federal standards in recognition of the markedly changed circumstances of clinical research."[185]

Other studies, however, sound a cautionary note in this regard. In a 2000 article, Dr. Mildred Cho and her coauthors report the results of a survey of the conflict of interest policies at nearly one hundred universities and conclude that most of them are vague about "the kinds of relationships that are permitted or prohibited."[186] Many university policies require only internal disclosures of conflicts of interest, with a slight majority of institutions also requiring disclosure to the public.[187] Only a small percentage (12 percent) of universities prohibit arrangements that might cause delays in publication, even though such clauses are common in industry contracts.[188] In a 2005 survey of medical school administrators, 24 percent reported that their schools permitted contracts that allowed sponsors to insert their own statistical analyses into manuscripts, 49 percent said they disallowed them, and 29 percent did not know. In the same survey, 50 percent said they permitted provisions allowing sponsors to draft the manuscript for publication, 40 percent said they disallowed it, and 11 percent did not know. Finally, 41 percent said they allowed contracts prohibiting investigators from sharing data with third parties after the study had been completed, 34 percent said they disallowed it, and 24 percent did not know.[189] As one Canadian scientist observed: "Many of the policies and procedures for the ethical oversight of research were put in place in an era when public funding was much more prominent than it is now."[190]

Numerous recent examples of continued ethical lapses on the part of university scientists also raise at least a modest concern about the effectiveness of these new ethical courses in the real world. For example, one of the nation's most prolific and highly regarded research psychiatrists resigned in 2006 from the board of editors of the journal *Neuropsychopharmacology* in the wake of news reports that he failed to disclose the financial ties he and his coauthors had with the manufacturer of a medical device they had favorably reviewed in that journal.[191] A large number of violations of the financial disclosure requirements adopted by *JAMA* in 1999 prompted the journal in 2006 to adopt even stronger disclosure requirements and to force some of the authors to publish letters of apology.[192] The journal's editor-in-chief concluded that deliberate omissions of conflict were "very rare" and that most failures to disclose conflicts were attributable to the fact that authors "just don't get it."[193] More disturbing are indications that some scientists regard the disclosure requirements as ill-considered impositions on their academic privileges.[194] On the theory that it may be impossible to teach an old dog new tricks, some degree of patience may be warranted before concluding that the newly implemented ethics courses are failing altogether. Still, some degree of skepticism is clearly in order.

The threat to science posed by financial conflicts of interest can even become a threat to the physical well-being of the members of the university community. After the University of Florida chemist Nicholas Bodor invented a chemical that he thought would be useful in delivering drugs directly to the human brain in the late 1980s, he created a company called Pharmatech to commercialize his idea.[195] Subsequently, one of Bodor's colleagues (and a member of the Pharmatech advisory board), Dr. Kenneth Sloan, learned that another compound with a similar chemical structure caused Parkinson-like symptoms in humans. Since Bodor was the chairperson of the department, Sloan asked their dean in confidence for a determination as to whether graduate students in the laboratories using the chemical should be warned of the possible risks.[196] When Bodor learned that Sloan had done this, he angrily responded, in two memos from Pharmatech to the dean, rebutting the claim. The dean, who was also a shareholder in the company, decided that a warning would not be necessary. The company went bankrupt soon after another scientist publicized Sloan's concerns at a congressional hearing, and Boder later gave Sloan a negative evaluation in a departmental academic performance review.[197]

The conflict of interest problem may be considerably worse at major research hospitals that are not associated with universities. These institutions are primarily in the business of serving patients, but they also conduct a great deal of clinical research on cutting-edge medical technologies. It is not at all uncommon for doctors at these facilities to be conducting clinical trials on medical devices while holding equity or royalty interests in the companies (often small start-ups) that are attempting to commercialize those devices.

CONFLICTS AT THE CLEVELAND CLINIC

At the time he was recruited to join the staff of the Cleveland Clinic, a nationally prominent research hospital, Dr. Jay Yadav had recently invented a special blood filter for use with carotid artery stents. The stents offered a promising alternative to more invasive carotid artery surgery. Both the filter and the stents were medical devices subject to FDA approval. The start-up company seeking approval for the filter device was founded by Dr. Yadav and several partners, some of whom were also doctors at the Cleveland Clinic. Although they sold their equity interests to a major device manufacturer for $40 million, Dr. Yadav continued to receive royalties from the manufacturer as the inventor of the device.[198] During his tenure at the clinic, Dr. Yadav played a prominent role in several clinical studies of the safety and efficacy of the blood filter and associated stents and in one major study comparing the safety and efficacy of stents to traditional carotid artery surgery. He did not, however, fully disclose his financial connections to the device manufacturer in several published articles prominently mentioning the filter.[199]

Another Cleveland Clinic researcher, Dr. Isador Lieberman, received valuable stock options in a company that was developing an alternative treatment involving a special bone cement for treating spinal fractures at the same time that he directed clinical studies concerning the safety and efficacy of the bone cement technique.[200] He did not disclose that relationship in several published articles, many of which the company cited in its promotional materials, though he did report that he served as a consultant to the company. When researchers at Johns Hopkins University published a study concluding that the bone cement technique was more closely associated with serious side effects than a far less expensive approach, Dr. Lieberman challenged the methodology of the study and the completeness of the reported data.[201] All the while, he maintained that his disclosures were fully consistent with the institution's conflict of interest policies, a matter about which he could speak with some authority, because he served on the Cleveland Clinic's conflict of interest committee.[202] The hospital responded to these revelations by severing its relationship with Dr. Yadav and by hosting a meeting in September 2006 of three hundred doctors, academics, industry representatives, and government officials to hear presentations on conflict of interest.[203] ❐

Codes of conduct and institutional conflict of interest policies are welcome indications that the scientific community itself has begun to recognize the ongoing shift in support for research from government to the private sector and the resulting threat that financial conflicts of interest will influence research outcomes. The ultimate measure of the effectiveness of such codes and policies, however, is in the degree to which violations result in meaningful sanctions. While reports of universities and research hospitals punishing scientists for ethical lapses involving fraud and failure to protect human subjects appear with some frequency in the media, reported instances of sanctions of the sort the Cleveland Clinic ultimately imposed on Dr. Yadav for violating conflict of interest restrictions and disclosure requirements are quite rare.[204] It is one thing to be forced to resign a position on the board of editors of a scientific journal and quite another to be sanctioned by one's home institution for unethical conduct. As things currently stand, it appears that university scientists who violate scientific norms in their collaborative pursuit of their sponsor's economic goals still face little risk of reproach from their colleagues and virtually no risk of sanctions from their institutions.

Limited Legal Constraints

A few legal constraints on research manipulation are available to fill in the gaps left open by universities' failure to discourage academics' attempts to shape science on behalf of private sector sponsors. Although the courts impose only the broad constraints of relevance and reliability on the research used in common law litigation, regulatory agencies administering premarket testing regimes have specified some relatively rigid protocols and requirements to govern the testing of drugs, pesticides, and toxic substances.[205] These "cookie cutter" research protocols, when applicable, make it difficult for a sponsor to shape science by altering research designs, because they leave the researcher with little discretion. Unfortunately, rulemaking processes effectively restrict the ability of the agency to prescribe these cookie cutter protocols in all of the situations where they might be useful. Regulations establishing such protocols are typically subject to an extended period of public "notice and comment," and they can become the subjects of lawsuits brought by parties who deem them to be arbitrary. As a result, agencies experience lengthy delays—sometimes for as long as a decade—in finalizing new protocols.[206]

For most health and environmental research used to inform regulation, then, the main shield against shaping is the limited oversight provided by agency scientists or, in court, the "*Daubert* hearing" on the admissibility of expert testimony. If these screening devices do not detect ends-oriented adjustments to that research, decision-makers are likely to treat the results as a reliable source of policy-relevant scientific information.[207]

Indeed, despite the obvious danger that sponsors will control research to produce predetermined results, regulatory agencies and courts do not appear to inquire into even some of the basic, publicly available circumstances surrounding privately commissioned research. The EPA, for example, does not formally require any conflict disclosures for research submitted in support of a registration to market a pesticide or in support of a permit to emit pollutants or handle hazardous wastes. The FDA has instituted a conflict of interest policy requiring financial disclosures for safety research conducted by private parties in support of an application to market a drug or food additive,[208] but these disclosures do not discriminate between sponsored research where the sponsor controls the design or reporting of the research and studies where the sponsor relinquishes control over the research process. Because the *Daubert* screen focuses primarily on whether expert testimony follows the scientific method in the abstract and is therefore "testable," courts need not identify or exclude biased research, and they rarely do so on their own.[209] Consequently, the judges typically send litigation science through the adversarial process with their fingers crossed that vigorous adversaries will do their job and identify and expose any underlying ends-oriented decisions.

Conclusion

The pervasiveness of the attempts to shape public health and environmental research documented in this chapter suggests that collective scientific knowledge does not always result from scientists dutifully applying the scientific method, but instead sometimes reflects successful efforts by advocates to influence researchers and research outcomes. Former *NEJM* editor Marcia Angell observes that the clinical trials used to evaluate the safety and effectiveness of pharmaceuticals "can be rigged in a dozen ways, and it happens all the time."[210] The possibility that some portion of our collective "scientific" knowledge about public health and the environment is the product of distorted methods and biasing interpretations should be quite disturbing to a public that depends on agencies and courts to protect it from the risks posed by novel products and modern industrial processes.

Perhaps even more disturbing, the increasingly pervasive practice of shaping science appears to be altering the trajectory of scientific knowledge. Known as the "funding effect," empirical studies in the biomedical arena reveal that private sponsorship makes a statistically significant difference (consistently favorable to the sponsor) on research outcomes.[211] A comprehensive review article summarizing 1,140 biomedical research studies concluded that "industry-sponsored studies were significantly more likely to reach conclusions that were favorable to the sponsor than were nonindustry studies."[212] The funding effect is also evident in food safety and environmental health research. The best predictor of the conclusions in published reviews assessing the health impacts of passive smoking, for example, is whether they are written by authors affiliated with the tobacco industry.[213] A 2005 survey of the endocrine disrupting effects of the heavily used plasticizer bisphenol A revealed that of the 115 studies published as of December 2004, none of the 11 industry-funded studies reported positive effects at low doses, but 94 of the 104 government-funded studies did.[214] A study of articles published in nutrition journals on the nutritional benefits and risks of soft drinks, juices, and milk during 2003 concluded that industry-funded studies were four to eight times more likely to reach results that were favorable to the sponsoring entity.[215] Other studies have detected the funding effect in studies commissioned by the oil industry examining the rate of cancer in workers and in studies commissioned by certain product manufacturers (alachlor, Atrazine, formaldehyde, and perchlorethylene) on the increased incidence of various chemical-related harms from long-term exposures.[216]

These developments are of growing concern to scientists, many of whom passionately oppose them. Letters to the editor identifying shaped science are becoming more common in the scientific literature.[217] Scientists are now convening conferences and symposia to shine a public light on the science-shaping techniques employed by savvy consultants and to condemn the scientists that go along with the ruse.[218] Scientific journals are waking up to the problem, and courageous past and present editors at the top journals, like *JAMA, Science,* and the *NEJM,* are speaking out against the practices and rallying others to prescribe and implement reforms. While these developments alone will not overcome the tremendous resource disparity that currently exists between the parties with an interest in bending science and the scientists who oppose their efforts, they are making headway. More important, they are paving the way for more effective reforms aimed at monitoring and controlling the privately funded research that too often dominates the highly contested and poorly supervised arena of policy-relevant science. We explore some of these suggestions in Chapter 10.

CHAPTER FIVE

Hiding Science

The Art of Concealing Unwelcome Information

When sponsored science produces unwelcome discoveries—despite possible efforts on the sponsor's part to shape the research in advance—it may still be possible to avoid any adverse consequences by hiding those discoveries from regulators, courts, and the public. Much private sector research already takes place in a culture of secrecy, inspired by the legitimate need to shield economically valuable information from competitors. Companies can sweep unrelated research, like internal studies indicating that their product causes unexpected harm, under the same conveniently broad umbrella of "confidential business information."[1] Even corporate-sponsored research conducted in public universities has been covered by contracts that give the sponsoring entity a right to deny the academic authors the right to publish their results, at least for a period of time sufficient to allow company scientists to prepare a response.[2] Yet, as Sheila Jasanoff explains, "science and secrecy do not sit comfortably together."[3] In the context of policy-relevant science, the tension between the two is often palpable.

THE SCIENTIST WHO WOULD NOT "BE BULLIED"

An expert in hereditary blood diseases, Dr. Nancy Olivieri began practicing medicine and conducting research at Toronto's Hospital for Sick Children in the mid-1980s.[4] Although widely feared by her underlings and regarded by many of her peers as a tireless self-promoter, she was greatly respected by her patients and praised by her mentors at Sick Children, which was affiliated with the University of Toronto's medical school, and the Harvard Medical School, where she did postgraduate work. She soon began to specialize in thalassemia, a disease in which children are unable to make proper hemoglobin to deliver oxygen from their lungs to the rest of their bodies. The only effective treatment for thalassemia is a monthly blood transfusion, but that in turn causes blood iron levels to increase to levels that will over time clog up vital organs. Fortunately, the drug Desferal can remove excess iron from a child's body. Unfortunately, very large quantities of the drug must be injected for hours at a time into patients several times per week. Some teenagers refuse the painful and unsightly injections, but few thalassemia victims who do not do something to remove the iron live past their twenty-fifth birthdays.

In 1988, Dr. Olivieri obtained funding from the Canadian Institutes of Health Research to undertake a pilot study for a long-term clinical trial on deferiprone, an experimental drug that could be taken orally and appeared to be effective in removing iron from thalassemia patients. When the Institutes of Health Research turned down her application for a grant to conduct a long-term follow-up study, she approached the Apotex Corporation, a highly successful generic drug manufacturer run by a billionaire alumnus of the university, for funding. Intrigued by the possibility of developing its first nongeneric drug, the company in the spring of 1993 agreed to fund the study for three years and to manufacture a sufficient quantity of the drug for use in the study. The contract with the company required Dr. Olivieri to submit all data to the company, and it further provided that the results could not be disclosed for up to three years without the company's consent. Desperate to obtain funding for her research, Dr. Olivieri did not object to the provisions.

Dr. Olivieri's small research team soon became overwhelmed by the study's intense informational demands, many of which were necessitated by the requirements of the regulatory agencies that would have to approve the drug for commercial use. To the consternation of company scientists, Dr. Olivieri departed from some of the study protocols and fell behind on the documentation obligations. By the end of 1994, the relationship between Apotex and Dr. Olivieri had begun to sour, and Apotex complained in writing to Dr. Olivieri that she was withholding data the company needed to monitor the progression of the trial in accordance with federal regulations. Her relationship with Sick Children also deteriorated, as her demands for additional resources to help with her burgeoning paperwork responsibilities went unmet.

Worst of all, the evidence was mounting that over time, the iron levels in some of the patients were beginning to rise, indicating that deferiprone was losing its efficacy. Dr. Olivieri was not regularly providing data to

Apotex at the time, and she did not mention the indications in a presentation she made at a December 1994 scientific meeting.[5] An article she published in *NEJM* in April 1995 concluded that deferiprone was not quite as effective at removing iron as Desferal, but was much easier for patients to take and had few side effects. The article did not mention, however, that it may have stopped working in a few patients. Some scientists, including her mentor at Harvard, were beginning to wonder whether Dr. Olivieri, whose work now depended on funding from Apotex, was becoming biased in favor of the drug. Bias was certainly possible, because the study, for sound ethical reasons, could not be blinded, and the researchers knew which patients were taking which drug.[6]

By the spring of 1995, Dr. Olivieri had reluctantly concluded that deferiprone was not working in some of the patients, and she wrote up protocols for a new study to probe why it worked in some but not in others. She then met with Dr. Michael Spino, the research director for Apotex, to share her concerns and request support for an additional five years under the new protocols. Surprised to learn of the problems, Dr. Spino reiterated his demand that she provide all of her raw data.[7] After conducting its own analysis of Dr. Olivieri's data, the company agreed that the drug was not working in some patients, but it had a different explanation (involving the proper dosage of the drug) for the cause of the problem, and it offered to support a follow-up study probing that theory. It also objected to her sharing her conclusions with her patients until they had agreed on a proper follow-up study.

Dr. Olivieri ignored the demand and gave her patients a new consent form stating that deferiprone was working in only a "minority" of the patients taking it. Apotex immediately terminated her contract and instructed her that all information from the study "whether written or not, obtained or generated by the Investigators" had to remain "secret and confidential." If she did not maintain that confidentiality, Apotex would "vigorously pursue all legal remedies" available to it.[8] Her Harvard mentor, however, advised her that she had a moral obligation to communicate her findings at the upcoming meeting of the American Society of Hematology (ASH). By then, she had received a letter from the company's lawyers threatening to sue her if she disclosed her results.[9]

Dr. Olivieri continued to study deferiprone surreptitiously by collecting the blood and biopsy results of her patients who continued to take it. Having learned that a similar drug had caused scarring of the livers and hearts of laboratory animals, she decided to examine slides from the liver biopsies of her patients. Although the original pathologist had not detected scarring at the time he examined the slides, Dr. Olivieri was confident that she did in fact detect liver scarring in some of the patients who were taking deferiprone. Unlike the original pathologist, however, Dr. Olivieri was not blinded as to whether the slides she was viewing came from the deferiprone treated patients or the patients receiving Desferal. She now concluded that deferiprone was so toxic that no patients should be taking it, and she received permission from the journal *Blood* to add a few lines to report this new finding in a review article she had written that was about to come out. She also began to share her findings with her present and former colleagues, one of whom (who was also a consultant to Apotex) cautioned her against claiming that the liver scarring she had observed was caused by deferiprone because it could also have been caused by a number of other things, like hepatitis, which was a common affliction of thalassemia patients who received frequent blood transfusions.

Dr. Olivieri presented her efficacy findings and her liver scarring hypothesis at the ASH meeting in December 1996. In a sharply worded exchange, Dr. Spino strongly objected to both conclusions and told the audience that she had not presented all of the relevant data. Unwilling after that to trust Apotex to inform the authorities, Dr. Olivieri wrote a letter relating her conclusions to the FDA. Dr. Spino complained to Dr. Olivieri's superior at the hospital, and he in turn demanded to know why she was still giving the drug to her patients if it was so toxic that it warranted a letter to the FDA. She said that she planned to raise the issue at the next patients' meeting.

Dr. Olivieri then arranged to have three prominent pathologists review the slides from the study in a blinded fashion for liver scarring. After writing up the results, she invited the three pathologists to be coauthors, but one of them declined on the ground that he would have reached a different conclusion. In early 1998, the article was accepted by *NEJM,* and Dr. Olivieri felt free to speak with the press about her findings. With the help of a public relations expert, who agreed to work with her on a pro bono basis, she prepared ten binders of documents on her travails for the press to review in the interim. The published article concluded that deferiprone "does not adequately control body iron burden in patients with thalassemia . . . and may worsen hepatic fibrosis."[10]

At this point, Dr. Olivieri became a celebrity. The reports that broadcast on the airwaves the night the article was published portrayed her as a dedicated scientist who had published the study despite the threat of a lawsuit from Canada's largest drug company because she was not "going to be bullied."[11] Doctors at Sick Children and prestigious scientists from around the world quickly lined up on both sides of the controversy, which ultimately wound up on *60 Minutes* in a segment that featured the CEO of Apotex referring under his breath to Dr. Olivieri as "nuts."[12] She also began to receive honors in both the United States and Canada for what some considered a courageous decision to stand up toa big drug company.[13] At the same time, the journal *Blood* published a series of letters to the editor criticizing her study on a number of grounds, and two subsequently published studies failed to find liver scarring in deferiprone-treated patients.[14] Neither Health Canada nor the U.S. FDA has approved the drug for commercial use. Apotex has, however, received approval to market deferiprone in twenty-five countries outside of North America, and its CEO has become one of the richest people in the world.[15] ❐

The Olivieri story demonstrates once again that competent scientists frequently disagree about how to analyze and interpret policy-relevant research, and it reinforces the message of Chapter 4 that control over the production, analysis, interpretation, and distribution of the underlying data matters a great deal in determining how scientists and policymakers perceive and use scientific studies. But it also illustrates another powerful reality: sponsors of scientific research who retain control over its dissemination and use have a strong incentive to prevent scientists who reach unwelcome conclusions from disseminating them to the scientific community or even communicating them to human research subjects who may be at risk. Whether or not she deserves a hero's mantle, Dr. Olivieri clearly paid a high price both personally and professionally when she decided to go public with her concerns. The media accounts of her story inspired the author John le Carre to write *The Constant Gardener,* but they also sent a clear message to other scientists engaged in sponsored research that they ignore sponsor demands to keep data confidential at their peril.

Of all the tools that affected parties employ to distort science, the tactic of hiding science is the most straightforward. With the possible exception of shaping science through overt scientific fraud, hiding science presents the most dramatic affront to both science and informed public policy-making. The numerous examples of suppression that follow suggest that it may also be the most frequently employed strategy for bending science. Given its very nature, it seems likely that many more accounts remain suppressed than are uncovered. From the very beginnings of modern science to the present day, sponsors of scientific research have attempted to hide results that have adverse implications for their economic, political, or ideological positions, and we have not identified any discernable historical trends.

Since the temptation to cover up inconvenient information is universal, the incidence of hiding will depend on the availability of opportunities to hide, and the evidence suggests that the opportunities to hide science are legion. Most of the examples recounted here involve private companies, for the simple reason that they sponsor much policy-relevant research and often stand to lose a great deal financially from the dissemination and use of information that casts their products and activities in a bad light. The government, too, funds a large proportion of the research used in regulation and litigation, and funding agencies have likewise attempted to hide policy-relevant research for ideological and political reasons. Plaintiffs in lawsuits do not fund a great deal of scientific research, but they have frequently cooperated with defendants in covering up company-generated science by agreeing to place litigation-generated documents under seal as part of settlement agreements.

Whatever the private advantages that flow from concealing adverse results, hiding science is socially costly. Suppressing adverse research retards scientific progress, keeps vital information out of the reach of regulators, and deprives those exposed to undisclosed risks of the option of avoiding them or suing responsible entities for compensation when those risks manifest themselves in real harm. An influential 2002 NAS report recognized the "general principle that publication of scientific information is intended to move science forward" and concluded that "[a]n author's obligation is not only to release data and materials to enable others to verify or replicate published findings . . . but also to provide them in a form on which other scientists can build with further research."[16] Despite virtual unanimity of opinion on the substantial dangers that concealment presents to both the scientific community and the public, existing laws have been ineffective in combating this strategy for bending science.

Why Hide Science?

Many socially beneficial products and the byproducts of socially useful activities pose nontrivial risks to public health and the environment. The companies that sponsor research into the nature and extent of these risks, both in their own laboratories and in universities and privately run facilities, occasionally find themselves in the precarious position of knowing that their products or activities are more dangerous than they appear. Public disclosure of this information not only runs the risk of increased regulatory burdens and future litigation costs but also raises the grim prospect of substantial reputational damage. Company officials sometimes conclude that concealing such damaging research is the safest business strategy, at least in the short run. Thus when a large manufacturer of polyvinyl chloride, a widely used plastic, discovered in 1966 that exposure to the precursor chemical, vinyl chloride, caused its workers to suffer from a rare disease called acroosteolysis that produced skin lesions, circulatory problems, and the destruction of bone at the ends of their fingers, it urged other companies in the industry to "use discretion in making the problem public" so as to "avoid exposés like *Silent Spring* and *Unsafe At Any Speed*."[17]

From a purely economic point of view, the polyvinyl chloride manufacturers' inclination to keep adverse research under wraps was financially prudent. Sharing preliminary research indicating that a product is harmful will usually increase the costs of doing business, often well beyond the profit margin. Workers will insist on greater protections in the workplace. Regulators may decide that the product should be manufactured or used in more restricted ways. Consumers will be less enthusiastic about purchasing a product if they know it is potentially toxic to them or their children. And perhaps most worrisome of all, plaintiffs' attorneys could learn of the research and file lawsuits on behalf of exposed individuals who exhibit suspicious symptoms, whether or not they were actually caused by the product.

The potential for tort liability may, in fact, provide the greatest incentive for companies to suppress adverse research results, even when those results are preliminary and equivocal. Tort law requires a private company to pay for harms to victims in many situations, but only if the alleged victims can prove that the company's product, pollutant, or activity caused their harm.[18] As we discuss in the next section, the potential for tort liability forces companies to walk a tightrope in determining whether to conduct or sponsor safety research on their products and byproducts. Although jurors appear particularly impatient with defendants that adopt a "see no evil" strategy with respect to the risks posed by their products and activities,[19] a lawsuit can proceed only as long as a plaintiff has *some* credible evidence, even though the results of more comprehensive but later-arriving research might ultimately exonerate it.[20] One obvious way out of this Catch-22 is for the company to conduct testing under tightly controlled conditions, and this requires a degree of suppression. Yet while suppression can be a winning strategy, it only works if any damaging evidence that results remains deeply hidden. Leaked internal acknowledgments of risk may be worse than publicizing the information widely or resisting testing altogether.[21]

One quite legitimate excuse for failing to disclose research results is the sponsor's reasonable determination that they remain preliminary or tentative and therefore need further verification and checking. Premature disclosure of research results may convey erroneous information and unnecessarily alarm regulators and the public.[22] In situations where public health may be a risk, however, public health goals—or erring on the side of protection—ordinarily trump general scientific principles that would entrust scientists and their sponsors with unfettered discretion to determine when research is sufficiently robust to be shared with others. While scientific norms counsel against the premature release of incompletely analyzed or vetted findings, in public health research this norm is countered by the worrisome possibility that leaving the decision of when to share preliminary evidence of adverse effects solely to sponsors or their researchers could lead to under-protective, self-interested decisions by manufacturers to "wait and see." In any event, concerns for the accuracy and completeness of the underlying data were not what motivated the actors in most of the instances of suppression detailed here. Instead, the sponsoring entities covered up or terminated credible research that indicated serious risks to public health because the results threatened potentially catastrophic economic consequences to the sponsors themselves.

Where Possible, Choose Ignorance

One way for companies to avoid the dilemma of deciding whether to suppress damaging research findings is to avoid testing altogether.[23] As the handwritten notes on the agenda for a pharmaceutical company staff meeting revealingly noted: "If FDA asks for bad news, we have to give it, but if we don't have it we can't give it to them."[24] The same is true for the information needed to resolve common law tort litigation. Huge gaps in scientific research relating to the impact of products and industrial activities on health and the environment serve as a testament to this tendency of many companies, consistent with rational choice theory, to choose ignorance.[25] Even as late as 1984, an NAS panel found that *no* toxicity testing existed for more than 80 percent of all toxic substances used in commerce,[26] and by 1998, at least one-third of the toxic chemicals produced in the highest volumes still failed to satisfy minimal testing standards recommended by an international expert commission.[27]

Anticipating the natural preference of some companies to resist safety research, Congress has enacted several product licensing statutes that require premarket health and safety testing as a condition to regulatory agency approval.[28] These statutes, however, apply only to pesticides, drugs, certain medical devices, and a small set of other suspect products.[29] For most other products and pollutants, affirmative testing requirements are effectively nonexistent.[30] Even when the authority does exist to require testing, the agencies generally need some limited evidence of harm before they can take meaningful postapproval regulatory action—a Catch-22 that reduces their ability to order more testing.[31]

The FDA, for example, has authority to require premarket testing of drugs and devices, but its postmarket testing authority has been quite limited until recently. A 2006 report of the NAS's Institute of Medicine noted that "[o]nce a drug is on the market, it can be difficult to compel sponsors or others to undertake appropriate comparative trials."[32] Even when manufacturers have agreed to undertake postmarket clinical trials, they have not always completed them. The FDA reported in March 2006 that 65 percent of the twelve hundred studies sponsors had agreed to undertake during several previous years had not even been initiated.[33] The result is that dangerous drugs might have remained on the market while their manufacturers chose ignorance. Examples include:

Prozac—After a 1990 epidemiological study raised serious questions about Prozac's potential to cause some patients to commit suicide, an FDA advisory committee strongly recommended that the manufacturer conduct a large clinical study to probe the suicide question. The company hired a prominent scientist who met with the author of the earlier study, drew up protocols for the recommended study, and approached investigators about participating in the study. By then public pressure had subsided, and the company never conducted the more ambitious follow-up study. Without the study, the FDA was reluctant to take action to protect potential victims, and it did not require warnings on the high suicide risks that Prozac and similar drugs may have posed to children and adolescents until New York's attorney general brought a highly publicized lawsuit in 2004.[34]

Dalkon Shield—After the Dalkon Shield intrauterine birth control device had been on the market for more than three years, a company scientist recommended testing to probe the effect of the device on the uterine environment. The company's management declined to do the testing, and hundreds of women subsequently suffered adverse health effects, including death in rare cases. Because the Dalkon Shield was "grandfathered in" to the medical device system, affirmative testing was not required unless the FDA demanded it. Since the company did not share their concerns with the FDA, the agency was kept in the dark about the potential risks of the Dalkon Shield.[35]

The EPA has somewhat similar authority to order manufacturers to test products that may pose an unreasonable risk to human health under the pesticide and toxic substance statutes, but it has exercised this power sparingly.[36] Particularly with respect to toxic substances, the EPA often finds it more expedient to negotiate with individual manufacturers regarding these testing obligations. These negotiations, which can drag on for years, do not always result in especially stringent testing requirements.[37]

This relaxed approach to government-required testing gives companies the upper hand in disputes with regulators over whether more testing is needed. For example, the petroleum industry went to considerable lengths during the 1980s to convince the EPA to remain optimistic about the unstudied, long-term health effects of and human exposures to the chemical methyl tert-butyl ether (MTBE), while at the same time dramatically increasing its use as a fuel additive to enhance octane levels and reduce air pollution. The industry assured the EPA that no additional exposure information was necessary because the EPA could trust the industry's "worst case" calculations and that no chronic toxicity testing was needed because the calculated exposures were well below the no-observed-effect level in short-term toxicity studies.[38] Later testing revealed that MTBE caused cancer in laboratory animals and moved very rapidly in groundwater from leaking underground gasoline storage tanks, but by then it had contaminated drinking water supplies throughout the country.[39]

One might expect that the prospect of catastrophic tort liability would bridge this disturbing gap in regulatory authority by frightening companies into safety testing. But tort law is a very blunt instrument, so blunt in fact that it probably encourages companies to "choose ignorance."[40]

Since plaintiffs bear the burden of proving that a company's products or activities caused damage, companies have little incentive to fund research that may demonstrate that their products and activities pose health or safety risks. Absent strong scientific evidence demonstrating clear and powerful scientific links between the product and a particular disease, like the associations of asbestos exposure with a rare cancer called mesothelioma and DES with clear cell adenocarcinoma, victims are effectively without recourse. Better to let sleeping dogs lie, at least until someone else conducts research that forces the issue.

Pervasive Hiding in the Private Sector

Choosing ignorance is not always a realistic option, and some companies that have decided to conduct and then suppress damaging information have found themselves under the public spotlight when internal documents emerge revealing their prior knowledge of potential harms resulting from their products or activities. Documents produced during litigation over the past twenty-five years have revealed that Johnson & Johnson (for ultraabsorbent tampons),[41] A. H. Robins (for the Dalkon Shield),[42] Dow Corning (for silicone gel breast implants),[43] and the asbestos,[44] vinyl chloride,[45] pesticides[46] and tobacco[47] industries have all concealed research about their products' adverse health impacts. Unfortunately, by the time the truth comes out, many innocent people may already have been injured so severely that no monetary award can make them whole.

Despite heavy oversight by the FDA, the pharmaceutical industry has over the years been especially aggressive in suppressing adverse scientific information. In the 1970s, several pharmaceutical companies were caught withholding adverse animal testing studies from the FDA.[48] A study of all of the pediatric clinical trials undertaken by drug manufacturers from 1998 through 2004—undertaken in order to receive a six-month additional market protection for pediatric drugs—revealed that only 45 percent were ultimately published in peer-reviewed journals.[49] Some of this unpublished research revealed an increased risk of suicide for children taking the drugs. The attorney general of New York in 2004 filed a civil action against the manufacturer of the antidepressant Paxil alleging that it had failed to disclose to doctors and patients, studies indicating that the drug was not effective as a treatment for serious depression in children and, worse, suggesting that it doubled the incidence of several behaviors associated with suicide.[50] In settling the case, the company agreed to post on its website the results of all clinical trials involving all of its drugs.

Sometimes timing is everything. If a company can suppress a study until after a critical decision point in the legal process, it can preserve the market for a product for months or even years. For example, at the time that the manufacturer of the pain reliever Celebrex published a six-month report on an ongoing clinical trial concluding that the drug was associated with a reduced incidence of gastrointestinal ulcer as compared to two other painkillers, it had obtained data for the full twelve months of the study showing that it did not in fact have an advantage over the other two drugs.[51] Although the FDA ultimately did not allow the company to claim that the drug reduced the incidence of ulcers, the market for the drug had already climbed by the time the full study data were published.[52] Similarly, the manufacturer of a drug designed to prevent bleeding during surgery failed to report to the FDA the preliminary results of a study that it had commissioned from an outside contractor indicating that the drug increased the risk of death, kidney damage, congestive heart failure, and strokes prior to a critical meeting of the FDA advisory committee that was considering whether to recommend that the FDA take action to limit the use of the drug in light of two other studies indicating that it caused similar effects.[53] The company turned the study over to the agency two days later, after one of the scientists who had worked on it blew the whistle to an FDA employee.[54] One of the members of the advisory committee later admitted that the study probably would have played a critical role in the committee's deliberations.[55]

Companies have also concealed information on the risks to neighbors and innocent bystanders from their polluting activities. For example, scientists working for the W. R. Grace Corporation were well aware of the high levels of tremolite asbestos in the dusty air that permeated its Libby, Montana, vermiculite plant and blew into the surrounding neighborhoods from the 1940s through the 1970s. Yet Grace concealed what it knew about "the tremolite problem" from this small community for twenty years while it continued to spread tremolite-containing mine tailings and dust throughout the town, including the local baseball diamond and high school track.[56] By the 1990s, hundreds of spouses and children of plant workers and other members of the community who had no connection to the plant had contracted asbestosis from the plant's air emissions and waste products.[57] The indictment of the company for this criminal behavior in 2005 was cold comfort to the devastated community.[58]

The individuals most directly and seriously affected by private sector suppression of health-related scientific findings are often workers who are exposed to chemicals in the workplace for extended periods of time and usually at much higher levels than are neighbors and consumers. The Johns Manville Company, an asbestos manufacturer, began to observe adverse health effects in its workers in the early part of the twentieth century, but its executives "developed a corporate policy of not informing sick employees of the precise nature of their health problems for fear of workers compensation claims and lawsuits."[59] When Dow Chemical Company's medical director concluded that exposure to benzene posed a cancer risk to exposed workers at the company's Freeport, Texas, chemical complex in the mid-1970s, upper-level management simply closed down his laboratory and turned it into a cafeteria.[60] Though one might wish that the foregoing examples were relics of a bygone age, companies continue to hide critically important scientific information from their workers, despite modern workplace hazard communication laws. In the 1990s, a textile manufacturing company—wielding a confidentiality agreement that a researcher had signed a year earlier—threatened to sue him if he published a study revealing adverse lung effects caused by workplace exposure to microscopic nylon fibers at the company's plant.[61] The researcher published the results anyway, and the Centers for Disease Control and Prevention (CDC) subsequently recognized "flock lung" as a new occupational disease caused by nylon fibers.[62]

Unfortunately, the documented cases of suppression may only reveal the tip of the iceberg. The extent to which manufacturers hide scientific data indicating that their products are harmful is largely unknown because the fact that a company has buried a study is rarely publicized until a whistle-blower or an aggressive litigant brings it to light.

Legal Protection of Business Secrets

If the law has historically provided an incentive to hide science, it has also provided effective tools for protecting hidden science. The most effective legal device for hiding science is a contractual provision forbidding scientists who sign the contract to disclose the results of scientific studies without the approval of the sponsor or employer.

Employment contracts and contracts with private testing companies typically contain such provisions. Although they are less common in research contracts with university scientists, Dr. Olivieri's experience shows that nondisclosure provisions do exist. A more broadly applicable legal vehicle for hiding science is the claim that the information is "proprietary," "trade secret," or "confidential commercial information."[63] Over a period of almost two centuries, the common law courts have fashioned a body of trade secrecy law in the context of faithless employees and industrial spies to protect information that, if revealed, might destroy a company's legitimately acquired competitive advantage. One purpose of the laws protecting trade secrets and confidential business information is to foster innovation and reward creativity.[64] These laws have evolved with little awareness of or regard for their potential to facilitate company suppression of important public health information.[65] Yet even after the onset of modern health and safety regulation, business secrecy claims continue unabated, with only limited legal restrictions on deliberate suppression of health and safety studies. For example, a majority of pharmaceutical companies until very recently claimed that all data underlying clinical trials used to secure and maintain FDA approval of drugs came within the broad ambit of "trade secrecy" and therefore should be unavailable to the public under the Freedom of Information Act.[66]

RISKING A BILLION-DOLLAR BUSINESS ON RATS PULLING A LEVER

In the early 1970s, the tobacco industry knew from its own research that nicotine, a powerful habit-forming drug, was the critical element of tobacco smoke that made it attractive to consumers. A document written by an RJR executive during that time period observed that the tobacco industry was "based upon design, manufacture, and sale of attractive dosage forms of nicotine."[67] To avoid the looming threat of federal regulation, however, the industry steadfastly maintained that nicotine was not addictive. Thus, while companies were anxious to research the pharmacological effects of nicotine for purposes of providing more "effective" doses than their competitors, it was also critical that the scientific community generally did not become aware of that research, especially insofar as the results could be interpreted to support the conclusion that nicotine was in fact addictive.[68]

During the early 1980s, the Philip Morris Company established a secret laboratory and hired several scientists to study the effects of nicotine and other smoke components on the central nervous system and to pursue research on possible nicotine substitutes. The experiments were so secret that even the laboratory animals were delivered surreptitiously. Among other things, the laboratory conducted experiments on self-administration, a strong indication of a drug's addictiveness.[69] In one of the experiments, rats "conditioned themselves to absorb nearly one hundred nicotine jolts in a twenty-four-hour period."[70] Victor DeNoble, the scientist in charge of the laboratory, sent a paper entitled "Nicotine as a Positive Reinforcer in Rats" containing these results to the scientific journal *Psychopharmacology* in January 1983.[71]

At roughly the same time, Rose Cipollone's son brought a lawsuit against the tobacco industry alleging that she had become addicted to cigarettes and they had caused her fatal lung cancer. Prompted by the lawsuit, Philip Morris executives flew DeNoble to New York City to brief them in person on the results of his research.[72] According to Dr. DeNoble, when he explained to Philip Morris executives what his laboratory had discovered about the addictive nature of cigarette smoking, one of them asked: "Why should I risk a billion-dollar [business] on rats pushing a lever to get nicotine?"[73] The company ordered DeNoble to withdraw the paper immediately, and he did so on August 30, 1983, explaining to the editor that it was "due to factors beyond my control."[74] Then, as Mrs. Cipollone's lawyers were preparing a discovery request for the company's research documents, DeNoble was told "to close down his laboratory, to kill the animals, to suspend all further investigation of possibly less toxic or harmful alternatives to nicotine, never to try to publish or discuss his work on addicting rats, and to find work elsewhere."[75] When DeNoble and a colleague resubmitted their paper eighteen months after they were fired, Philip Morris threatened legal action against them under the confidentiality clause in their contracts.[76] Unable to afford lengthy and expensive litigation, they again withdrew their paper.[77] The paper was never published.[78] ❐

The unfortunate history of Dr. DeNoble's tobacco research shows how companies can take advantage of contract law to hide scientific information about their products from the scientific community and ultimately from the regulators and the general public. Companies like Philip Morris that establish their own laboratories are, of course, free to keep them secret, shut them down, and fire the employees who run them. In his Pulitzer Prize-winning book *Ashes to Ashes,* Richard Kluger describes many other cases in which "industry-funded researchers who came too close to the truth for their patrons' comfort . . . were abruptly dropped."[79] What is perhaps less well understood is the fact that companies can invoke confidentiality clauses in employee contracts to keep the former employees, employees of other companies, and even academic researchers, like Dr. Olivieri, from revealing the contents of the secret research.

In the business world, a standard nondisclosure clause in an employment contract is both an acceptable and a common way to ensure that employees do not share business secrets with competitors or otherwise damage the good name of the company by publicizing its internal dirty laundry.[80] Because they typically provide for prior company approval of any dissemination of research results, confidentiality clauses give a company the power to control the scientific information produced under its direction. Superseding scientific norms and the ethical commitments of employee-scientists, these business contracts create a control structure that gives corporate executives the ultimate right to decide whether the broader scientific community will have access to health-related scientific information.

Companies also routinely employ confidentiality clauses in commissioning outside researchers to conduct sponsored research.[81] In most cases, the risk of termination of the contract as well as the loss of any prospect of future business is sufficient to prevent independent contractors from disclosing potentially damaging scientific research without the company's consent.[82] Universities are in a somewhat better bargaining position, but the degree to which they use that power to limit or eliminate restrictive confidentiality clauses is unclear. In a 1996 national survey of executives of biomedical companies, more than half reported that their contracts with universities placed some restrictions on the ability of the researchers to communicate the results.[83] The most straightforward way to place these restrictions on publication is through "property rights" provisions in contracts that allow the manufacturer to retain ownership of the data or "prepublication review" clauses that allow it to bar or control how the research is reported in the literature.[84] A 2002 survey of 108 of the nation's 122 medical schools reported that while 98 percent of them included a provision in contracts with sponsors preventing confidentiality clauses from restricting researchers' publication rights altogether, none of them included a commitment to publish the results, and the median duration of confidentiality clauses was five years between completion of the research and publication.[85] Although some prestigious institutions like the Cleveland Clinic and the Duke Clinical Research Institute are able to negotiate so-called gag clauses out of contracts with sponsors, others have reported that they have felt powerless in contract negotiations with sponsors over confidentiality clauses.[86]

The voluntary nature of the employee-employer and contractor-researcher arrangements leads courts to treat confidentiality clauses as presumptively legitimate.[87] Courts can, however, refuse to enforce such contracts when they violate public policy and when they are invoked to cover up criminal activity or serious misconduct.[88] Even in these cases, however, the courts tend to strike the balance in favor of upholding the contract.[89]

On this commodious legal terrain, the cost to a sponsor of insisting on overly broad, vague and even contradictory confidentiality clauses is very low. Dr. Peter Shank, the associate dean for research at Brown University, observes that "[t]hese contracts frequently contain contradictory statements, such as the scientist 'has the right to publish' but the companies 'own all the data generated in the study.'"[90] The worst-case scenario, from the company's perspective, is that a researcher will test the limits of a confidentiality clause by disclosing research results and successfully challenge its validity when the company sues him. Even then, the employer is no worse off for having tried. The advantages of using such contracts to silence employees and contractors, by contrast, can be significant. Most employees and contract researchers will either adhere to the secrecy requirements or will be very reluctant to run the risk of unsuccessfully challenging the confidentiality clause. Win or lose, employees who disclose trade secret information must generally shoulder the considerable costs of defending the lawsuit. The *in terrorem* effect is usually enough to keep adverse studies behind the company's closed doors.

Not surprisingly, confidentiality clauses have played a significant contributing role in delaying or preventing public dissemination of privately sponsored health and safety research. During the 1930s, the asbestos industry contracted with Saranac Laboratories in upstate New York, one of the country's leading pulmonary disease research facilities, to conduct a number of tests on various aspects of asbestos toxicity. The contracts usually contained clauses ceding ownership of the data to the sponsors and giving them the right to approve studies prior to publication.[91] Although asbestos caused pulmonary disease in laboratory animals in most of the experiments, the results were rarely published in the scientific literature.[92] The industry's cover-ups continued into the late twentieth century when, according to a federal indictment, W. R. Grace contracted with Dr. William Smith of Farleigh Dickinson University to study the health effects in hamsters of tremolite asbestos and vermiculite. A consultant hired by W. R. Grace then revised a draft report concluding that ten hamsters had died of the rare cancer mesothelioma to remove the reference to carcinogenicity, and W. R. Grace exercised its contractual right to prevent Dr. Smith from publishing the true results of the study.[93]

The pharmaceutical industry also uses nondisclosure clauses in contracts with private consultants and university-affiliated medical schools to suppress the results of clinical drug trials.[94] For example, the manufacturers of the over-the-counter cold medicine phenylpropanolamine (PPA) contracted with the Yale Medical School to conduct a postmarketing study on the potential of PPA to cause hemorrhagic stroke after the FDA threatened to take regulatory action. The study was completed on October 17, 1999, and approved by a company-selected oversight committee, which urged the researchers to notify the FDA of the results immediately. Soon after the researchers complied, an attorney for one of the companies informed them that they had committed a "serious breach" of the contract by disclosing the results of the study.[95] When the FDA then scheduled a public advisory committee meeting to obtain the input of the scientific community, the Yale researchers told the agency staff that they needed more time to refine and analyze their data, and the meeting was put off for another ten months. The scientific community and millions of PPA consumers remained in the dark until the advisory committee finally met on October 19, 2000, a year after the researchers had completed the study. Less than a month after that, the companies voluntary withdrew the products containing PPA and reformulated them using alternative active ingredients.[96]

Even when a company does not have a contract with the person who threatens to disclose its internal research, it can frequently argue that any disclosure would violate its legal right to keep the information confidential. The legal definitions of "trade secret" and "confidential business information" are nebulous and have historically included such diverse items as client lists, business plans, and pricing information. Since all businesses are engaged in trade, the courts tend to construe the terms expansively to encompass anything a company wishes to keep secret.[97] Thus a company can sweep under the "trade secret" or "confidential business information" rug virtually any type of information that it is not otherwise required by law to disclose, including damaging health and safety research. For example, when CBS officials learned of the confidentiality contract Jeffrey Wigand had signed with the tobacco company Brown & Williamson, his former employer, they canceled plans to air a *60 Minutes* interview with him for fear of being sued by the company for tortuous interference with the confidentiality agreement.[98] In these settings, the third party making the disclosure—usually a journal or a media company, but sometimes even a regulator—is positioned essentially as an accomplice to the scientist who is attempting to publicize information subject to a confidentiality clause. Under federal law, government employees can even be charged criminally, imprisoned for up to one year, and terminated from their positions if they disclose information that companies legitimately claim to be trade secret or confidential business information.[99]

PUBLISH SECRETS AND PERISH

Concerned about the adverse potential health effects of overprescription of the AneuRx stent, an FDA-approved medical device, for treating heart aneurysms, three FDA scientists and a professor at the University of Michigan School of Medicine prepared a paper comparing the death rates in patients using the devices to the death rates resulting from the surgical procedure previously used. Relying on adverse event reporting data in FDA files, much of which had been provided by the device's manufacturer, the paper combined long-term and short-term mortality rates and concluded that the overall mortality rate was higher for patients using the device. After the peer-reviewed *Journal of Vascular Surgery* had accepted the article and posted a prepublication preview of the study on its website, an attorney for the manufacturer threatened to sue the publisher if it published the article or continued to mention it on the website. The manufacturer claimed that the reports it had submitted to the FDA could not be used in the study because they were proprietary trade secrets that could not be disclosed or used without the company's permission. The article was formally withdrawn by the FDA researchers' superiors, who claimed that the agency (not the scientists) owned the study. Although the University of Michigan coauthor disagreed with this decision, the article was never published.[100] ❐

Like the contract-bound researcher, scientists who have access to data that are not public and attempt to publicize damaging information that a company regards as confidential are subject to liability for any lost profits and competitive damage that results to the company if the confidentiality claim is valid. So long as a company has an arguable legal claim that the information forming the basis for a published report or statement is confidential "proprietary" information, it can also intimidate journals into declining to publish the results. For example, the peer-reviewed journal *Clinics in Occupational and Environmental Medicine* decided not to publish a previously accepted article written by a Boston University epidemiologist and his coauthor after attorneys for IBM Corporation threatened legal action because the data used in the article had been produced in litigation pursuant to a protective order. Even though the scientists, who had prepared the study for plaintiffs' attorneys in litigation against IBM, received legal advice that the information was in the public domain, the journal still refused to publish it. In an unusual expression of solidarity among the scientific community, the authors of nine other studies that were scheduled to be published in the same issue protested this action by withdrawing their submissions as well.[101] Two years later, IBM withdrew its objections, and the study was published in an online journal.[102] Scientific freedom notwithstanding, it is difficult to fault scientific journals for declining to subject themselves to the possibility of expensive litigation in such situations, even if they are likely to prevail in the end.

Even when nondisclosure contracts are not practical, companies can still use the nebulous "trade secret" claim to protect from public view some of the information they must nevertheless share with regulators, like basic information on chemical structure or even some toxicity information. In most EPA programs, classifying information as trade secret protected is easy to do: a company need only stamp the information "confidential business information" and the claim automatically applies, unless an outside party seeks the information under the FOIA.[103] In fact, even if the agency discovers that a claim is wholly without basis—usually in the course of responding to an FOIA request—there is no penalty.[104] As a result, from a company's perspective, there is little to lose and a great deal to be gained from routinely claiming trade secret protection, since the classified information is at least temporarily removed from public view. A number of companies in fact appear to be taking full advantage of this permissive program. Investigations conducted by the EPA, the General Accounting Office, and an outside consulting group all found a significant amount of unjustified overclaiming.[105] Some regulated parties have candidly conceded that they sometimes illegitimately claim trade secret protection. Furthermore, they defend the EPA's overly protective approach on the ground that any agency-imposed requirement to justify the numerous trade secret claims already in existence would violate the Regulatory Flexibility Act because of the enormous work burden it would create for small manufacturers to sort out which of their trade secret claims are actually valid and which are not.[106]

Legal Devices to Combat Suppression

Anticipating efforts by regulated entities to conceal adverse information and research results, several of the major environmental laws require companies to report health and safety studies and other evidence of adverse effects caused by their products to the relevant regulatory agencies.[107] For example, Congress in 1976 enacted section 8(e) of the Toxic Substances Control Act, which requires manufacturers and processors of chemicals to maintain records of "significant adverse reactions to health or the environment . . . alleged to have been caused by" the chemical and to report to the EPA any "information which reasonably supports the conclusion" that the chemical "presents a substantial risk of injury to health or the environment."[108] The FDA requires drug and device manufacturers to report adverse effects to the agency within twenty-four hours.[109] In the mid-1980s, OSHA also imposed an affirmative obligation on employers to alert workers to the risks of chemicals in the workplace through frequently updated material safety data sheets (MSDSs).[110] The MSDSs must be passed along to purchasers of the chemicals and made available to their workers as well. It is therefore now illegal for an employer to hide from its workers relevant science about "significant risks" arising in the workplace.

These laws recognize that when the results of hidden research indicate adverse effects on public health and the environment, the culture of secrecy operates to public detriment.[111] This recognition is consistent with economic theory. A free market functions efficiently only when workers and consumers are fully informed about the risks as well as the potential benefits of a job or a product. Yet some regulatees persist in suppressing policy-relevant science despite these well-intended provisions.

THE TOXIC SECRET BEHIND BUTTER-FLAVORED POPCORN

Diacetyl is a naturally occurring chemical that provides the artificial butter taste in products like microwave popcorn, but it can also be lethal to workers. After an attorney for a worker at a popcorn plant who contracted a fatal lung disease called *bronchiolitis obliterans* (now commonly referred to as "popcorn lung") reported the case to the Missouri Department of Health, the department investigated and found a suspicious cluster of workers at the same plant suffering from the same extremely rare disease. This in turn lead the National Institute for Occupational Safety and Health (NIOSH) to conduct a high-exposure animal inhalation test that produced what the NIOSH pathologist called "the most dramatic case of cell death I've ever seen."[112] This was not, however, the first time diacetyl had been tested in laboratory animals. A similar inhalation study a German chemical company had conducted a decade earlier had left the rats gasping for breath within four hours and killed half of them before the end of the day. The animals' lungs had shown effects similar to those identified in the NIOSH study.[113]

Although a summary of the industry study was included in a publicly accessible database maintained by the Flavor and Extract Manufacturers Association (along with dozens of other studies on diacetyl), it was never published or submitted to any U.S. government agency.[114] More important, the information was not included on the MSDSs for diacetyl that must by law be available for inspection by all workers exposed to chemicals in the workplace. It is not clear whether the company's failure to report this information was willful or simply careless, but the consequences were substantial. A number of injuries and some fatalities have been linked to worker exposure to the toxic chemical, many or perhaps all of which could have been avoided if the employers had carefully complied with the disclosure laws.[115] ❐

The preventable deaths and injuries from popcorn lung are not the only tragedies that could have been avoided if companies had complied with existing disclosure requirements, yet some blame is also due the regulators for failing to draft clear reporting requirements about what the law requires. Under at least some of the environmental and health statutes, adverse effect reporting requirements are needlessly ambiguous and leave the regulated industry "in charge of citing what information it would like to disclose and what analyses it would like to do, presenting ample opportunities for industry-funded researchers to keep underlying data and discrepancies confidential and to make strategic decisions as to whether to submit research studies for regulatory consideration."[116] For example, the Toxic Substances Control Act requires manufacturers only to report what they believe to be "substantial risks" arising from their products.[117] The high level of risk needed to trigger this reporting obligation, coupled with the ambiguities necessarily entailed in determining when research actually meets the "substantiality" threshold, provide manufacturers with great discretion to decide whether or not to report ambiguous testing or other information on nonfatal health effects.[118] For many other types of harms—like risks to groundwater, air, and other resources—there are no federal reporting requirements at all, unless the polluter releases a "reportable quantity" of a specifically listed pollutant in a twenty-four-hour period.[119] Moreover, companies are generally not required to document or otherwise explain the calculations or estimations on which they base their determination that their activities have or have not triggered the reporting requirements.[120] This ambiguity is not, however, a regulatory inevitability: in at least one other licensing program, the EPA has proved quite adept at crafting clear and aggressive rules for disclosure of adverse effects information.[121] In the adverse effects reporting requirements governing pesticides, the EPA requires companies to report "opinion" evidence by reliable experts;[122]

discontinued studies;[123] and anything on a lengthy and specific list of major and relatively minor effects.[124]

Further undercutting the effectiveness of adverse effects reporting requirements are the basic realities of enforcement that make suppression difficult to detect, thereby lowering the risk that violators will be caught.[125] Instances of noncompliance are likely to come to the agency's attention only when an employee or other insider blows the whistle or an outsider manages to penetrate industry files, usually through a tort lawsuit. These conditions are not common. A strongly worded nondisclosure contract will probably silence all but the most courageous whistle-blowers, and documents turn up in lawsuits only years after damage that might have been prevented has already occurred. The low priority agencies typically assign to overseeing private compliance with reporting requirements in allocating their limited resources further decreases the odds of detection.[126]

At the same time that enforcement risks are low, fastidious attention to these reporting obligations brings a company few rewards. Filing adverse reports does not immunize the company from additional regulatory action or protect it from private tort suits. Indeed, complying with these requirements seems much more likely to increase the likelihood that an agency will promulgate burdensome new regulations and that perceived victims will bring lawsuits.

Faced with an internal cost-benefit assessment that will sometimes counsel a company against disclosure of adverse research results, it is thus not surprising that a number of companies have a less than stellar compliance record with agency reporting requirements.[127] Dr. Jerome P. Kassirer, a former editor of *NEJM,* observes that adverse effects reporting to the FDA in clinical drug trials is "uneven, sometimes grossly flawed, and occasionally overtly deceptive." Examples of pharmaceutical company failures to comply with FDA's adverse effects reporting requirements include:

Fen-phen—A manufacturer of the weight loss drug fen-phen concealed from the FDA a number of reports it received of pulmonary hypertension allegedly caused by the drug.[128]

Ancure Endograft System—The manufacturer of a medical device designed to treat weakened blood vessels in the abdomen failed to tell the FDA about thousands of reports that the delivery mechanism for the device was breaking off in the bodies of the patients and twelve reports of deaths resulting from the malfunction.[129]

SangStat—After failing to demonstrate the bioequivalency of its generic drug with the FDA-approved version using apple juice as the vehicle, the manufacturer demonstrated bioequivalence using chocolate milk as the vehicle, but failed to tell the FDA about the apple juice test. The generic drug, which was not bioequivalent, later caused several hospital emergencies.[130]

Compliance with the EPA's adverse effects reporting requirements by the chemical manufacturing community may even be worse. For example, when EPA announced an "amnesty" program under which it assessed greatly reduced penalties for a limited time to encourage compliance with the substantial risk notification requirement of the Toxic Substances Control Act, regulated entities volunteered eleven thousand previously unsubmitted studies, four times the number of studies submitted during the fifteen years the requirement had been in effect.[131] Although adverse effect reporting has now become more routine,[132] the vast bulk of the reports filed with the EPA disclose uneventful and often benign findings.[133] The extent of possible undercompliance is underscored even more powerfully by the EPA's administrative enforcement action in 2005 against a highly sophisticated company, the DuPont Corporation, for failing to report substantial risks associated with a chemical known as perfluorooctanoic acid (PFOA) that it used in making Teflon. Documents disclosed in a lawsuit filed by neighbors whose groundwater had been contaminated revealed that DuPont had neglected to report in 1981 that it had discovered PFOA in umbilical cord blood fluid from one baby and in the blood of another baby born to female workers at its plant in Parkersburg, West Virginia.[134] Three years later, DuPont failed to report that it had discovered PFOA in a public drinking water well adjacent to its Parkersburg plant[135] and in tap water in an Ohio town located immediately across the Ohio River.[136] A decade after that, DuPont failed to report the results of a 2004 study in which it discovered high levels of PFOA in serum from twelve persons residing near the plant.[137] The $16.5 million fine DuPont ultimately paid was the steepest fine the EPA had ever administered for a reporting violation.[138] Even in hindsight, however, violating the reporting requirement might well have been worth the risk had it allowed the company to avoid the $107 million in damages it ultimately paid to settle the neighbors' lawsuit.[139]

According to a February 2005 federal indictment, the W. R. Grace Corporation suppressed a study it had commissioned from Richard R.

Monson of the Harvard School of Public Health in 1982 of worker mortality at the company's Libby, Montana, plant from 1961 to 1981. Although Dr. Monson concluded that an excessive number of employees at the Libby facility had died of respiratory cancer, the company never submitted the report to the EPA, as required by the EPA's regulations, nor did it add the information to the OSHA-required MSDSs for the Libby facility. In a private memo to upper-level management, Grace's health and safety director noted that it was "no surprise" to learn that "[o]ur major problem is death from respiratory cancer."[140] Yet management also decided that this "unsurprising" incriminating report needed to remain inside the company.[141] By this point, liability risks from already-exposed workers had become so enormous that it seemed more prudent for Grace to suppress adverse information than share it.

The ambiguous and narrow reporting criteria in most of the disclosure statutes, coupled with modest sanctions and a low probability of detection—in congruence with rational choice theory—lead to a prediction that companies will report adverse discoveries only when they are likely to be disclosed anyway or when the effects are not especially consequential. Although the DuPont example suggests that companies can miscalculate the costs and benefits of failing to report, the high volume of minor-consequence reports and the existing, albeit limited evidence of failures to report serious adverse effects caution against optimism about the effectiveness of existing statutory adverse effects reporting requirements.

Plaintiffs as Accomplices

So far, our tour of strategies for hiding science has visited only regulated companies that have been unwilling to share damaging internal information with agencies and the public. Yet the same companies sometimes enlist victims and their attorneys as willing accomplices in the cover-up. In these situations, defendant companies offer a bonus payment or other inducement to plaintiffs if they agree to seal (that is, legally suppress) damaging internal documents divulged in litigation so that regulators, future victims, and the media do not have access to them.

TAKE THE PAYMENT AND "SHUT UP"

When plaintiffs' attorneys in the early 1980s discovered internal company memoranda revealing that the manufacturer of the arthritis medicine Zomax knew of its serious allergenicity risks while continuing to market it without a suitable warning, the manufacturer swiftly responded by entering into secret settlements with each of the plaintiff-victims.[142] In exchange for bonus payments, the plaintiffs agreed to keep the company's secrets to themselves. Other unwitting patients continued to take Zomax until the FDA arranged a recall in 1985. At that point, according to FDA estimates, Zomax had been linked to 14 deaths and 403 life-threatening allergic reactions. One plaintiff's attorney summarized his own dilemma in agreeing to the settlement terms: "The problem is that they have a gun to your head. The client is concerned about being compensated in full. The lawyer must abide by the concerns and wishes of his client . . . not the fact that other victims may be injured."[143] Another plaintiff's attorney conceded that the company "paid my clients a ton of money for me to shut up."[144] ❐

Newspaper reports and legal periodicals reveal a number of other disturbing accounts of companies inducing plaintiffs and their attorneys to suppress information about product hazards revealed during discovery for litigation even as they continued to market the product to other consumers.[145] Such accounts date back to very early litigation in 1933 against the asbestos companies by eleven workers who filed claims for asbestosis.[146] Although some commentators have suggested that the incidence of cover-up confidential settlements is on the rise,[147] these accounts are not easily verified empirically because of practical obstacles to collecting records on even the number of such confidential settlements, much less their significance for public health policy.[148] For the same reason, the magnitude of human suffering that has resulted from hiding information produced during litigation is difficult to quantify, but it may well be substantial.[149] We must at least acknowledge the very real possibility that the tort system, which is the one legal institution that has been particularly successful at forcing suppressed information into the open, may currently be playing a much more limited role than it could be because key documents can be sealed in gag settlements.

There are, of course, countervailing considerations that warrant nondisclosure in some cases. For example, a discovery request might yield personal information on the subjects of a clinical trial or epidemiological study that the researcher agreed to keep confidential, or a defendant might use a discovery request to probe into a plaintiff's personal habits (with an implicit threat to make the information public in the future) to intimidate the plaintiff into dropping the case.[150] Businesses involved in litigation also have a legitimate interest in keeping sensitive financial and trade secret information out of the hands of their competitors.[151] Finally, nondisclosure agreements facilitate the free exchange of information among the parties during the pretrial discovery stage, thereby relieving busy judges of the obligation to resolve hundreds of individual discovery disputes that would invariably break out if produced documents were presumptively available to the public.[152]

While a balancing approach—under which the courts balance the legitimate interests of the parties in maintaining confidentiality against the public interest in disclosure of policy-relevant scientific information—is an obvious solution for resolving these conflicting objectives, the courts' review of all sealed settlements and protective orders would be extremely resource-intensive and, in some settings, procedurally difficult.[153] When sealing the record is a stipulated condition of the settlement, a judge is required to approve the condition before entering a judgment based on the settlement. Local rules, however, vary tremendously with respect to a judge's discretion to preside over the details of these settlements when they do retain the ultimate sign-off power.[154] Less amenable to judicial control are settlements in which the plaintiff agrees, usually in a "sidebar" contract with the defendant, to withdraw the lawsuit and seal the documents in return for a sum of money.[155] Parties' ability to agree or stipulate to broad protective orders preventing disclosure of internal documents produced during litigation limits still further the courts' ability to prevent them from suppressing health and safety information.[156] Since settlements are a much less resource-intensive way to keep judicial dockets manageable than full-blown trials, the judges themselves may have only limited incentives to inquire into their propensity to hide critical policy-relevant research.

Further Evidence of Hiding: Suppression by the Government

The government not only faces a number of challenges in overseeing private compliance with reporting laws but also must confront its own conflicts of interest in deciding whether to release damaging findings. The incentives motivating government to suppress adverse information are generally quite different from those that affect the private sector. In many instances military secrets or national security, rather than competitive advantage, provide the impetus for concealing information. Sometimes the incentive, however, is a much less commendable desire to hide bad news from the electorate. Regardless of motive, the resulting suppression can look surprisingly similar from the perspectives of the scientific community and the general public. And to the extent that the government's research can be categorized as preliminary or "predecisional," suppression is often too easily justified.

THE AIR IS SAFE TO BREATHE

When the World Trade Center towers were destroyed in the tragic attack of September 11, 2001, the EPA staff knew full well that some constituents of the enormous cloud of dust generated by the collapsed buildings, including asbestos, lead, glass fibers, and concrete dust, could present health hazards.[157] In fact, the asbestos-laden insulation used in almost forty of the lower floors in the North Tower had come from the vermiculite mines in Libby, Montana that EPA was at that moment requiring W. R. Grace to clean up at a cost of millions of dollars.[158] Yet because this information would undoubtedly frighten the already terrified residents of New York City, top-level EPA officials experienced substantial political pressure to downplay or even conceal the potential public health risks.[159] As a result, the EPA capitulated to efforts by the White House Council for Environmental Quality to soften its press announcements and tell residents that the "air is safe to breathe and their water is safe."[160] Two years later, however, an exhaustive report by the agency's inspector general concluded that the statement was not scientifically supported at the time the agency made it and that the Council on Environmental Quality had "convinced EPA to add reassuring statements and delete cautionary ones" to its press releases at the time.[161]

At the same time it was reassuring residents, however, EPA was using a reliable outside monitoring system to detect high levels of a variety of toxic contaminants in the city's air. Relying on modern asbestos detection technology, civilian testing companies working for the government determined that the dust contained asbestos at levels comparable to the dangerous levels found in the homes in Libby.[162] Although New Yorkers' exposure to the risk was much briefer than that of most Libby residents, the scientists quickly posted the test results on the website of the American Industrial Hygiene Association (AIHA) to put residents on notice of that risk, and EPA staff immediately relayed the information to the Federal Emergency Management Agency and the White House. The information flow, however, stopped there.[163] The report of health risks was removed from the AIHA website within hours of going up, and the two scientists who were responsible for uploading it were told that their services would no longer be needed at Ground Zero.[164] Government employees in charge of the cleanup effort and the people doing the actual work were not informed about the contents of the dust, and EPA headquarters continued to issue reassuring reports.[165]

A subsequent study concluded that exposure to air contaminants by Fire Department employees who were present at the site produced a "substantial reduction" in pulmonary function during the year following the collapse.[166] The loss of lung capacity during that single year was equivalent to twelve years' worth of normal age-related decline.[167] A report by the Mount Sinai School of Medicine team charged with implementing a congressionally authorized screening program for workers at the site reported that 2.5 years after the collapse, lung abnormalities continued to plague World Trade Center responders. Almost 70 percent of the more than ten thousand workers in the program reported new or worsened respiratory conditions, and almost 30 percent of the nonsmokers suffered from objectively diagnosed breathing impairments.[168] The study's director reported: "Our patients are sick, and they will need ongoing care for the rest of their lives."[169] ❐

Government suppression manifests itself in many ways. In some cases, the suppression involves simply barring government scientists from communicating their research to the outside world without prior approval. For example, a twenty-four-year old political appointee at the National Aeronautics and Space Administration (NASA) attempted for several years to prevent the NASA climate scientist James Hansen from speaking publicly about his conclusion that global warming had contributed to growth in hurricane intensity.[170] In April 2004, a high-level Department of Health and Human Services (HHS) official demanded that the World Health Organization (WHO) route any requests for participation by HHS employees at WHO meetings through the HHS Secretary's office to allow it to "assist" the WHO in selecting the most "qualified" participants.[171] In March 2007, the Department of Commerce issued an administrative order requiring employees to submit all written and audiovisual materials prepared in connection with a "Fundamental Research Communication" to the head of the relevant agency "for approval in a timely manner."[172] Agencies subject to the order included the National Oceanic and Atmospheric Administration, the National Weather Service, and the National Marine Fisheries Service.

Some incidents of government suppression arise from the desire of upper-level government officials to protect favored interest groups. The Salmon Recovery Science Review Panel assembled in 2000 by the National Marine Fisheries Service (NMFS) and approved by the NAS found strong scientific support for distinguishing between wild salmon and hatchery salmon in making endangered species designations and recommended a separate category for wild salmon. But the NMFS deleted this recommendation from the final version of the panel's report, apparently at the insistence of an attorney in the Office of the General Counsel who had previously, as an attorney for the timber industry, advocated including hatchery fish with wild fish to avoid Endangered Species Act restrictions on harvesting trees in national forests.[173] The recommendations only became public when six members of the Science Review Panel published their conclusions in *Science* in defiance of the agency.[174] Senior FDA officials in 2004 prevented an FDA staff scientist in charge of evaluating the risks posed by newly emerging antidepressant drugs to children and adolescents from presenting to an FDA advisory committee the results of a meta-analysis he had undertaken of both the published data and data the drug companies had submitted to FDA but refused to make public. The meta-analysis, the results of which were later confirmed by an FDA-commissioned team of scientists at the Columbia University School of Medicine, showed that children given the drugs were almost twice as likely to become suicidal as children on placebos.[175]

Although it is vastly more troubling to learn that the government has concealed important health and safety information from the public without adequate justification, the good news is that this practice seems less likely to succeed in government agencies than in the business world. Unlike most private sector employees, government employees are not bound by confidentiality clauses in their employment contracts and need not fear expensive lawsuits for lost profits if they blow the whistle, though they do risk termination from their jobs and associated reputational damage. Since agency employees are more inclined to view their government employment as a civic service, they may also be less tolerant of efforts to hide important health and environmental information from the public and more inclined to present it at scientific meetings and even leak it to the media than scientists working in the private sector.

Conclusion

Private sector entities that conduct health and safety research on their products and activities often face a dilemma. If they release preliminary scientific research that suggests the possibility of long-term adverse effects, those disclosures may drive away potential customers, stimulate government regulation, and generate expensive tort litigation. If they do not disclose that research, activists, plaintiffs' lawyers, or even government prosecutors may accuse them of covering up information needed to protect human health and the environment. Although government suppression presents a far more complex and individualized cost-benefit balance, agency officials face a similar dilemma. Making the public aware of preliminary scientific information indicating that commercially important products and activities may pose public health risks could unnecessarily frighten the public. In the case of products like drugs that have public health benefits, disclosure may do more harm than good. Failing to disclose the information may, however, subject the agency to subsequent accusations that it was involved in a politically inspired or even unlawful cover-up to protect powerful economic interests.

Current legal protections do very little to combat the natural instinct on the part of both private sector and government actors to suppress economically or politically damaging bad news for as long as possible. To the contrary, contract law and the common law of "trade secrecy" reinforce private sector instincts to suppress. Through employment contract nondisclosure clauses, employer-sponsors not only can threaten to terminate employees who do not agree to suppress damaging findings but also can sue them for damages that result from the disclosure if the information turns out to be erroneous and damages the company's reputation. The sanctity that trade secrets enjoy in U.S. law contributes to the carefully maintained illusion that strict controls over the dissemination of internal research are essential to the vitality of private intellectual property and therefore socially beneficial. Because countervailing public policy considerations played no role at all in the evolution of trade secrecy law, they are too often given short shrift in a world in which full public access to policy-relevant scientific information is necessary to reduce serious threats to public health and the environment.[176]

Modern health and environmental statutes have made some inroads into these sweeping common law entitlements, but the current legal regime is not capable of counteracting the powerful private sector incentives to suppress relevant scientific information. Legal reform is imperative to clarify and strengthen existing reporting and disclosure obligations, and in Chapter 10 we provide some suggestions for appropriate changes to the current legal regime. Until Congress acts, however, companies will continue to wrestle with the difficult choice between disclosing public health risks that may make them vulnerable to liability and regulatory costs on the one hand and concealing adverse effects to protect their profits on the other. The many examples of past concealment described in the preceding pages suggest that the public should not be sanguine about how companies will resolve that dilemma.

CHAPTER SIX

Attacking Science

The Art of Turning Reliable Research into "Junk"

Not all policy-relevant research is commissioned by advocates with an eye to how it will affect their interests when it enters the legal system. When an advocate encounters a study produced by an independent scientist or commissioned by another entity that is inconsistent with the advocate's view of the relevant issue, hiding the research is obviously not an option. The next best solution may be to attack the research as "unscientific" or otherwise "fatally flawed." The primary aim when attacking science is to cause others to lose confidence in the reliability of the challenged research. In the realm of policy, advocates can accomplish this goal by raising multiple esoteric complaints that only a few highly specialized scientists can understand or evaluate. The attacks may come from advocates or their clients, but they just as frequently come from hired surrogates or friendly think tanks. The attackers typically raise hypothetical concerns about methods, statistical competence, and unaddressed alternative explanations for the results, but they rarely point to properly conducted studies that reach different results or offer to conduct better studies with their own resources. The attacks, in other words, are offered not in the spirit of advancing collaborative scientific dialogue, but in the hope of throwing the research into doubt and, if possible, discrediting it. Some advocates even go beyond attacking the science to harassing the scientists who conduct unwelcome research, an even more disturbing strategy that we address in the next chapter.

Since skepticism is a familiar and critical feature of the realm of science, scientists generally presume that criticisms—even vigorous, cutthroat criticisms—are offered in the spirit of advancing scientific understanding and without hidden agendas. The scientists who generated the research are, of course, free to respond to the attacks, and we shall see that they usually do, often quite convincingly. In the policy realm, however, sophisticated attackers can easily dismiss perfectly valid responses as self-interested rationalizations. Policy-makers and the general public are in no position to distinguish genuine critiques of science from the illegitimate ones generated by advocates. Independent scientists who are prepared to review both the studies and the critiques are in the best position to rebut unwarranted attacks, but most of them are too preoccupied with their own research agendas to devote time and resources to setting the record straight. The illegitimate attacks therefore tend to have considerable staying power in the public policy arena, even after scientists have shrugged them off.

While the primary aim of illegitimate attacks is usually to reduce the likelihood that the relevant legal actors will rely on the unwanted research, the advocate may also succeed in delaying public acceptance of research that ultimately survives the attacks for years or even decades.[1] Any delay, in turn, can be extremely valuable in forestalling compliance costs and putting off liability judgments. Accordingly, attacks on science remain one of the most widely used strategies for bending science. A tobacco company put it well when it boasted that "[d]oubt is our product since it is the best means of competing with the 'body of fact' that exists in the mind of the general public."[2] The strategy of sowing doubt by attacking science has spread far beyond the tobacco industry and has in fact become so routine that scientists have coined the term "manufacturing uncertainty"[3] for the tactic. Modern public relations firms frequently recommend this strategy as the client's first defense against unwanted research.[4]

This chapter surveys the range and force of illegitimate attacks on science. After introducing several features that make attacking science different from run-of-the-mill scientific skepticism, we detail some of the more creative techniques advocates have employed to launch attacks, including targeting scientific journals before and after publication. The chapter closes with a survey of attacks that are aimed not so much at convincing the scientific community that research is flawed as at achieving the more modest goal of persuading policy-makers, judges, and the general public that the science is unreliable.

How "Attacking Science" Differs from "Healthy Scientific Skepticism"

Many scientists may be skeptical of the position that scientific criticism can ever be a bad thing. "If a critique is unjustified, then it will be easy to dismiss," they might assume. "Better to find too much wrong with research than too little." But the attacks employed to bend science are far more ruthless and damaging than most scientists might imagine. When the basic rules of scientific research—honesty, disinterestedness, balance, collaborative and constructive criticism—are suspended, even the most cautious expert can be caught off guard. Consider the case of the Japanese epidemiologist Dr. Takeshi Hirayama.

HIRAYAMA IN THE CROSS-HAIRS

In January 1981, Dr. Takeshi Hirayama published in the *British Medical Journal* an epidemiological analysis of data on the incidence of lung cancer in more than ninety thousand nonsmoking wives of Japanese smokers from 1966 to 1979.[5] Hirayama concluded that the wives' risk of contracting lung cancer was between one-third and one-half the relative risk of a smoker.[6] To Hirayama these results indicated "the possible importance of passive or indirect smoking as one of the causal factors of lung cancer."[7] Hirayama's suggestion that ETS caused lung cancer was extremely threatening to the tobacco industry, because it would provide strong ammunition to activists seeking to ban smoking in public places. The U.S. tobacco industry immediately prepared critiques of the Hirayama paper and commissioned reviews by its stable of consultants.[8] As the medical journals began to publish letters criticizing the study from paid tobacco industry consultants, who never identified themselves as such,[9] Hirayama patiently responded to each one.[10]

An analysis of Hirayama's paper prepared by Peter N. Lee, a statistician retained by the British Tobacco Advisory Council, identified several weaknesses in the study, but concluded that it was "difficult to argue that the study was fundamentally flawed."[11] When he was later asked to comment on a statistical critique prepared by an RJR statistician with the help of Dr. Marvin Kastenbaum, the U.S. Tobacco Institute's chief statistician,[12] Dr. Lee disputed their calculations and provided detailed calculations of his own to rebut them.[13]

Unwilling to leave it at that, Dr. Kastenbaum revisited the article and came to the conclusion that the Hirayama paper contained a "fundamental error."[14] Kastenbaum thought that Hirayama had forgotten to take the square root of a critical number and had therefore reported a chi-squared value, rather than the proper chi value. He then hired Nathan Mantel, the creator of the statistical test Hirayama had employed, "to seek an independent evaluation of the results."[15] Mantel agreed that the article at least raised the question "whether Hirayama has conducted a more refined analysis about which he is giving us no clues, or he has mistakenly interpreted his chi-square [*sic*] value as a chi value."[16] Although a "[m]uch more careful analysis of the data would be needed" to know for certain, Mantel was at least willing to entertain the possibility that Hirayama had made a critical statistical mistake.[17]

Without attempting to contact Dr. Hirayama for an explanation or conduct further research, the Tobacco Institute initiated an aggressive public relations blitz using the asserted statistical error "to discredit Dr. Hirayama's" paper.[18] On June 10, the institute's chairman sent a telegram to Hirayama's boss notifying him of the "very grave" error Dr. Kastenbaum had discovered.[19] Copies of the telegram were sent to Dr. Hirayama, the *British Medical Journal,* and the press.[20] A simultaneously issued press release announced that Dr. Mantel had "confirmed the existence of a fundamental mathematical error" in the Hirayama study.[21] The institute's public relations department also distributed a package on the Mantel critique to a field force it had assembled with orders to "personally visit more than 200 news organizations nationwide" on Friday, June 12.[22] About one hundred video news clips were sent to local television stations, and four hundred audio cassettes were sent to large radio stations.[23]

At this point, events took an unexpected turn. The Tobacco Institute's own paid consultants privately broke ranks. Five days after the press blitz, Lee provided a four-page detailed response to the institute's critique of his earlier analysis.[24] Another statistician employed by the Brown and Williamson Tobacco Company also examined the Hirayama data in light of Mantel's critique and related that his findings "agree[d] with Hirayama and Lee."[25] Franz Adlkofer, a German tobacco industry scientist, reported that he also agreed with Lee and believed that Mantel and Kastenbaum were wrong.[26] Adlkofer later took the position that "Hirayama was correct, that the TI [Tobacco Institute] knew it and . . . TI published its statement about Hirayama knowing that the work was correct."[27]

Despite the limited scientific basis for its attack, the Tobacco Institute's public relations blitz had its desired effect. Stories appeared in major newspapers across the country with titles like "Miscalculation Reported in Study on Cancer in Wives of Smokers,"[28] "Math Mistake Claimed in Smoking-Peril Study,"[29] and "Tobacco Unit Raps Smoking Study."[30] Local newscasts devoted brief segments to the news about the

Mantel letter and the Tobacco Institute's unequivocal statement that "an arithmetical error invalidates this study."[31] Hirayama later demonstrated in a letter to the editor that he had in fact used the correct statistical test, and Mantel allowed that the institute had "put words in my mouth."[32] Yet the institute did not recant, and it in fact steadfastly maintained for the next twenty years that the study had been the subject of many "highly critical" letters to scientific journals and was therefore "scientifically questionable."[33] ❐

Many sociologists and philosophers of science agree that vigorous skepticism is one of the most critical ingredients of rigorous science.[34] To ensure that science is advancing with the greatest possible objectivity, scientists have a responsibility to scrutinize each other's work and act as collective gatekeepers over the resulting body of research. The corollary to this proposition is that scientists must likewise be prepared to defend their methods, data, assumptions, and conclusions equally vigorously when colleagues challenge their work.

It is one thing, however, to encourage a genuine debate about the quality of scientific studies and the validity of the implications drawn therefrom and quite another to allow entities with economic or ideological agendas to launch attacks on policy-relevant scientific research solely for the purpose of "derail[ing] the use of sound science in the regulatory process."[35] One scientist that has been on the receiving end of such attacks has warned other scientists who publish policy-relevant research to "be prepared for the sort of review that can take years of your life, keep you from doing other work, and may never be fully resolved." Even if the work "is published in the best journals," is "widely acclaimed as ground breaking and important," and is "replicated by others," once a federal agency uses it to support a significant regulatory action, "the accountants in the OMB can still declare it junk science" and "some industry-financed body with an Orwellian name like Center for Regulatory Effectiveness will be cheering them on."[36]

While illegitimate attacks undermine science by siphoning limited resources from useful research and by creating doubts about the value of the enterprise in the public mind, distinguishing the illegitimate attacks from normal and productive scrutiny is not always easy. Indeed, the most insidious feature of the attack strategy is that it so closely resembles legitimate scientific debate. Yet a more careful look reveals that the attacks are consistently different in nature, duration, and quality from the vital skepticism and scrutiny that accompanies healthy science.[37] The following discussion compares three distinguishing characteristics of illegitimate attacks on science with the corresponding characteristics of legitimate scientific dialogue.

Obfuscation versus Helpful Criticism

In a productive critique of research, a reviewer will identify one or more significant weaknesses in the study and offer suggestions on how to improve those particular features. For example, epidemiologists will criticize research that is not careful about separating exposed workers from the unexposed or control group and suggest better sorting techniques. Toxicologists will criticize an animal carcinogenicity study that terminates the exposures too soon and suggest a more appropriate duration, or they may critique an acute toxicity study that employs too few animals to detect an effect or suggest a more appropriate dosing regime. These criticisms identify features of a study that deviate from what the critic believes to be the better research practice and suggest how the researchers can bring them into line.

Two qualities of illegitimate, obfuscatory attacks on science help distinguish them from the legitimate criticism that is essential to scientific advancement. First, the illegitimate attacks typically include challenges that are scientifically inappropriate, either because they are unsubstantiated or because they are careful to take issue with unresolvable questions about research design. Because unsubstantiated attacks on the merits of a study are likely to fail in the long run, as in the case of the Tobacco Institute's attacks on Dr. Hirayama's study, illegitimate attackers often find it more productive to deconstruct research by raising hypothetical charges about research design that are not supportable, but are likewise not easily refuted.[38] A deconstructing attack might posit, for example, that the standard-sized aquarium used to house an aquatic test species was too small for the particular experiment. Even if the criticism seems trivial, the researcher cannot experimentally prove that the aquarium was precisely the right size. The sociologist H. M. Collins coined the term "experimenter's regress" to describe the ability of those who attempt and fail to replicate an experiment to generate "a potentially endless series of questions designed to probe every aspect of the 'failed' experiment, from the reliability of the instruments used to the honesty of the researchers."[39] A scientist whose research is the target of a deconstructing attack faces an equally endless series of questions, selected precisely because they cannot be answered, from scientists who have not attempted to replicate the experiment and have no intention of doing so.[40]

Second, illegitimate attacks are rarely straightforward and comprehensible to laypersons or even to scientists who are unfamiliar with the particular subject matter of the research. Indeed, such critiques are often consciously structured in very sophisticated ways so as to require a great deal of time and energy from the evaluator, thereby forcing supporters of the challenged research to divert disproportionate amounts of their scarce resources to its defense. By demanding much from would-be defenders, an attack that raises highly esoteric criticisms can enjoy a longer life span. Even if the appropriate specialists in the field ultimately dismiss the attacks as unfounded, the march of time and the complexity of the issues may lead scientists outside the field to assume that the attacker was misguided but well intentioned. Consequently, scientists who launch such illegitimate attacks have little reason to fear that the abuses will adversely affect their own careers.

Endless Challenges versus Consensus-Building

In healthy scientific debates, close scrutiny of research is traditionally provided gratis by scientists who voluntarily contribute their services for the purpose of advancing science. A scientist who chooses to dedicate time to critiquing another scientist's work incurs significant opportunity costs because that time will be unavailable for writing her own grant applications, conducting research, and drafting articles for publication. As a result, peer scrutiny of published scientific research outside of the routine peer review process administered by the journals and granting institutions tends to be undertaken sparingly. When a scientist engages in this activity, it is usually because she believes that her input will make a valuable contribution to science.[41]

Illegitimate attacks on policy-relevant science are thus unlikely to be initiated by busy, unpaid scientists who must carefully pick their battles, but instead appear more often to be liberally financed by parties with an economic stake in the outcome of a pending or anticipated legal proceeding or public policy debate. Scientists who assist in sponsored attacks may receive handsome remuneration—often more than they would receive from grant applications and salary rewards for conducting meaningful research. More important, when one scientist tires of mounting a challenge or loses confidence in its legitimacy and returns to his lab, the sponsor can usually find other scientists willing to take over the attacks. As long as there is money to mount the challenge, there will be someone available to mount it. The attacks therefore go on and on, even as the scientific community draws closer and closer to consensus about the legitimacy of the research under attack. The petroleum industry's long-standing challenge to the hypothesis that greenhouse gasses cause global warming, discussed below, is an excellent example of such an endless challenge.

Challenges financed and managed by advocates are also not aimed at facilitating the emergence of a scientific consensus around an established scientific truth. To the contrary, they are meant to prolong the perception of dissensus both within and outside of the scientific community. Thus, on the heels of the famous 1964 surgeon general's report concluding that "[c]igarette smoking is causally related to lung cancer in men,"[42] the tobacco industry assembled a team of lawyers called the Committee of Counsel to fund critiques and obfuscatory research through consulting contracts, rather than grants, with the costs being split up among the companies on a market share basis.[43] The secret process lacked even the pretense of peer review, and contracts were awarded not on the basis of scientific merit, but on the anticipated impact that resulting critiques would have on evolving public policy. Scientists whose projects proved useful to the lawyers could expect future funding. For example, after receiving an initial contract for $54,000 in 1968,[44] the statistician Theodor Sterling received millions of dollars in a continuing flow of largess from the committee's special account for the next twenty years. In 1989 Sterling's projects alone drew over $600,000 from the secret fund.[45] One participant later related to the FDA head David Kessler that the Committee of Counsel had so much money to spend on scientists like Dr. Sterling because they "protected the paradigm."[46]

Although the Committee of Counsel's thirty-year effort set the gold standard for covertly funding attacks on science, other industries have created similar, if less well-endowed, strategic support operations over the years to fund equally illegitimate attacks on science. If anything, the incidence of such institutions has increased during recent years. A few of them are highlighted here:

Formaldehyde—When scientific reports began to appear in the late 1970s and early 1980s indicating that formaldehyde was carcinogenic in rats, a coalition of companies in the wood products industry devised a four-part plan to: attack the rat study and any future studies indicating risk; fund a rat study that would be designed to minimize the likelihood of a "false positive" result; and aggressively steer research in directions that would downplay the risks of formaldehyde.[47]

Ephedra—A 1997 FDA proposal to regulate dietary supplements containing more than 8 mg ephedra as drugs inspired the dietary supplement industry to create the Council for Responsible Nutrition, which contracted with several toxicologists to reevaluate the adverse event reports on which the FDA had based its proposal and conduct an "independent" risk assessment that differed from the FDA's assessment.[48]

Fine Particulates—When the EPA relied on a highly influential epidemiology study known as the "Six Cities Study" to revise its air quality standard for particulate matter in the mid-1990s, more than six hundred potentially affected companies from the petroleum, automobile, and other business sectors formed the Air Quality Standards Coalition and contributed $5,000 to $100,000 apiece to support a coordinated attack on the Six Cities Study and related epidemiological studies.[49]

Whether launched by single companies or coalitions, these attacks on science far exceed the normal vetting processes commonly encountered in science. When the chromium industry attacked EPA's comprehensive study of the effects of chromium-6 on workers in the workplace, it hired a consulting company to analyze a "simulated cohort" of workers and to obtain and reanalyze the raw data from EPA's study. The industry then used both of the costly critiques to challenge the validity of the EPA analysis.[50] In a similar effort to resist the emerging scientific consensus that aspirin caused Reye's syndrome, scientists working for the pharmaceutical industry identified seventeen separate flaws in the four studies that the CDC relied on in 1980 to alert the medical community to that risk.[51] Likewise, industry-funded scientists have over the years challenged the science used to assess a number of substances identified by the International Agency for Research on Cancer as carcinogenic in humans.[52] Although the relevant agencies usually reject these prolonged challenges, they sometimes succeed in buying time by manufacturing uncertainty.

Unbalanced Attacks versus Meaningful Collaboration

Healthy scientific scrutiny usually involves multiperspective input from a broad group of peers reflecting many points of view.[53] According to one sociologist of science, "[s]tructuring the community to include multiple perspectives and values will do more to advance the aims in relation to which value-free science was an ideal—impartiality and universality—than appeals to narrow methodology ever could."[54] In contrast to this healthy skepticism, illegitimate attacks on science are typically mounted by a narrow slice of the scientific community. Indeed, some attacks are initially choreographed by lawyers and public relations firms, and the scientists do not become involved until near the end of the process. In setting priorities for the Tobacco Institute–sponsored scientific journal *Tobacco and Health Research,* for example, it was the institute's public relations firm that recommended the criteria for publication: "The most important type of story is that which casts doubt on the cause-and-effect theory of disease and smoking."[55] One should not be surprised to find that this publication did not solicit or attract papers reflecting a wide variety of perspectives on the smoking and health controversy.

Another hallmark of legitimate criticism is its constructive nature. When a scientist takes issue with research, it is his or her obligation to "show, constructively, that [he or she] has the potential to find solutions to the predicaments" revealed in the criticized study.[56] Because the goal of illegitimate criticism is to tear down threatening research, rather than to build up scientific knowledge, a signature characteristic of illegitimate attacks is their negativity. Thus while Dr. Lee's statistical reassessment of the Hirayama data identified several apparent weaknesses in Dr. Hirayama's analysis that might have contributed to a legitimate scientific dialogue and resulted in improvements in future publications of the ongoing study, Dr. Kastenbaum did not include Lee's analysis in the critique of the Hirayama study that he sent to the medical journals because it ultimately supported Dr. Hirayama's conclusions. Healthy and productive skepticism involves thoughtful commentary and suggestions for improvement, but illegitimate attacks offer only destructive criticisms that usually lack rigorous support.

While illegitimate attacks are neither balanced in nature nor collaborative in spirit, their sponsors sometimes present them as if they were. The industry's Tobacco Strategic Attack Plan against government attempts to regulate tobacco in the 1990s highlighted the value to the industry of "allied attacks where friendly third parties are engaged on our side but without direct or obvious connection to the industry."[57] Indeed an attack originating from an "independent" think tank or university-affiliated center is particularly useful to sponsors because such institutions have no apparent economic interest in the outcome of any particular decision-making process. Even if it cannot exercise the degree of control over a think tank or university center that it can over a hired consultant, it behooves an industry to keep sympathetic think tanks well nourished in the hope they will launch allied attacks on scientific studies and reports that put the industry's economic well-being at risk.

Battle Plan

Once an advocate is poised for action, an attack on an otherwise credible study generally follows a familiar pattern. Consistent with its leadership role in advancing the art of bending science, the tobacco industry has provided a sophisticated, multi-faceted battle plan for attacking science.

A POLITICAL CAMPAIGN MENTALITY

Soon after the EPA published its long-awaited risk assessment on ETS, Tom Humber of the public relations firm Burson-Marsteller sent a memorandum to his clients at Philip Morris suggesting a multifocused strategy for addressing the regulatory and litigative consequences of the anticipated public reaction to that document.[58] The memorandum shines an unusually powerful light on science-based strategies for attacking unwelcome research. Burson-Marsteller was, and still is, one of the two or three dominant public relations firms in the world, and its clientele included many of the major actors in most of the important industries affected by the same regulatory and litigative forces as those that threatened the tobacco industry. The strategies it outlines are no doubt indicative of the advice it has provided to many of its other clients.

Humber recommended that the tobacco industry adopt a "political campaign mentality" in an all-out effort to "[d]iscredit the EPA report on ETS," to "[b]lunt the thrust of employer and manufacturer liability suits," and generally to "protect the franchise."[59] Humber's first recommendation was "Sue the bastards!" In his view, "no other action . . . will accomplish as much across *all* fronts as effectively." Not only would lawsuits "establish both focus and direction" but also they would "at the very least, delay or cloud precipitous actions against us."[60]

Second, Humber recommended that the industry seek out stories that demonstrated "the scientific weaknesses of the EPA conclusions." This could be accomplished in part by pointing to "EPA excesses and mistakes unrelated to tobacco," by demonstrating EPA "corruption," and by "[s]timulat[ing] non-tobacco industry, anti-regulation groups and others to provide their own perspectives in order to portray EPA as an agency correctly under siege."[61]

The third task would be to elicit a corps of scientists to attack the EPA report and belittle the risks posed by ETS. Thus far, the industry's scientific effort had "been conducted under industry aegis, and the results—from a public relations perspective—have been less than successful."[62] It was "[a]bsolutely critical" that the industry "call upon the scientific experts" already in the industry's stable "for public service" and "expand their number along a variety of fronts, some of which will not necessarily require discussion of smoking or ETS."[63]

Fourth, the tobacco industry should build coalitions with other industries and mobilize third parties to attack what they all viewed as "junk science." In this regard a small nonprofit think tank called the Institute for Regulatory Policy, headed by the Reagan administration OSHA director Thorne Auchter, offered an "existing mechanism that currently is in the best position to assemble and mobilize a wide variety of business groups, corporations, local governments, and other parties concerned about or victimized by EPA excesses."[64] Another entity, Citizens for a Sound Economy, headed by former White House aide C. Boyden Gray, already had a "track record of activity," including an August 1992 conference in Columbus, Ohio, featuring Vice President Dan Quayle, and was "well-positioned to undertake a number of coalition-building activities."[65] ❒

The Burson-Marsteller strategy suggests that successful attacks on science should be aimed simultaneously at three distinct target audiences. The first audience is the scientific community. Convincing scientists that a study is badly flawed is the ultimate goal; if the scientific community itself begins to question the reliability of research, convincing the public and policy-makers to reject it becomes a much easier task. Even if scientists are not ultimately persuaded, it may be enough that outsiders are persuaded that the reliability of the research is a matter of genuine debate within the scientific community. That level of doubt in the public mind will greatly enhance the likelihood that the attacks aimed at the other two targets will succeed.

The attacker must simultaneously focus public relations efforts directly on lay audiences. Since the public and policy-makers generally lack the capability to evaluate the scientific legitimacy of attacks, this second audience can often be persuaded by incomprehensible technical critiques and related smoke and mirror tactics. As long as the scientific community stays out of the debate or appears to be divided on the question, the attacker is likely to succeed in undermining public confidence in the research, especially when the attacker coordinates its efforts with sympathetic commentators in the mass media (a strategy we take up in Chapter 9).

Finally, the attacker is well advised to target legal decision-makers in mounting challenges to unwelcome research, particularly when agencies or other advocates attempt to use the research in litigation or administrative proceedings. Even if the claims are meritless, the legal system lends technical challenges an air of authenticity. Lawsuits also have a way of attracting public attention, thereby reinforcing simultaneous efforts to undermine the legitimacy of the research in the public mind. And because the judiciary and the upper levels of regulatory agencies are staffed largely by nonscientists, errors are common in the science-based conclusions drawn by courts and sometimes even agencies. Lawmakers, like the general public, may erroneously conclude that reliable research is junk science that should be excluded from their consideration as well.

Together, simultaneous attacks on inconvenient scientific research in these overlapping venues offer a chance of at least delaying the time when policy-makers are comfortable addressing the implications of that research in legislation, regulation, or litigation. Even in situations where it is clear that an advocate has commissioned the attack, credible-sounding critiques cannot be ignored. By at least postponing the day of reckoning, an industry can continue to accumulate profits, and that may well make the effort worth the trouble.[66] In the following discussion we elaborate on each of the target audiences and the corresponding battle plan.

Bull's-Eye: The Scientific Community

The most sophisticated attacks take place within the realm of science, and scientific journals are the preferred battleground. Although scientists frequently interact and debate with one another at conferences and meetings, the journals are the most visible and accepted vehicles for scientific communication. Since journal editors endeavor to screen articles and commentary through collaborative-minded and disinterested specialists, scientists assume that the work printed in most journals has been carefully scrutinized by peers. The scientific literature is thus the most logical forum for attackers to use in mounting their attacks.

Prepublication Attacks

The dream attack, from the advocate's standpoint, is one that convinces a scientific journal in the throes of considering a piece for publication that it should not publish it. If a damaging study remains out of the literature altogether, it is essentially a dead letter in the scientific community. It may still be considered by courts and agencies, but both institutions prefer the patina of peer review, and a showing that a journal has considered and rejected a study can be the kiss of death. Dr. Neil Pearce experienced precisely this kind of attack when he attempted to publish in a prestigious medical journal a study that had damaging implications for a widely used asthma medication.

LOBBYING THE "LANCET"

Dr. Pearce and his colleagues suspected that a dramatically increased death rate among asthmatics in New Zealand was due to fenoterol, a drug that had been introduced in that country in 1976 and heavily marketed for the following three years. Dr. Pearce's group first conducted clinical studies demonstrating that repeated use of fenoterol resulted in more cardiac side effects than alternative drugs, and other research groups subsequently confirmed these studies. Dr. Pearce and his team then conducted a case-control study comparing patients who died of asthma with patients who were afflicted with asthma of similar severity but did not die.[67] The results were powerful and strongly suggested that fenoterol caused these deaths.

The New Zealand Department of Health assembled an advisory panel to review the "striking" results of the case-control study. Although it had some criticisms, the panel concluded that the study was sufficiently well supported "to justify public health action."[68] Encouraged by this review, Dr. Pearce and his coauthors submitted a paper containing the results to the prestigious British medical journal the *Lancet*. Following peer review and some minor changes to the text, the journal accepted the article for publication on February 20, 1989. At the same time, the authors provided a copy of the article to the Department of Health with strict orders not to share it with the drug's manufacturer. At the urging of the department's director general, however, Dr. Pearce allowed him to send a copy of the manuscript to the company, on the explicit written condition that the company would make no attempt to prevent its publication.

The company ignored the condition, and the *Lancet* soon began to receive a barrage of negative reviews from scientists that the company had hired to critique the study. Within a week, the journal had rescinded its offer to publish the article. Outraged, Dr. Pearce and his coauthors informed the journal of the origin of the criticisms and offered to answer any of the charges. Dr. Pearce and his colleagues ultimately demonstrated to the editors of the *Lancet* that the critiques were without merit, and the journal agreed to publish the article after all. Had Dr. Pearce not been aware of the provenance of the attacks, it could easily have been otherwise. ❒

As the fenoterol experience suggests, advocates are often willing to go to considerable lengths to prevent publication of studies that might cause them economic grief or undermine their ideological or political positions.[69] Stifling publication ensures that few scientists will learn of the results and incorporate them into their own research agendas and scientific viewpoints. Perhaps more important, agencies and courts increasingly view publication in a peer-reviewed journal as a proxy for reliability.[70] Hence depriving a troublesome study of this exalted status is highly desirable from an advocate's perspective. Even if publication cannot ultimately be prevented, prepublication knowledge of the contents of an adverse scientific study can give a company and its public relations consultants a head start on preparing scientific critiques and press releases for rapid response teams to use when the study does come out (a strategy we explore in more detail in Chapter 9).

Advocates can ascertain the contents of prepublication studies in several ways. The researcher or an associate may present some or all of the results at a scientific conference attended by consultants or scientists otherwise sympathetic to the advocate's position. For example, when a University of Washington medical researcher, Dr. Bruce Psaty, returned from a scientific meeting at which he presented the results of a study finding that new antihypertensive drugs called calcium channel blockers were associated with a higher risk of heart attacks than older treatments, he learned that his dean had already received a fax from the drug's manufacturer complaining about his presentation and demanding to have access to all of the underlying data. Soon thereafter, a state legislator called the dean to complain about the presentation, and company officials called another dean to suggest that the study should not be published.[71] Dr. Psaty did publish the paper in *JAMA*.[72] A meta-analysis of clinical trials involving the same drug subsequently confirmed the study's conclusions, and the FDA ultimately required the manufacturer to add a warning to the drug's label.[73]

In some cases, journals submit studies to employees or consultants of interested parties for peer review because they are knowledgeable about the subject matter. Although the peer review process is supposed to be confidential, scientists have on occasion leaked the drafts to their employers or sponsors. For example, the Tobacco Institute in the late 1960s learned of a National Cancer Institute–funded study by the pathologist Oscar Auerbach on the effects of tobacco smoke on the lungs of beagle dogs from industry consultants who were asked to peer review it for *JAMA*.[74] An institute publicist and an outside attorney told the president of the American Medical Association (AMA) that several of the peer reviewers had serious reservations about the paper and unsuccessfully urged him to cancel Auerbach's upcoming presentation at the AMA's annual convention. The Tobacco Institute then released a press release declaring that the still-unpublished Auerbach study "may be one of the great scientific hoaxes of our time."[75] Because of the institute's pressure, the journal editors insisted on a number of changes to the study as a condition to publishing it in *JAMA*. Auerbach declined and opted to publish the unrevised, original version of his paper elsewhere.[76] The now famous article provided one of the first indications that smoking caused cancer in animals and therefore posed a cancer risk to humans.

Postpublication Damage Control

When prior censorship fails, another time-honored technique is to attack the study in the scientific literature. A commissioned critique published as a letter to the editor in a scientific journal may persuade some fellow scientists that the study is at least partly flawed, but even if that fails, a published critique tends to cause at least some policy-makers and the public to think the challenge has merit. Ultimately, these manipulations can snowball to the point where the journals themselves become unwitting accomplices in the attacks, as Dr. Ignacio Chapela's recent experience suggests.

UNDERCUTTING CHAPELA

In the summer of 2000, Dr. Ignacio Chapela of the University of California–Berkeley confirmed with standardized testing protocols that some of the native corn his graduate student David Quist had gathered in Mexico contained remnants of the cauliflower mosaic virus, a vector frequently used in genetically modifying plants.[77] This was an alarming discovery, because the Mexican corn is the genetic precursor of all modern corn crops.[78] Furthermore, data derived from a different detection technique suggested to Chapela and Quist that modified genes were moving around in the plants' DNA over several generations. This discovery, if true, was even more alarming because it suggested that genetically modified (GM) corn in the wild was not as stable as it appeared to be in laboratory tests.[79] This discovery in turn could have enormous repercussions for agricultural biotechnology in general because of its potential to undermine the industry's strong assurances that modified genes could not spread in the wild.[80]

After Chapela and Quist submitted their study to the prestigious scientific journal *Nature,* its editors sent the paper to three anonymous peer reviewers for their assessments in early 2001. Two of the reviewers immediately approved the paper, but the third wanted to see data from tests done on corn collected from the same area prior to the time that any corn was genetically modified. Chapela and Quist complied with this request, and all three reviewers approved the slightly revised version of the paper.[81] By that time, however, the chair of Mexico's Commission on Biosafety had reported the study's results at a public subcommittee meeting of an international food safety organization, and an environmental group had demanded action to protect the genetic integrity of Mexican corn.[82] The editors decided to send the revised paper to a second set of peer reviewers, explaining to Chapela and Quist that the "intense media interest" required that the paper be given "close scrutiny."[83] After the second set of reviewers also approved the paper, it was published on November 29, 2001.[84]

Within two weeks, *Nature* received two letters to the editor accusing Chapela and Quist of sloppy research—from two graduate students (one present and one former) in another department at Berkeley that had recently received a $25 million grant from a major biotechnology company.[85] The letters first challenged the authors' conclusion that local Mexican corn had become contaminated with DNA from genetically engineered corn.[86] (Subsequent research was mixed, although some later studies conducted by the Mexican government found that local corn had become contaminated, and some even contained genes from the ill-fated Starlink strain of GM corn that had been banned from U.S. food because of its allergenicity risk.)[87] The letters also attacked the article's conclusion that the DNA was not stable, arguing that the tests Chapela and Quist had used were not capable of supporting that conclusion and that the researchers had in any event committed a common error by erroneously interpreting "background noise" as evidence of the presence of a foreign gene. In addition, the paper did not identify the source of the "control" corn or state how many times the experiment had been repeated to verify the results.[88] Tacitly acknowledging that the debate extended beyond the realm of science into the realm of policy, one of the letters noted that it was "important for information about genetically modified organisms to be reliable and accurate, as important policy decisions are at stake."[89]

At the same time the letters to the editor were arriving in *Nature*'s mailbox, the internet was abuzz with similar criticisms of the article cast in "far more loutish language."[90] The internet critiques, which were reported in *Science,* also criticized *Nature* for publishing the piece and demanded that *Nature* force the authors to retract the paper.[91] Unbeknownst to anyone at the time, much of the early internet criticism had in fact been generated by two apparently fictitious scientists whose transmissions originated from a Washington, D.C., public relations firm that was working for a major biotechnology company.[92]

The barrage of criticism motivated *Nature*'s editors to demand that Chapela and Quist return to the laboratory and use a different gene detection technique to validate the conclusions of the already published paper.[93] Strangely, the editors further insisted that this be accomplished within four weeks and that the response be limited to three hundred words.[94] The researchers complied with the request but were able to provide only preliminary results by the deadline. Their response admitted that they had misidentified two sequences (a factor that could undermine the original paper's instability thesis), but they rejected the argument that they had mistakenly identified foreign DNA in the local Mexican corn.[95] The editors then had the Chapela and Quist response peer-reviewed by three additional scientists. Although all three of the scientists disputed the original paper's instability conclusion, two of the three were persuaded that the new data supported the conclusion that foreign genes were present in local corn.[96]

The editors of *Nature* published the Chapela and Quist response,[97] but they also took the entirely unprecedented step of writing an editorial in which they agreed with the internet critics that the journal should not have published the original paper.[98] Yet the editors never asked Chapela and Quist for a retraction, and it remains a fully citable part of the scientific literature to this day.[99] A team of scientists from Ohio State University and Mexico found no transgenes in seeds from hundreds of corn plants that they sampled in the same area three to four years later. The authors suggested that either the foreign genes were never present or that they did not fare well in nature over a three year period.[100] Time will tell whether the Chapela and Quist findings remain resilient to further research and investigation by others. What is clear from the controversy, however, is that neither the scientists nor the journal editors were fully prepared for unidentified outsiders' unrelenting attempts to deconstruct the published research. ❒

Once a study with negative implications for an industry's economic well-being is published in a peer-reviewed scientific journal, the industry typically moves to damage control mode. As Dr. Chapela's experience demonstrates, a well-orchestrated assault on a published study may persuade the journal to express public regret at having published it in the first place. Even if the original authors can convince the journal that the attacks are ill-founded attempts to bend science, the necessity of responding to such attacks can easily wear them down. In her book *When Smoke Ran Like Water,* the epidemiologist Devra Davis tells of an urgent fax she received from the *Lancet* soon after it published her study on the direct health impacts of reducing power plant emissions to meet greenhouse gas targets. The fax contained "a very serious set of charges" claiming that the data were "flawed" and the analyses "wrong."[101] On further examination, she discovered that the critic, an employee of the industry-funded Citizens for a Sound Economy (the same think tank the Humber memo recommended to the tobacco industry), had never published a peer-reviewed article in the field of environmental health, and his most recent work consisted primarily of letters to the editor and other postpublication attacks on similar public health estimates. Nevertheless, Davis said she "spent more time rebutting what he wrote and persuading the *Lancet* that this was not a serious scientific critique than my team had spent in responding to the original peer reviewers' substantive comments."[102]

Even when retraction is not a serious option, letters to the editor pointing out "fundamental flaws" in the data or analysis of an unwelcome study can still be a useful attack strategy.[103] The tobacco industry alluded to the critical letters to the editor that it had generated in response to the publication of the Hirayama study (discussed earlier) a decade later in arguing to the EPA that the study had long ago been discredited.[104] And the industry continued to employ the commissioned letters to the editor as a damage-control strategy throughout the 1990s.[105] A study of letters to the editor published in English-language journals in response to Dr. Psaty's calcium channel blocker study (discussed earlier) found that twenty-three of the twenty-four authors writing letters criticizing the study had financial ties to manufacturers of those drugs.[106] In another study, researchers found that authors with financial ties to the food industry were about four times more likely to write letters to the editor, commentary, or review articles that were favorable to the fat substitute olestra than authors without these ties.[107]

The Next Target: Policy-Makers and the Public

Backed by credentialed scientific consultants, a good public relations firm, and allies in industry-funded think tanks, an industry can make a legitimate scientific study or report look like junk science in the eyes of lay observers. If information is relatively complicated, like the typical research publication, then its credibility can be reduced simply through strategic and sustained attacks on the methods, experimental design, or integrity of the researcher—all of which may be flatly wrong.

A "NOVEL" APPROACH TO GLOBAL WARMING

Faced with growing consensus within the scientific community that climate change is real and attributable in part to fossil fuel consumption, industries hoping to avoid vigorous regulatory requirements developed multifaceted strategies for undermining public acceptance of the phenomenon. The industries pooled their resources to support a number of think tanks to manufacture uncertainty about the validity of climate change research on their behalf. In 2003 the Competitive Enterprise Institute petitioned several federal agencies under the newly enacted Information Quality Act to "cease dissemination of the National Assessment on Climate Change" and demanded that it be removed from all federal databases and websites.[108] The George C. Marshall Institute published several highly publicized attacks on scientific reports prepared by governmental and international panels and financed promotional efforts for articles published by "senior scientists" associated with the institute whose expertise in climatology was at best questionable.[109] The American Enterprise Institute (AEI) even hosted a public lecture, entitled "Science Policy in the Twenty-First Century," by Michael Crichton, the popular author, who had written a fictional attack on environmental groups and the scientists who have concluded that global warming poses a threat to the future of the planet. In welcoming Crichton to the podium, AEI president Chris DeMuth praised him for conveying "serious science with a sense of drama to a popular audience."[110] Crichton told the sympathetic listeners that he had studied the science underlying the global warming debates and found it "shockingly flawed and unsubstantiated."[111] Not surprisingly, climate change scientists had reached exactly the same conclusion—only in their case with considerable scientific support—about Crichton's book.[112]

These accumulated attacks on climate change science apparently succeeded, at least in the short term. Polls conducted in 2004 suggested that about 50 percent of the public worried "only a little" or "not at all" about climate change, despite the fact that there was by that time strong consensus within the scientific community about both its existence and the need for social action to reduce greenhouse gas emissions.[113] While this gap between the views of the public and the scientists can be attributed to many causes, some of the divergence no doubt reflected the success of the clever, industry-funded strategists in manufacturing uncertainty in the minds of the public and policy-makers. ❐

Scientific attacks for the purpose of influencing lay audiences are relatively easy to orchestrate, and they are often successful, at least in the short term. Indeed, it is no longer necessary to mine the academic literature for professors who might be willing to manufacture uncertainty for a fee—one need only consult the yellow pages. According to David

Michaels of the George Washington University Medical School, "[e]mphasizing uncertainty on behalf of big business has become a big business in itself," as various "product-defense firms have become experienced and successful consultants in epidemiology, biostatistics and toxicology."[114] An eager scientist hired to mount a challenge can usually come up with questions, however trivial, about the validity of protocols governing a study. Careful inspection of the underlying data may reveal minor measurement mistakes, or a reanalysis of the data using a different statistical approach may produce different results. Critical scientists in the employ of an advocate can question the authors' interpretations of the results and offer different interpretations. Unless lawmakers or members of the public have the time and wherewithal to investigate the source and nature of sponsored attacks, they will at the very least create uncertainty about the validity of the challenged study and its suitability for use in the decision-making process.[115]

The master of the reverse Midas touch, the tobacco industry, turned valuable science into junk for a half century. Among statistician Theodor Sterling's many lucrative projects for the Committee of Counsel was his June 1967 critique and reanalysis of the seminal "mortality study" the National Center for Health Statistics (NCHS) did on the incidence of diseases other than lung cancer in smokers.[116] After a Brown & Williamson scientist expressed doubts about relying on Sterling's possibly biased analysis, a colleague responded: "we must use every possible method to cast some doubt on the statements and data presented by our opponents."[117] The industry was "making every effort to buy time and . . . any doubt that can be placed in the minds of the jury (or the public) is better than no doubt at all."[118] On April 24, 1969, Sterling fulfilled those expectations when he testified in congressional hearings that the NCHS morbidity study contained "a number of major flaws" in the data and analyses that were "so serious that clear-cut conclusions should not be drawn."[119] Committee member Richardson Preyer (D-NC) delighted the tobacco industry when he announced that Sterling's testimony had "blown the morbidity study right out of the water."[120] A Brown & Williamson pamphlet quoted Sterling liberally in a discussion entitled "How Eminent Men of Medicine and Science Challenged the Smoking-and-Health Theory during Recent Hearings in the U.S. Congress."[121]

Other advocates have followed the tobacco industry's lead to this day, as the aforementioned attacks on climate change research indicate. In the early 1980s, the formaldehyde industry persuaded the Consumer Product Safety Commission to delay a possible ban of formaldehyde-based insulation while it considered the claims of an industry consultant that the cancer study supporting the proposed action had employed "weak rats" that were susceptible to viral infections and known for spontaneous tumors.[122] The manufacturer of the herbicide Atrazine commissioned critiques of research suggesting that it was a potent endocrine disrupter, in the hope of persuading EPA program managers that "some inadequacies in the study design, the carrying out of the study, and the researcher's interpretation" warranted ignoring the research altogether.[123] As the states in the late 1980s began to rely on the EPA's 1985 risk assessment for dioxin to establish extremely stringent state water quality standards, the Chlorine Chemistry Council (an industry trade association) and its allies in think tanks like the Competitive Enterprise Institute launched a sustained attack on the risk assessment that went on for more than a decade as the EPA unsuccessfully struggled to revise it.[124]

Scientific wizards who prove especially adept at converting science to "junk" can depend on a steady flow of income from consulting work for entities that have a need for that expertise. As noted earlier, Theodor Sterling built a very lucrative moonlighting career out of critiquing epidemiological studies for the tobacco industry. The Louisiana Chemical Association hired Dr. Otto Wong in 1989 to evaluate the epidemiological evidence on carcinogenicity of the known human carcinogen vinyl chloride after a Tulane University epidemiologist concluded that southern Louisiana residents living within a mile of a chemical plant or a refinery were four times more likely to die of cancer than people living 2–4 miles away from the facilities.[125] Dr. Wong concluded that the increased incidence in cancer was more likely attributable to the lifestyle of area residents, who in his view tended to "smoke more, eat low amounts of fresh fruits and vegetables, and work in high-risk industries associated with lung cancer," than to their exposure to vinyl chloride emissions.[126] Impressed with his work, the American Petroleum Institute hired Dr. Wong to head up a $27 million epidemiological study in Shanghai, China, after a 1997 study funded by the National Cancer Institute concluded that workers exposed to benzene were at risk of developing non-Hodgkin's lymphoma.[127]

The Final Target: Summoning the Legal System to Attack Credible Science

Having savaged a policy-relevant scientific study in both the scientific and policy arenas, the final step is to use available legal tools to undercut its credibility even further or, better still, to exclude it altogether from the relevant policy-making processes. To up the odds of success for this strategy, advocates have sponsored efforts by sympathetic think tanks to create more effective legal tools for undermining reliable science.

TOZZI'S LAW

Like many of the strategies recounted in this chapter, the Information Quality Act—a statute that allows citizens to file complaints against information disseminated by federal agencies—had its origins in the efforts of the tobacco industry to avoid restrictions on ETS.[128] The statute itself was the brainchild of a Philip Morris consultant named Jim Tozzi, who parlayed an intimate knowledge of the regulatory process gained during twenty years as a high-level federal bureaucrat into a multimillion-dollar conglomerate of consulting firms and tax-exempt nonprofit think tanks.[129] With the help of Philip Morris, Tozzi founded several nonprofit organizations that he then used to coordinate utility, paper, and diesel industry groups who were similarly threatened by EPA risk assessments "to support a new administration policy on risk assessment which would prevent agency disseminations of health risks which cannot be supported by objective science."[130] A common theme running through Tozzi's activities was the broader overall goal of imposing external standards on regulatory agencies for the quality of the scientific information they used and disseminated.

Philip Morris supported Tozzi's organizations generously, presumably in the expectation that they would provide the tobacco industry with new legal tools to isolate and undermine credible research that threatened greater liability or regulation, particularly in the ETS arena. Philip Morris's records indicate that the company paid one of Tozzi's organizations $65,000 per month, up to $780,000 in 1998, for improving "data integrity."[131] An agenda for a meeting held by one of Tozzi's groups on June 30, 1998, in New York City entitled "Data Integrity and Data Sharing" suggests various strategies for reaching industry goals with respect to data access and data quality.[132] These coalitions enjoyed some degree of success in October 1998, when Congress enacted the Shelby Amendment, a two-sentence rider to the Treasury and General Government Appropriations Act for fiscal year 1999, which gave any citizen access to the data underlying federally funded (but not privately funded) research relied on by federal agencies.[133]

Yet the data access provisions did not go far enough, and the industries under Tozzi's tutelage were eager to gain additional tools that would allow them to challenge the quality of agency science. Tozzi first attempted to persuade congressional allies to include data-quality language in an appropriations rider for fiscal year 1999. Athough language to that effect was included in the House report on the bill, the provision did not appear in the bill itself.[134] One year later, Tozzi's persistence paid off. He persuaded Representative Jo Ann Emerson (R-Mo.) to sandwich almost identical language into the text of the 2001 appropriations bill, between a property acquisition for the Gerald R. Ford Museum and a provision relating to non–foreign area cost-of-living allowances.[135] Congress held no hearings on the new statute, and it was not mentioned when the House and Senate debated contents of the appropriations bill. Legal scholars were later shocked to learn that this hidden rider provided an entirely new administrative mechanism for challenging agency information.[136]

Under the several-sentence-long Information Quality Act, advocates now have a potentially powerful and largely unconstrained procedural weapon for challenging the quality of science (and other information) that agencies rely on in implementing their statutory missions.[137] Any person can file a petition demanding "correction" of information "disseminated" by an agency, including scientific studies, if that person believes the information is not reliable, is not objective, lacks utility, or is biased.[138] The Act and its implementing regulations, promulgated by the White House OMB, provide no mechanism for compensating the agency for processing complaints, and they place no limits on the type or nature of complaints that can be filed. Professor David Michels concludes that the law "gives corporations an established procedure for killing or altering government documents with which they do not agree."[139] During its first five years, it was used by groups bankrolled by the oil industry to discredit the National Assessment on Climate Change; by food industry interests to attack the WHO's dietary guidelines; by marijuana activists to get cannabis removed from the federal government's list of illicit drugs; and by the Salt Institute to challenge the advice of the NIH that Americans should reduce their salt consumption.[140] Advocates have petitioned the EPA in particular to exclude or withdraw from public databases path-breaking research on the endocrine disruption properties of the herbicide Atrazine; brochures warning auto mechanics about risks of exposure to asbestos from brake linings; and technical reviews, including EPA's risk assessment of a chemical used in plastics.[141] ❒

Combined with several other tools available in the judicial and administrative arenas, the Information Quality Act gives advocates an opportunity to argue that science must be "perfect" before it can inform policy.

When deployed in strategic and nonscientific ways, these tools can undermine credible research, confuse policy-makers and the public at large, and alienate respected scientists.[142] Their emergence also marks a sharp detour from the approach legal institutions have historically taken toward facts and evidence. Courts, for example, never demand perfection in the nonscientific evidence on which they routinely rely, such as the testimony of "percipient" witnesses and documents that may or may not have been forged. Like the recollection of an eyewitness, science is a human endeavor and is therefore subject to human error. The fact-finder is expected to weigh the available evidence and draw conclusions about the existence or nonexistence of facts on the basis of relevant legal tests ("preponderance of the evidence" in civil trials and "beyond a reasonable doubt" in criminal trials).

In this emerging new world of administrative law, where the science underlying public health and environmental protection must be "perfect," credible research can routinely encounter two kinds of legal attacks. The first category consists of frontal attacks arguing that a study is unreliable and therefore unworthy of consideration. These attacks closely resemble the previously discussed strategies that advocates have employed to exclude studies from scientific journals and public forums. The second category consists of "corpuscular" attacks that challenge technical features of many of the individual studies underlying broader conclusions about the risks posed by regulated products and activities in an effort to undermine these more general and often well-accepted weight-of-the-evidence risk judgments. The more ambitious goal of these attacks is thus to alter the ways legal institutions approach a body of evidence regarding risky products or activities.

Frontal Attacks

Both the courts and the regulatory agencies allow advocates to launch frontal attacks by arguing that a particular study is so unreliable that it should be excluded from the decision-making process altogether. The Supreme Court's well-known *Daubert* test provides courts with a screening mechanism to ensure that expert testimony meets minimal indicia of reliability before that testimony can be submitted to a jury or otherwise entered into evidence.[143] Under *Daubert,* trial judges are to determine reliability by reference to its "scientific validity" when measured against the methods and procedures of science.[144] The IQA mimics *Daubert* by providing advocates with a similar legal vehicle for arguing that information used or disseminated by an agency is flawed and should be corrected or else removed from the public domain.[145] The agency then has the burden of defending its decision to use or disseminate the information.

These legal mechanisms offer two advantages to advocates who are determined to undermine the credibility of an unwanted study. First, they now have a legal vehicle for padding the public record with voluminous technical critiques of the study's design, methodology, execution, and statistical analysis, all in support of their underlying contention that the study is scientifically unreliable. Even if the court or the agency ultimately rejects the challenge, mounting it in such a formal way in a legal proceeding bolsters the advocate's contention that something may be amiss with the research. With any luck, the challenge will attract the attention of the media, which will in turn convey the message beyond the particular legal proceeding to the general public. Second, there is always the possibility that the policy-maker will be persuaded by the challenge, either because she lacks sufficient scientific competence to evaluate it properly or because she is inclined to exclude the evidence for other reasons and the challenge offers a convenient excuse. When an advocate can convince a court or agency formally to "exclude" a scientific study from its consideration, it may have gone a long way toward staving off liability or regulation. A victory in an agency or a court may also help to convince the public and other, less attentive scientists that the study was "fatally flawed."

An all-out victory of this sort is by no means a hypothetical possibility. In his book *Toxic Torts: Science, Law, and the Possibility of Justice,* Carl Cranor provides a number of examples of scientifically misleading attacks that have successfully persuaded judges to exclude valid scientific research from jury consideration in important cases involving real victims.[146] For example, several courts have refused to allow experts to testify about human health risks of chemicals to the extent that their conclusions are based on extrapolating from animal studies.[147] These rulings ignore the fact that scientists regularly use and generally depend on animal evidence in assessing human health risks because direct human testing and epidemiological studies are often unavailable.[148]

Law professor Sidney Shapiro has observed that Information Quality Act administrative appeals can "become part of the litigation strategy of regulated entities to slow, or even stop, the government from disseminating information that is legally or politically inconvenient for them";[149] and the epidemiologist Devra Davis worries that "science used in regulatory actions . . . must now be pleasing to the political appointees at OMB [who implement the Act], few of whom have much expertise with, or sympathy for, real science."[150] By contrast, most information that private sector entities submit to the agencies is protected from challenge under the Act by one or more exclusions provided in the OMB's implementation guidelines. The guidelines exempt information submitted by private entities that is "trade secret" or "confidential business information," an expansive category that can include health and safety studies, or at least portions of those studies.[151] Information submitted as "public filings" or used in "adjudications," categories that appear to include pollution permit applications, industry discharge and emissions monitoring reports, and pesticide registrations, are also exempt under the guidelines.[152] Thus, as Professor Sheila Jasanoff observes, the law "authorizes challenges to publicly generated information on the basis of claims that are not themselves subject to equivalent standards of openness."[153]

Corpuscularize the Evidence

Advocates have also employed the same legal tools—*Daubert* in the courts and the Information Quality Act in the agencies—in clever ways to alter the way courts and agencies approach the entire body of scientific evidence surrounding the safety of an activity or product. Ideally, scientists consider all of the credible scientific evidence bearing on a public health question as a whole. Under this judgment-laden, weight-of-the-evidence approach, the scientist or, more frequently a committee composed of scientists drawn from the relevant disciplines, considers all of the proffered studies and determines the weight to be afforded to each study on the basis of its identified strengths and weaknesses.[154] Some studies are so poorly designed or executed that they are entitled to no weight at all, but many studies that are otherwise flawed in one or more aspects may appropriately be considered to the extent that they add to or detract from conclusions based on studies in which the scientist or committee is inclined to place more confidence.[155]

In an ambitious effort to preclude experts from providing these weight-of-the-evidence judgments when they are particularly unwelcome, advocates have devised strategies for corpuscularizing the evidence and thereby undermining the experts' overall assessments. A corpuscular attack on scientific information focuses the policy-maker's attention on flaws underlying each individual study, rather than on the scientific reliability of the expert's overall conclusions. Only after each study has survived the advocates' frontal attacks can it then form part of the body of evidence that an expert or expert body considers in determining whether the weight of the evidence supports particular conclusions.[156]

Since 1992, defendants in toxic tort and products liability litigation have been remarkably successful in persuading the courts to adopt this corpuscular approach to decision-making.[157] They argue that each study must be flawless before it can inform an expert's overall conclusion about cause and effect, and the courts occasionally agree.[158] A good example is a 2001 Georgia case in which two plaintiffs alleged that the drug Parlodel, a medication they had taken to suppress postpartum lactation, caused them to have hemorrhagic strokes.[159] Even though a number of studies and reports supported a clinical hypothesis of an increased risk of at least some types of strokes, the district court held that no individual study, standing alone, passed *Daubert*'s reliability test when applied to the legal question of causation-in-fact. The studies addressed more limited questions, such as the effects of the drug if it was immediately withdrawn, the scientific implications of the adverse effects reports filed with the FDA, and so forth.[160] Having excluded the plaintiffs' experts' weight-of-the-evidence testimony, the court then dismissed the entire case without allowing a jury trial, because the plaintiffs had no remaining scientifically reliable evidence to support their causation claims.

The corpuscular approach is especially disruptive in decision-making contexts in which epidemiological studies play an instrumental role. Scientists can rarely express the conclusions of individual epidemiological studies with a high degree of certainty, because epidemiologists encounter great difficulties in designing and executing studies in a world in which health and mortality records are notoriously bad; data frequently come from human recollections; and potential confounding factors are always present.[161] The corpuscular approach invites the entity seeking to exclude epidemiological evidence to search every detail of each epidemiological study for possible flaws in the statistical analysis and to speculate at great length about potential confounding factors and other possible sources of bias in an effort to persuade the decision-maker to eliminate it from consideration. Given the practical impossibility of conducting a perfect epidemiological study, such searches are nearly always fruitful. For example, when the staff of the National Toxicology Program recommended placing trichloroethylene (TCE) on a list of human carcinogens, they based their conclusions on a number of human and animal studies, as well as a meta-analysis of several epidemiological studies. The Halogenated Solvents Industry Alliance (HSIA) hired two well-known epidemiologists to determine whether each of the studies, standing alone, demonstrated that TCE caused cancer, and the scientists predictably answered that question in the negative. In insisting that "only faultless studies merited inclusion in any summary of the work," HSIA advocated an impossible test that may have also helped them achieve their desired result—the National Toxicology Program ultimately elected not to place TCE on the list.[162]

Conclusion

A good scientist maintains a healthy skepticism about the work of other scientists and is hesitant to take at face value another scientist's interpretation of scientific data. The scientific norm of reproducibility even encourages scientists to be skeptical about the reliability of scientific data until they have been sufficiently validated by additional studies. This makes it very difficult for a layperson to distinguish a scientist's legitimate skepticism about another scientist's work from an economically or ideologically motivated attack on a study.

One helpful approach is to "follow the money."[163] If a strongly negative comment by a peer reviewer, a letter to the editor criticizing a published study, or a demand that it be retracted comes from a consultant for an entity with an interest in how the study might be used in the policy-making process, then the scientific community, the regulatory agencies, and the general public should be skeptical of the skeptic. If the critic is employed by a think tank, then the criticism should be discounted to the extent that the think tank is supported by entities with an economic or ideological stake in the outcome of a policy-making process to which the study or report is relevant.[164]

Regulators and judges should also be aware of advocates' attempts to exclude research through strategic corpuscular attacks. Agencies and especially courts must inoculate themselves against the tendency to demand perfection from each individual study before allowing the study to be considered in a larger body of scientific evidence used to assess causation. Epidemiological studies, in particular, are notoriously plagued with practical roadblocks to an accurate depiction of the examined relationships. Yet epidemiological studies are often the only available evidence that courts or agencies are willing to consider on the critical issue of whether human exposure to a substance causes a particular disease. As the epidemiologist Devra Davis observes, in science, as in other human activities, "the perfect must forever remain the enemy of the good."[165]

CHAPTER SEVEN

Harassing Scientists

The Art of Bullying Scientists Who Produce Damaging Research

Desperate to undermine damaging research, some advocates have found that harassing or defaming the scientist who is responsible for the unwelcome findings can be a very effective, if somewhat unsavory strategy.[1] The strategy of harassing scientists differs from the attacking strategy in that it focuses on the scientist and not the science. Attacks on science consist primarily of critiques of scientific papers, challenges to scientific presentations, and similar attempts to manufacture uncertainty by casting doubt on the results and interpretations of scientific research. As such, illegitimate attacks are often difficult to distinguish from the legitimate skepticism scientists expect of their colleagues. The harassment strategy consists of personal attacks on the target scientist backed by advocates whose primary goal is to undermine the entire body of that scientist's work in the relevant area. The tools can range from unreasonable demands for underlying data to defamation suits and even scientific misconduct charges. Harassment is much rarer and, consequently, easier to identify than attacking, although advocates sometimes deploy both tactics simultaneously by commissioning critiques of individual studies at the same time they issue overbroad subpoenas and file misconduct charges against the scientist.

Like attacks on science, the harassment of scientists is not a recent phenomenon. When Dr. Irving Selikoff began to publicize his now famous research on the health effects of asbestos in 1964, an executive for a large manufacturer of asbestos-containing products wrote in a confidential memorandum: "Our present concern is to find some way of preventing Dr. Selikoff from creating problems and affecting sales."[2] The industry met this concern with a multiyear attack on Dr. Selikoff and his research.[3] To his credit, he endured this sustained industry-generated harassment, and the scientific truth ultimately prevailed. Asbestos is no longer a component of modern consumer products, and tens of thousands of victims of asbestos exposure have been compensated for at least some of their losses.

Dr. Selikoff's success has by no means discouraged advocates from harassing scientists, however. If anything, the frequency of harassment has increased in recent years as the law and the scientific community have crafted new legal and institutional tools for managing scientific misconduct and as readily available vehicles for public attacks on a scientist's reputation have emerged on the internet and in the media. Fifty percent of the epidemiologists responding to a 2001 survey of members of the International Society for Environmental Epidemiology reported that they had been harassed by advocates following the publication of their research.[4] A similar 2005 poll of government scientists reported that between 30 and 40 percent of the scientists engaged in environmental and public health research perceived the possibility of retaliation from within the agency if they voiced their positions on controversial issues.[5] Drawing on his experience as director of the NSF and head of the White House Office of Science and Technology Policy, Dr. Neal Lane has warned that "powerful groups will not hesitate to attack scientists whose findings they don't like."[6] Understandably, not all scientists have been as persistent as Dr. Selikoff in shrugging off unfair harassment.

Harassment is a particularly useful tool for bending science because it can impugn the researchers' integrity while at the same time hampering their ability to continue their potentially damaging research. Even wholly unsupported allegations of scientific dishonesty, for example, may have a lasting impact on a scientist's reputation and the perceived validity of his research that can be rectified only over many years as other researchers replicate the suspect work. The targeted scientists must also divert their time and attention away from their ongoing research, and the resulting delays in or even termination of the challenged research also benefit the harassing party. Legal scholar Robert Kuehn argues that these tactics "raise the most serious concerns about scientific freedom" because they are designed either "to prevent the creation of certain unwelcome data or theories or, alternatively, to deter or block the dissemination of unwelcome data or theories that already exist."[7]

In this chapter we explore the diverse tools advocates use to harass scientists. They include demanding all documents underlying ongoing research under the threat of contempt of court, suing scientists for libel for speaking out against a product on the basis of their research, convening lopsided congressional hearings to question a scientist's research, diverting funding from unwelcome research projects, threatening a scientist's future employment, and filing unsupported charges of scientific misconduct. These tactics have little or nothing to do with protecting the legal rights of the entities that finance the attacks. Yet the legal system generally condones them and rarely penalizes their abuse. Indeed, the fact that advocates can mount these personal attacks through legitimate processes using available legal tools only makes the resultant abuses that much more demoralizing to the scientists who are at the receiving end. These serious intrusions into the practice of policy-relevant science happen because the law, like science, errs on the side of tolerating open debate and criticism. When the stakes are high, however, permissive legal rules and open participatory processes are vulnerable to exploitation. Parties can use litigation and related legal procedures not just to uncover the truth but also to alter it to their advantage.

Impugning a Scientist's Reputation

The most devastating and therefore potentially most effective strategy for discrediting bothersome research is to challenge the researcher's professional integrity. Once a scientist's reputation for doing honest research becomes tainted, her peers are less likely to take her conclusions seriously for a long time to come. The reason is simple—in most areas of policy-relevant research, the only way to be absolutely sure that a study is reliable is to replicate it and reproduce the same results. But that requires substantial resources and is therefore not an efficient process for scientific discovery. The primary alternative to replicating research is to trust researchers who have reputations for doing reliable work. The work of scientists who earn reputations for conducting high quality research thus enjoys an imprimatur of trustworthiness without the added assurance of protracted replication.

Nobel laureate George Akerlof has supplied a convincing explanation from the economics of information for why reputational attacks are especially effective.[8] Using as one of his minilaboratories the used car market, where sellers have much better information than buyers, Akerlof concluded that the reputation of the seller becomes an important, indeed vital proxy for the credibility of the information. A dealer who is respected will find himself earning higher returns than other dealers, even though the buyers have no additional information about the cars. The sociologist Robert Merton likewise has shown that reputation can become such an important proxy for quality that a highly respected reputation can trump substantive merit, and well-respected scientists can publish papers in the best journals and earn prestigious awards at the expense of less well-known but more deserving colleagues.[9]

Advocates eager to discredit damaging research understand the importance of reputation as a proxy for scientific credibility and at the same time appreciate its value to the scientist. In one survey, 60 percent of scientists who were ultimately exonerated of misconduct charges reported that they nevertheless experienced reputational damage as a result of the proceedings against them.[10] Because an academic scientist's reputation is his most valuable asset, reputational attacks can also intimidate scientists and may cause them either to halt controversial research projects or avoid controversial policy-relevant research altogether.[11] Once a scientist's reputation becomes the focal point of an attack, the stakes become very high for him. Less obvious but no less important, the scientific community and society as a whole also have a great deal to lose when the reputations of good scientists are unfairly impugned.

Scientific Misconduct Complaints

An especially brutal strategy for bending science is to generate a complaint against a scientist who is producing unwanted research, formally charging him or her with scientific misconduct. Even if misconduct claims are largely unsupported or badly exaggerated, the stigma associated with the mere allegation can have a lasting effect.[12] That does not, however, discourage some desperate parties from resorting to this tactic in order to rid themselves of unwanted science and scientists who produce it.

DR. NEEDLEMAN'S NEEDLESS TRAVAILS

Dr. Herbert Needleman was an associate professor of psychiatry at Harvard Medical School when he became the target of a demeaning personal attack financed in part by a powerful industry with a long history of dominance over the relevant research agenda. Focusing his research on the health effects of lead on children in the 1970s, when tetraethyl lead was still a ubiquitous gasoline additive, Dr. Needleman and a coauthor published an important epidemiological study in *NEJM* demonstrating an association between lead exposure and lowered IQ and poor school performance in children.[13]

Needleman's research was instrumental in setting tighter lead standards for products like gasoline and more stringent cleanup levels for lead-tainted Superfund sites. As a result, he became quite unpopular with the lead companies, who paid the increased costs resulting from these regulatory actions and faced potentially ruinous liability in lawsuits brought by cities and injured individual plaintiffs. As part of their larger effort to attack the science—sometimes legitimately, sometimes not—the companies hired two academic scientists to scour Professor Needleman's work for possible flaws.[14] Although he initially resisted their demands, he ultimately allowed the two critics to spend two days in his laboratory examining his notebooks.[15] On the basis of their visit, they filed a formal scientific misconduct complaint against him with the federal government's Office of Scientific Integrity (OSI) charging him with manipulating his data, failing to take into account the effects of "poor housekeeping and parenting" and other factors that might have accounted for the lower IQs, and generally failing to cooperate with them as fellow scientists.[16]

At the OSI's request, Needleman's employer, now the University of Pittsburgh, agreed to convene a special inquiry into the charges.[17] While his research and statistical practices were not impeccable, the lengthy hearings ultimately revealed that the scientific misconduct charges against him were not supportable. Yet as a result of the serious charges, his research agenda was set back almost a decade, and his reputation was thrown into question. For their efforts, the two scientists who brought the failed misconduct charges received generous remuneration from the lead industry.[18]

A number of investigators have since validated Needleman's original conclusions in their own research.[19] In the 1980s, he published a follow-up study of the same children showing that the adverse effects of lead continued into their adolescence.[20] In response to this generally well-regarded research, the CDC has gradually lowered the "safe" standard for childhood blood lead concentrations.[21] In recent years, a number of health organizations have honored Dr. Needleman for his path-breaking and courageous lifetime of work on lead toxicity.[22] ❐

Dr. Needleman was subjected to the precursor of what is now an established national program, created by Congress in 1985, to investigate and punish scientific misconduct in federally funded research.[23] To deter fraudulent practices, Congress created an agency—now called the Office of Research Integrity—to investigate claims that federally funded researchers have engaged in scientific misconduct.[24] An allegation of misconduct, defined as "fabrication, falsification, [or] plagiarism" of data,[25] must be filed with the accused researcher's home institution, and serious allegations ordinarily precipitate a hearing before a board of peers. Open-ended rules typically govern such important procedural questions as the threshold evidentiary showing for initiating a hearing and the extent to which accused scientists have an opportunity to confront their accusers. If a researcher is found guilty of misconduct, the home institution must determine the proper internal punishment and decide "whether law enforcement agencies, professional societies, professional licensing boards, editors of journals in which falsified reports may have been published, collaborators of the respondent in the work, or other relevant parties should be notified of the outcome of the case."[26]

As Dr. Needleman learned firsthand, this salutary process for protecting the public from research fraud is subject to abuse by those intent on bending science. Because every penny of the home institution's federal research funding is theoretically at stake if it performs this function inadequately, it has a strong incentive to appear receptive to all types of misconduct complaints and willing to initiate hearings to demonstrate to federal authorities that it is not trying to cover up bad behavior. Thus a survey by the Office of Research Integrity in 2000 of scientific misconduct programs at responding universities found that only 11 percent of 156 responding institutions required an accuser to describe the misconduct in the complaint and only 10 percent expected supporting documentation to initiate the process.[27] Fair-minded university scientists may sometimes be persuaded by a vigorous critic that irregularities in lab notebooks or substandard statistics are telltale signs of research fraud, and they may not probe deeply enough into the critic's bona fides before initiating a full-fledged hearing.

The federal regulations also contain no provisions for punishing or otherwise discouraging paid consultants from filing false or harassing misconduct claims against innocent scientists.[28] While the regulations do not prevent or impede home institutions from administering sanctions to the extent that they have any power over the accusers, the relevant literature contains no evidence that any university has ever done so.[29] Ironically, then, a legal authority that Congress created to prevent fraud can become a risk-free vehicle for unscrupulous accusers to perpetrate it. Because it tends to err on the side of protecting the accuser more than the accused, even when the accuser's motives may be suspect, professional and scientific societies during the 1990s expressed skepticism about the program's current design.[30]

In stark contrast to the ease with which critics may pursue scientific misconduct allegations, defending against them can be very resource-intensive. Because even meritless charges of scientific misconduct can significantly damage an academic career, targets of such charges must vigorously defend themselves. As a practical matter, the targeted scientists are pulled away from the research they would otherwise be conducting to prepare and present their defenses. The claims may also have a wider negative effect on an entire field of research. In addition to providing a direct warning to funders and superiors to keep a close eye on the targeted researcher's work, scientific misconduct allegations also send an indirect message to prospective researchers to avoid policy-relevant research that might arouse the ire of powerful economic or ideological interests.

Since the Office of Research Integrity program applies only to institutions that rely on public funding to support their research, the tool has a somewhat more limited range than other harassing techniques. Still, both the tobacco and lead industries have launched scientific misconduct campaigns against federally funded researchers whose studies produced unwelcome results. In each case, the researchers' universities determined that the charges were unfounded, often after devoting a great deal of time and effort to the proceedings.[31] Plaintiffs' attorneys have also abused scientific misconduct proceedings. After researchers at the University of Washington published a study questioning the reliability of immunodiagnostic tests for chemical sensitivity, the researchers were subjected to a thirteen-month-long scientific misconduct hearing initiated by attorneys representing plaintiffs in chemical sensitivity cases. The complaints continued even after the researchers were formally exonerated.[32] The science policy literature contains evidence of other abuses of scientific misconduct claims in areas outside of policy-relevant science.[33] Because the procedures do not apply to research in private settings, however, the tactic is generally unavailable to public interest groups that might want to challenge the integrity of scientists, like those doing clinical drug trials for CROs, who may be actively involved in shaping policy-relevant science in the private sector.

Publicized Attacks

Coordinated attacks in the media or other public forums can also be a very effective tool for harassing and intimidating scientists. Public disparagement of a scientist who produces unwelcome results may be unseemly and even unfair, however, like the negative political campaign advertisement, the strategy often works. As in the political arena, public relations firms specializing in generating negative press are available for those who can afford their hefty fees.

SKEWERING DR. SANTER

In late May 1996, Dr. Benjamin Santer, a specialist on climate modeling at the Lawrence Livermore National Laboratory, gave a preview of the findings of the soon-to-be-issued report of the Intergovernmental Panel on Climate Change (IPCC). Since the IPCC report was, in the view of some, "a scientific equivalent of a smoking gun," Dr. Santer's presentation was of great interest to representatives of the energy industry and associated think tanks who attended the presentation.[34] During the question-and-answer session that followed, members of the audience accused Dr. Santer of secretly altering the previous year's IPCC report, suppressing dissenting views of other scientists who contributed to the report, and eliminating all references to scientific uncertainties. Things got worse as stories began to appear in the *Washington Times* and *Energy Daily,* an energy industry newsletter, accusing Santer of making unauthorized and politically motivated changes to the text of the IPCC report.

Within days, the unsubstantiated allegations made their way into the mainstream press. Dr. Fred Seitz, a particularly prolific climate change skeptic, complained in an op-ed piece in the *Wall Street Journal* that he had "never witnessed a more disturbing corruption of the peer-review process than the events that led to this IPCC report."[35] The public relations specialists for the Global Climate Coalition, an industry trade association, even coined the term "scientific cleansing" for industry critics to use in framing their accusations.[36] Dr. Santer wrote to all the scientists who had assisted in preparing the report to warn them that he had become the subject of "some very serious allegations" that "impugn my own scientific integrity, the integrity of the other Lead Authors . . . and the integrity of the IPCC itself." On a personal note, he reported that "the last couple of weeks—both for me and my family—have been the most difficult of my entire professional career."[37]

Forty IPCC scientists signed a letter to the *Wall Street Journal* responding to these extreme allegations and defending the science of the report and Dr. Santer's role in particular, but the newspaper allowed only Dr. Santer's name to appear on it.[38] In a letter to Santer, the scientists indicated their support for him and their disgust with Dr. Seitz's op-ed piece because it "step[ed] over the boundary from disagreeing with the science to attacking the honesty and integrity of a particular scientist."[39] Another letter to the *Wall Street Journal* signed by the IPCC chair, Robert Bolin, and Sir John Houghton and Louise Galvan Meira Filho, the chairs of the IPCC's working group on science, stated: "No one could have been more thorough and honest" than Santer in incorporating the final changes suggested by peer reviewers into the text.[40] ❐

As Dr. Santer's travails suggest, carefully orchestrated publicity campaigns can portray attacks on the veracity of researchers and the validity of their research as legitimate scientific disagreements. In late 2006, the public interest group People for the Ethical Treatment of Animals (PETA) launched an internet campaign against researchers at Oregon Health and Science University who were studying the effects of hormones on the sexual preferences of rams, approximately 5 percent of which are homosexual. The purpose of the experiments, which had been quietly proceeding for several years, was simply to learn more about the basic biology of sexual development. The PETA website, however, reported that the researchers were "killing homosexual sheep and cutting open their brains" in an attempt "to 'cure' humans of their homosexual tendencies." The website accusations, coupled with a story in the London *Sunday Times* erroneously reporting that the scientists had successfully isolated a homosexuality hormone, generated tens of thousands of email complaints to the university, and the tennis star Martina Navratilova wrote a highly publicized letter to its president. The scientists spent weeks responding to the misinformed accusations.[41]

A more disturbing venue for a publicized assault on a scientist and his work is a congressional hearing, where an invited scientist can be peppered with politically motivated questions or positioned against an unrepresentatively large number of scientist-critics.[42] These hearings can be held at the whim of the chair of a congressional committee, who also controls the selection of a majority of the invited witnesses. Thus if a company or trade association has sufficient influence to persuade the chair to hold a hearing on a scientific issue, it can work with the committee staff to orchestrate the hearing so as to cast public doubt on what may actually be a solid scientific consensus. Onlookers might believe that the hearing provides balanced information that has scientific merit, but impartiality is by no means guaranteed. As Dr. Rodney Nichols observes, such hearings may "ask the wrong questions, put key social issues on a pseudo-technical basis, and thereby undermine the confidence of everyone."[43]

One of the most notorious uses of a congressional hearing to launch a highly publicized attack on respected scientists occurred in 1995 at a series of hearings entitled "Scientific Integrity and Public Trust," convened by Representative Dana Rohrabacher, the chair of the Subcommittee on Energy and Environment of the House Committee on Science.[44] The official goal of the hearings was to take a closer look at the quality of the science underlying public policy approaches to ozone depletion, global warming, and the EPA's risk assessment on dioxin.[45] But the identities of the invited experts revealed that the hearings were in fact designed to provide a forum for marginally credentialed skeptics to criticize mainstream scientists and federal scientific agencies. Representative Tom DeLay also attempted to embarrass the respected scientists who were the targets of the hearings by charging them with writing "[t]he conclusion . . . before the study is even done."[46] After the hearings had been completed, Rohrabacher opined that "money that goes into this global warming research is really money right down a rat hole."[47] The hearings were so one-sided and unreliable that they inspired a lengthy report by the ranking minority member of the committee, entitled "Environmental Science under Siege," that offered a detailed rebuttal to many of the intemperate charges made during the hearings.[48]

Although less dramatic than a full-fledged hearing, a letter from a powerful congressperson to a federally funded scientist or funding agency demanding information and answers to specific questions (sometimes suggested by the advocates behind the attack) is also an effective harassment tool.[49] These letters are the equivalent of a mini-congressional investigation, even when initiated by a lone congressperson, because they can be backed up by the full subpoena power of Congress. Indeed, subpoenas coming from Congress can be far more onerous than parallel inquiries in court because there is no higher authority, like a judge, to appeal to in cases of unreasonable or abusive demands. Chair Joe Barton's letter seeking information underlying Professor Michael Mann's "hockey stick" model for global warming, an episode discussed in more detail in Chapter 11, provides an apt illustration.[50] Barton not only demanded all of Mann's underlying data and analyses, but insisted that he respond in "detail" and in "a narrative fashion" to a multifaceted critique published by an economist and businessman in a nonscientific journal.[51] Mann's response to Barton's inquiry took weeks of his time and that of his pro bono counsel,[52] even though all of the underlying data had long ago been posted at more than thirty websites and much of it had already been replicated.[53] Both the American Association for the Advancement of Science (AAAS) and the NAS sent letters to Barton protesting that his demands set a very dangerous precedent of replacing scientific peer review with scientific review run by politically motivated congresspersons.[54]

Advocates have also employed public attacks to discourage scientists from pursuing important research projects that might run counter to their economic or ideological interests. That is how the W. R. Grace Corporation dealt with a study of lung disease that a newly arrived physician and county health officer, Richard Irons, proposed to undertake after he discovered an unusually high incidence of lung disease among his patients in Libby, Montana, the site of Grace's huge vermiculite mine and processing facility. Soon after he announced his intentions, in 1979, one of the county commissioners told Irons to "stop making waves." When he persisted, the county hospital accused him of illegally using drugs, and it revoked his hospital privileges, which is the kiss of death for a small-town doctor. He had in fact been taking a pain medication for his back over many years. Irons decided that he could not fight the company and left town, leaving behind an environmental disaster that over the next twenty years caused more than half the town's population to contract asbestos-related lung diseases.[55]

The internet offers an even more accessible way to insinuate reputational attacks into the public policy-making arena and even into debates within the scientific community. We saw in Chapter 6 how internet-based critiques of Dr. Ignacio Chapela's study on the changed genetic composition of indigenous Mexican corn apparently originated from two fictitious scientists working for a public relations firm hired by the biotechnology industry. That attack played a role in *Nature*'s decision to state publicly that it should not have published the piece in the first place. The *Nature* controversy in turn took a toll on Dr. Chapela's career.[56] In December 2003, he was informed that his university had denied his application for tenure, despite the fact that only one member of his department's thirty-six-member faculty had opposed it.[57] After he filed a lawsuit against the university and 145 professors signed a petition in his support, the university offered Chapela tenure with back pay in the spring of 2005.[58]

Draining Resources

Advocates can also waylay bothersome research by draining a scientist's energy and resources through onerous lawsuits or burdensome demands for data and documents. In some cases, such abusive use of the necessary tools of the civil justice system have caused researchers to give up entirely on research that threatened powerful economic interests.

Suing the Scientists

In the 1970s, strategic litigation against public participation (SLAPP) became a popular way to silence activists who were speaking out against undesirable business practices.[59] The companies would sue the citizen groups or other "rabble-rousers" for harming their reputations and interfering with their business relationships. A variant type of lawsuit, called strategic litigation against unwelcome research (SLAUR) has been evolving since the early 1990s in the context of policy-relevant science. These suits allege that scientists have unreasonably tainted the good name of a company or product by making public statements about product hazards on the basis of their research. The comparatively low costs associated with employing litigation as a harassment tool, coupled with the compelling benefits that can flow from filing the litigation, give companies that wish to cast doubt on damaging research a powerful tool for bending science.

DEFAMING HALCION

The sleeping pill Halcion had been on the U.S. market for five years when the FDA ordered a comprehensive review of a disturbingly large number of adverse reaction reports the agency had received from U.S. doctors. Two FDA reviews concluded that Halcion continued to generate from eight to thirty times the number of adverse event reports as its nearest competitor, but the drug remained on the market.[60] Civil litigation against the Upjohn Corporation, the manufacturer of Halcion, mounting adverse effects reports, and a decision by the British government to ban Halcion converged to focus attention on the company. Its executives decided that it needed to do something to protect a billion-dollar product for which doctors were still writing seven million prescriptions per year.[61] An internal company memorandum suggested that the company might simply sue somebody. Although the executives were "not in a position to assess the legal ramifications of such action," they could certainly "assess the business ramifications": a lawsuit "would publicize our intent to defend Halcion against unjust action" and thereby "encourage . . . physicians to continue writing Halcion prescriptions."[62]

As luck would have it, a plausible defendant appeared a month later in the person of Dr. Ian Oswald, an emeritus professor of psychiatry at the University of Edinburgh who had conducted a clinical trial of Halcion in the late 1980s and found an association between Halcion and anxiety, stress, and weight loss.[63] Oswald had recently appeared in a BBC documentary in which he accused Upjohn of covering up a clinical study of its own that had reached similar results. He repeated this accusation in a January 20, 1992, story in the *New York Times,* where he was quoted as saying that the Halcion marketing program was "one long fraud."[64]

Dr. Oswald's accusation had a considerable basis in fact. Because he had been retained as an expert witness in the earlier civil litigation, he had reviewed the sealed documents in that case, as well as the actual data underlying the original 1972 Upjohn clinical study, called Protocol 321.[65] That study had been particularly problematic for Upjohn, because its report to the FDA had mentioned only 6 percent of the observed cases of nervousness or anxiety and only 29 percent of the observed cases of paranoia in the twenty-eight participants. The company later admitted these misrepresentations and reported them to the FDA as innocent "transcription errors." A contemporaneous internal company memorandum referred less charitably to the "manipulation of data by one of our people."[66] The report of an FDA criminal investigation into the incident had concluded that the evidence "strongly suggests that the firm was aware that the report [on Protocol 321] was inaccurate and incomplete but continued to withhold notification to the FDA that the report was flawed."[67] Even more damning, the committee advising the agency that banned Halcion in Britain concluded that if the information in the study "had been presented completely and correctly" when the drug was originally licensed, it was "highly unlikely" that the committee would have made a favorable recommendation.[68]

Faced with this incriminating evidence, the company decided that a lawsuit against Oswald would provide an opportunity to argue in a public forum not only that Halcion was safe but also that complaints against the manufacturer's actions were untrue. Even more conveniently, Dr. Oswald's British citizenship allowed the company to sue him in England, where there is no First Amendment protecting free speech. Oswald would therefore bear the legal burden of proving that his statements were accurate, and he would have no right to a jury trial.[69] At the same time that it launched the lawsuit, the company accused Oswald in the press of purveying "junk science."[70]

The litigation strategy paid off when a British judge found that Oswald had libeled Upjohn and ordered him to pay the company £25,000. It also ordered him to pay the Upjohn executive Royston Drucker £75,000.

The essence of the court's judgment was that although the company had made "serious errors and omissions" in its report on Protocol 321, they fell short of actual fraud.[71] On Oswald's countersuit for libel, based on the "junk science" allegation, the judge ordered the company to pay him £50,000. Oswald came out the overall loser, however, when the judge ordered him to pay the company's litigation costs, which were expected to amount to at least £2 million. ❐

In the Halcion case, Upjohn invested nearly $4 million in litigation costs to pursue a libel claim worth roughly one-twentieth of that amount. Even if Upjohn thought it could win a large judgment, it could not have expected a retired scientist to pay that sum. Assuming that the company is a rational economic actor, there must be an explanation for why it would pursue a lawsuit that would cost far more than it could realistically bring in. The power of such a lawsuit is not in the amount of the potential verdict, but in its effect on public perceptions of the credibility of the science being questioned.

The prospect of defending a lawsuit is also a severe threat to the settled life of an academic. The cost of defending a defamation suit, even putting aside the possibility of an adverse verdict, can easily bankrupt a university scientist.[72] A suggestion, however subtly conveyed, that a scientist might be sued for publishing or publicly summarizing legitimately conducted scientific research can therefore be a powerful weapon in the hands of an entity with sufficient resources to make the threat a reality. Thus, in traditional SLAPP fashion, the SLAUR can successfully intimidate scientists and discourage them from speaking out about events or problems that they know to be true, even in the United States, where there is a First Amendment right of free speech. It is a rare scientist who so highly values her right to speak out about research that she will risk financial ruin to exercise it.

While lawsuits are obviously extreme measures, they are not uncommon. The manufacturer of the AIDS drug Remune sued the scientist who was in charge of a large company-sponsored clinical trial after he submitted a paper, over the company's objections, to *JAMA* concluding that the drug was ineffective.[73] One of the strategies outlined in the Tobacco Strategic Attack Plan developed by the tobacco industry was to launch "surgical strikes," consisting of lawsuits against people bearing bad news for the industry. The memo noted that well-timed lawsuits gave the industry "tremendous media opportunities, which we are continuing to exploit," and cited as an example a single lawsuit against the

American Broadcasting Corporation that succeeded in eliminating the word "spiking" (spraying concentrated nicotine on processed tobacco leaves) from "the lexicon of the anti-tobacco crowd" because of fear of being sued. The memo concluded: "if this is all the suit ever does, it will have been worth it."[74]

As the tobacco industry's plan suggested, an entity with sufficient resources can strategically employ lawsuits against scientists to reframe facts and research results in ways designed to advance its interests in the public policy arena.[75] The mere existence of the litigation implies that a company is so outraged by an inaccurate statement that it has gone to the trouble of filing a lawsuit. The company is thus able to portray itself as the victim rather than the perpetrator of bullying and harassment. Litigation also has a good chance of making the news, especially if it is relatively simple and "sexy."[76] A company attempting to discredit a scientist's unflattering research may have a hard time getting its rebuttal published or even acknowledged in the mainstream media through traditional devices like press releases and phone calls to reporters. Suing the scientist, by contrast, can generate media attention by providing the added drama of formal public conflict and by offering numerous opportunities for staged media events.[77]

Even if the affected entity is reluctant to invest in a full-fledged lawsuit, the mere threat of litigation can have a serious deterrent effect on scientists of modest means who cannot necessarily depend on their home institutions to assist in the litigation. For example, after serious concerns were raised in the mid-1990s over whether a proposed privately run low-level radioactive waste disposal facility would leak radioactive waste into groundwater, the Department of the Interior, which owned the proposed site, hired the Lawrence Livermore National Laboratory and two scientists who had been members of a NAS panel to study the site in detail. Soon thereafter a private company that would have profited from the disposal facility sent letters to the two scientists warning them that if they participated in the research project, the company would "seek compensation from any persons or entities whose conduct wrongfully injures its interests in this manner."[78] Although the Deputy Secretary of the Interior denounced the letters as "raw intimidation tactics," the scientists decided not to risk a lawsuit, and the study was halted.[79] Threats of litigation have also reportedly dissuaded editors from publishing controversial books exposing scientific misconduct by the tobacco industry; articles and letters containing sharp criticisms of research practices by biomedical companies; and research findings suggesting that certain pesticides may be more risky to public health than expected.[80]

Of course, not all defamation lawsuits against scientists are wholly lacking in merit. Perhaps Oswald did go too far in asserting that Upjohn had engaged in "one long fraud," and other scientists have also crossed the line by making unsupported factual assertions that caused economic harm to companies. Yet generous pleading rules, even when a scientist has done nothing wrong,[81] often allow harassing lawsuits to proceed for an extended period before they are finally dismissed or otherwise resolved. Federal rules allow courts to penalize those who file frivolous litigation, but judges concerned about preserving the right to a fair trial tend to administer the penalty sparingly, thus allowing virtually all claims to go forward except those that are almost completely lacking in legal and factual support.[82] At most, the penalties for filing the litigation involve paying for the opposing party's attorneys and litigation costs.[83] State laws aimed at discouraging SLAPP suits in theory might provide added sanctions for this type of litigation abuse, but these statutes typically provide relief only to citizens who become targets of lawsuits as a result of actively engaging in public hearings, protests, or litigation and are generally inapplicable to SLAUR suits.[84]

Subpoenas and Public Records Requests

Advocates can also overwhelm researchers with abusive subpoenas and open records requests demanding documentation underlying every aspect of a research project, including documents containing confidential information of a highly personal nature obtained from human subjects and other information not normally shared, even among scientists. In the same way that they can file SLAUR suits, advocates can strategically employ the basic commitments to open government embodied in these legal tools to harass scientists and chill scientific activity that produces unwelcome research. Moreover, the tools are curiously asymmetrical, in that far more of them are available to advocates seeking information from government scientists and academics working on government grants than to those seeking similar information from scientists working for or funded by the private sector.

JOE CAMEL GOES TO COURT

Dr. Paul Fischer's consumer marketing research on the appeal of Joe Camel, RJR's tobacco mascot, revealed that young children identified very positively with "Ole Joe" and that the identification was so strong that it was probably more than an unhappy coincidence.[85] Soon after his research was published in *JAMA*, Fischer was served with subpoenas filed by RJR in a lawsuit in which Dr. Fischer played no role whatsoever. As a defendant in litigation brought by a smoker, RJR invoked the court's power to subpoena relevant information produced by third parties to request Fischer's data. Rather than simply seeking information that Fischer would provide in the normal course of scientific data sharing, however, RJR requested all of the data underlying Fischer's ongoing Joe Camel research. The subpoena thus requested all lab notebooks and other confidential information concerning the children who participated, including their home phone numbers and addresses.

When the federal court refused to enforce RJR's exceedingly broad subpoena, the company filed a similar request under a state open records statute that allowed citizens to access state government documents. Since Dr. Fischer worked as a tenured professor at the State Medical College of Georgia, his research records were considered government documents under the state's broad law and therefore subject to disclosure unless they came within narrow statutory exceptions. Despite Fischer's strong contention that the confidential records were not appropriate for release under the state statute, a judge ordered him to share all the documents. Faced with a court order and very little support from his university, Fischer turned the documents over to RJR. There is no evidence that the company ever used this information (it ultimately agreed to have the confidential data redacted), but Dr. Fischer devoted a great deal of time and effort to fighting both the subpoena and the state open-records litigation. The tobacco company's subsequent attempts to replicate Fischer's data verified his results, and Joe Camel is now a historical curiosity. Unfortunately, the incident left such a sour taste in Dr. Fischer's mouth that he resigned his tenured post in disgust and took up a family medical practice in rural Georgia. ❐

Burdening a scientist with unreasonable document requests does nothing to advance peer scrutiny of the research. At the same time, such requests effectively undercut scientific freedom by overwhelming scientists with sanctions-backed demands for documentation and, in some cases, by intimidating scientists with the threat of further legal proceedings after they produce the documents.[86] At the very least, scientists who are subjected to these harassments will have to seek legal counsel, sometimes at their own expense. While sharing data and records is a tenet of good scientific practice as well as good government, a scientist carrying out a review in good faith does not need access to all research records, including those giving rise to reasonable expectations of privacy, to evaluate the reliability of a study. Clear limits are required to prevent abuse, but current legal restrictions on subpoenas and open records requests do not always provide them.

With limited exceptions, state freedom-of-information laws generally require all written information in government files to be made available to any citizen who requests it in writing. All records related to studies conducted by researchers in public universities, including data tables, lab notes, and the like, may therefore be subject to such requests.[87] By contrast, only a party in active litigation may issue a subpoena, but subpoenas can seek documents not only from parties and experts in the litigation but also from third parties with relevant information, and the exceptions are much narrower.[88] The subpoenas, like freedom-of-information requests, routinely seek all underlying data, all notes of any kind (including lab notes), and all prior drafts of any written testimony.

These legal tools typically establish no threshold standards of reasonableness for document requests, relying instead on the good faith of requestors to limit their demands for documents that are not within the defined exceptions. Consequently, when a scientist receives such a request, the burden is on her to justify denying documents she thinks should come within the typical exceptions for confidential records and ongoing research.[89] Judges' appreciation for the importance of protecting scientists and their subjects varies. Although federal courts do routinely limit unreasonable subpoenas after the recipients challenge them in formal legal proceedings, they generally will not limit subpoenas in advance. From the scientist's perspective this modest protection may be too little, too late. Without the assistance of an able attorney, a scientist may wrongly assume that full compliance with a subpoena is required even if the request is patently unreasonable. At the same time, universities that are sensitive to the political environment may not always come to the aid of scientists who are receiving information demands from powerful political actors. As a result, unreasonable private efforts to obtain underlying scientific information have some chance of actually succeeding.

Sanctions for overly aggressive use of these tools are either limited or nonexistent. A court can penalize subpoena abuse, but the punishment consists, at most, of lost wages and attorney fees.[90] A well-endowed private party may well decide that it is worth risking a modest penalty to achieve the substantial benefits that may result, so long as it does not also generate bad publicity. In addition to overwhelming the targeted scientist, who typically lacks legal or even paralegal assistance, subpoenas and freedom-of-information requests have the potential to dredge up information that may later be used in attacks on his research. Indeed, the specter of future harassment and adverse publicity may be sufficiently intimidating to cause a scientist to rethink future research projects that could implicate a private party's product or activity.

The impropriety of burdening scientists with harassing document requests is a matter of growing concern to the scientific community.[91] The Washington University neurologist Brad A. Racette points out: "When investigators see the difficulties faced by other researchers, their natural reaction is to find something else to do."[92] He was served with ten subpoenas after he published a study suggesting that exposure to manganese fumes from welding rods caused Parkinson's disease in welders. The university came to his defense, to the tune of $200,000 in legal expenses, but the records were ultimately turned over to the defendants in the subsequent class action lawsuit.[93] The Exxon Corporation, threatened by scientists' efforts to assess environmental damages resulting from the Valdez oil spill, served harassing subpoenas on one of them seeking all records, data, and ongoing research. That researcher complained that complying with the subpoenas "permanently disrupted" his research project "due to the constant need to respond to motions and affidavits."[94] Lawyers for the North Carolina pork industry invoked that state's open-records law in an attempt to require a University of North Carolina epidemiologist to produce copies of the completed surveys he and his colleagues had administered to residents living near industrial swine operations as part of their study of the health impacts of such operations. The researcher objected to the request because he had promised the residents, some of whom worked at the plant, that their responses would remain confidential.[95]

Harassment through document requests can extend to others involved in data gathering, analysis, and publication. For example, the defendants in a lawsuit the state of Rhode Island brought seeking damages for harm to children caused by lead-based paint subpoenaed twenty-five years' worth of raw and even unpublished data from three scientists who had agreed to serve as experts for the state, another scientist who had declined to serve as an expert after receiving the defendants' request for data, and still another scientist who had merely agreed to share a data set with one of the three testifying experts for a pooled analysis of IQ and blood lead levels.[96] After the historian Gerald Markowitz testified in a Louisiana lawsuit about vinyl chloride manufacturers' historical attempts to avoid responsibility for harm to their workers, he included research gleaned in part from documents obtained in that litigation in two chapters of a book published by the University of California Press.[97] When he served again as an expert witness in a second case, the chemical company defendants subpoenaed and deposed all of the book's peer reviewers. One of them, the well-known biographer of Eleanor Roosevelt, Blanche Wiesen Cook, called the process "harassment to silence independent research" and an attempt to have a "chilling effect on folks who tell the truth."[98]

Scientists have gradually begun to speak out as a group against these serious intrusions on their research. The largest collective statement by scientists in opposition to legal tools that could be used in harassing ways was triggered by the enactment of the Data Access Act. That statute requires federally funded scientists to make all "data needed to validate a federally funded study" that is used to inform or support regulation available to requesting parties through the FOIA.[99] Because this type of information sharing could require disclosure of virtually all data and records underlying research used for regulation, including ongoing research, the AAAS, the NAS, and a variety of other scientific organizations filed vigorous objections to proposed rules implementing the law.[100] In response to the scientific uproar, the federal agency responsible for the regulations—the OMB—narrowed the reach of the Act to provide access only to certain types of standard data underlying *completed* federally funded research and also to establish rules requiring the requestor to compensate the researcher for time and copying costs.[101]

Targeting Funders and Employers

Modern scientific research is normally quite expensive. Successful scientists must therefore be not only good investigators but also good fund-raisers.[102] One way to ensure that scientific research does not undermine an advocate's position is to make sure that the research never gets funded or that the scientist's financial support is reduced sufficiently to slow down or halt his research. When an advocate is the source of the funding, this strategy is especially easy to deploy. However, advocates have on occasion successfully persuaded other employers to discharge or demote troublesome researchers, deny them promotions, benefits, and transfers, and otherwise make life difficult for them.[103]

MUZZLING WYNDER

One of the earliest and most serious threats to the tobacco industry came from mouse studies conducted in the early 1950s by Ernest Wynder, a scientist employed by the Sloan-Kettering Institute.[104] Wynder had already conducted a small epidemiological study suggesting that tobacco smoke caused cancer, and he attempted to verify that statistical association experimentally by conducting a two-year study of the capacity of tobacco smoke distillate to cause cancers in mice. The results clearly demonstrated that the distillate caused skin cancers in the mice.[105] Several tobacco companies, in what might at first glance appear to be an ill-considered move, then began to make large donations to the institute. Before long, according to an internal tobacco company memo, the institute's director and other officials began "subjecting Wynder to more rigorous screening procedures before letting him speak in the name of the institute," and this "had a proper and pleasing effect." In the end, "[t]he deductible contribution to Sloan-Kettering [was] probably the most effective of all health research contributions" because it diverted Wynder from his self-appointed role as an industry critic.[106] Wynder himself became a consultant to the tobacco industry in the late 1980s during the debates over passive smoking.[107] ❒

When the bothersome science is coming from a scientist employed by someone else, advocates can still harass the scientist by complaining to her employer that she is biased or otherwise incompetent in the hope that the complaint will generate indirect pressure on her to modify or even retract her damaging conclusions.[108] Research facilities that rely heavily on grants and contracts from regulated industries are especially susceptible to such overt attempts to influence research outcomes. For example, after Dr. Arpad Pusztai published in the *Lancet* a study in which he concluded that rats fed GM potatoes suffered various health problems and expressed to reporters the worry that the food industry was treating consumers like "unwitting guinea pigs,"[109] representatives of the biotechnology industry complained bitterly to his employer, a British research institute. The institute, which had recently received a $224,000 grant from a large biotechnology company, then fired Pusztai, broke up his research group, terminated six ongoing projects, and requested that the British Royal Society convene a commission to review his research.[110]

This tactic has even proven effective in persuading public sector employers to prevent government researchers from doing studies that might be damaging to powerful private sector actors. After lawyers for the tobacco industry learned from industry consultants that Jack Henningfield, a scientist at the National Institute on Drug Abuse (NIDA), was conducting research on the addictive properties of nicotine, it lobbied its allies in Congress and the Reagan administration to put a stop to his threatening research. The administration's drug czar then visited the acting director of NIDA and reportedly suggested that "if NIDA had so much money to waste studying the addictive properties of nicotine, then maybe NIDA did not really need its budget."[111] When Henningfield received word from the acting director that not only his budget but also the funding of his colleagues at Johns Hopkins University was at risk, he abandoned the research, thus delaying independent research on tobacco addiction for years. The action also had a predictable *in terrorem* impact on Henningfield's colleagues, who did not attempt to carry on their research.[112]

When advocates direct similar campaigns against researchers at universities, where the tradition of academic freedom is better established, however, these institutions sometimes stand behind their scientists. When the Indiana State University microbiologist Dr. Kathleen Donnelly published a study concluding that Bausch & Lomb's ReNu MultiPlus contact lens cleaner did not kill bacteria as well as other solutions on the market, the company wrote a strong letter to upper-level officials at the university complaining that the study's methodology was flawed and demanding a published retraction, but the university stood behind Dr. Donnelly.[113] Years later, the company had to remove the product from the market after scientists at the CDC concluded that it was linked to more than one hundred serious corneal fungus infections.[114] Similarly, a highly regarded professor of medicine at Stanford University rebuffed an angry phone call from the medical director of Merck Pharmaceutical complaining about negative statements that an adjunct professor was making in medical seminars about Merck's refusal to give him data on the potential adverse effects of its blockbuster drug Vioxx. When the professor responded with a letter to Merck's CEO complaining of the attempted intrusion, the company backed off.[115]

Conclusion

Sadly, the reports of scientific harassment outlined in this chapter may represent only the tip of a much larger iceberg. Frederick Anderson, an attorney with a firm that represents mostly corporate clients, concluded in 2000 that "[w]ell-organized campaigns against certain types of research and the researchers who conduct them do appear to be on the rise."[116] Instead of providing needed protections for researchers who go about their work competently and transparently, the legal system subjects them to abuse that would never be tolerated within the scientific community and forces them to fend for themselves.

Perhaps even more tragic, those at the receiving end of these unprincipled yet legally sanctioned attacks are generally not the scientists whose work yields little of value to society. They are instead the scientists who produce information and discoveries that have great significance for social problem-solving. After his ten-year ordeal defending himself from unfounded scientific misconduct charges, Dr. Needleman observed that "the federal investigative process can be rather easily exploited by commercial interests to cloud the consensus of a toxicant's dangers, can slow the regulatory pace, can damage an investigator's credibility, and can keep him tied up almost to the exclusion of any scientific output for long stretches of time, while defending himself."[117]

Although it could potentially provide enormous moral suasion in this area, the scientific community has seemed unduly complacent or else uninformed about many of the incidents of harassment detailed in this chapter. The public record contains only a few accounts of scientists banding together to support the victim against an unfounded attempt to besmirch his reputation or to launch a counterattack on the entities responsible for the harassment. In this regard, the commendable outpouring of protests from the scientific community against the energy industry's attempts to skewer Dr. Santer and Representative Barton's harassing informational demands on Dr. Mann, both of which came in the highly publicized area of climate change science, are the exceptions that prove the rule. Perhaps scientists believe that any attack on reputation is a serious enough matter that it must have some merit because bona fide members of the community of science would never stoop so low otherwise. Or scientists may be justifiably concerned about the potential for soiling their own reputations if they come to the aid of a colleague in a battle against adversaries who are demonstrably willing to sling mud.[118]

At this intersection of policy-relevant science and the law, where both the costs of abuse and the incentives to commit it are high, structural legal change is imperative. In Chapter 11 of this book we suggest specific parameters for a positive restructuring of the legal rules to discourage abuse while allowing legitimate demands for information and accountability to proceed. Higher evidentiary burdens for filing challenges, greater oversight over the charges, and much more vigorous sanctions for engaging in abuse will render these tools less attractive for advocates attempting to bend science.

Without reform, the future for scientists who become the targets of these abusive tactics is bleak. Advocates will continue to misappropriate legal processes initially designed to advance good science and provide effective civil justice, making unreasonable demands on scientists and jeopardizing their reputations. Scientists will succumb to the legal attacks either because they think they must or because they find it impossible to mount successful counterattacks against adversaries with greater economic resources and legal expertise. Like Dr. Fischer, they may decide that the game is no longer worth the candle. The resulting exodus of scientific talent, however, is most certainly not a positive development for society, nor is it something the scientific community can tolerate in the long run.

CHAPTER EIGHT

Packaging Science

The Art of Assembling an Expert Group to Advance a Favored Outcome

Purchasing pronouncements from one or more handpicked experts on what the "science says" is an extremely effective way to mold science to fit one's needs. An advocate with a lot of resources and a need to bend an emerging body of science utters the right incantation, and a group of scientists magically appears in a high-ceilinged conference room in Washington, D.C., or better still, at a well-appointed conference center at some other commodious location to produce a "consensus" statement that is "independent" and "expert," but in truth diverges significantly from mainstream scientific views in a direction that favors the economic interests or political needs of the sponsor. A wave of the magic wand, and a book or review article authored by an expert in the field graces the literature with a "definitive" encapsulation of existing scientific information on a topic near and dear to the same sponsor. Yet unraveling the origins of these faux "consensus statements" and "authoritative works" is so difficult and so rarely done that most scientists, policy-makers, and judges seem to take them as legitimate summaries of the relevant scientific information bearing on one or more questions that frequently arise in regulation and litigation.

In this chapter we look under the hoods of these sleek but powerful vehicles for manipulating the perceptions of scientists, regulators, and judges to shed light on how advocates manage to package science in misleading ways with little fear of detection. We also try to identify the conditions that are most likely to produce these sponsor-manufactured "consensus statements" and review articles and to understand when these highly choreographed efforts to bend policy-relevant science are likely to succeed. Before embarking on this undertaking, however, we should take a moment to underscore the brilliantly nefarious nature of this particular tactic. As we discussed in Chapter 3, one of the annoyances agencies and courts encounter in their attempts to access and use policy-relevant science is the difficulty they face in isolating the mainstream views of the scientific community from the largely unsupported or extreme views they hear from some of the scientists and consultants working for the advocates. This conundrum is a direct result of the diffuse nature of scientific governance. Scientists vet each other's work through conferences and peer review of individual journal articles, but they much less frequently take on the task of bringing several strands of recent research together in a single review article, and they only very rarely attempt on their own to congregate to articulate in a single document "what the science is" on any given issue. A consensus statement thus fills this large void.

When produced discreetly, the manufactured "consensus statement" or sponsored review article is also useful to advocates attempting to influence courts and agencies, because it is extremely difficult for other scientists to evaluate or challenge. Since independent scientists are not well organized, they rarely engage in the potentially futile exercise of challenging panel statements, and they almost never do so collectively. As a result, criticisms of packaged panels generally come from individual scientists or parties whose views are not adequately represented on the panels. Yet many onlookers may discount criticisms by one or two scientists when compared to the negotiated statement of a "distinguished" group of carefully selected experts; indeed, they are likely to dismiss these isolated scientists' criticisms as "sour grapes." The formality of the packaging exercise and the number of scientists contributing to it provide it with an imprimatur of credibility that can effectively insulate the resulting "consensus statements" from meaningful review or challenge.

If there is any good news in this story, it is that this highly effective vehicle gets driven only on rare occasions. It tends to be quite expensive to operate and is therefore employed only as a last resort by advocates in the private sector who have lots of available resources or a huge ideological stake in the outcome. Although government agencies can assemble ad hoc advisory committees somewhat less expensively than private sector entities, it is still a very resource-intensive and time-consuming process, and time is often a resource in short supply for agencies. Hence public and private scientific advisory panels tend to make the scene in cases where adept packaging is needed to resolve very high-stakes issues like climate change, ETS, or catastrophic risks of lucrative and widely used products. Furthermore, packaging works best when the scientific community has not gelled around a clear consensus position or at least when such consensus as does exist is not readily apparent to most players in the policy arena. When these three conditions—high stakes, a government agency or a private sector advocate with adequate resources, and malleable science—are present, the stage is set for effective packaging.

In this chapter we explore the three main approaches that have evolved for packaging science when the conditions are ripe. The first involves *stacking* or skewing federal science advisory panels, a feat that is difficult to accomplish from outside the agency that convenes them, but easy enough when an agency is inclined to do so for political or ideological reasons. Under the second and much more costly approach, a private sector entity carefully assembles a panel, symposium, or team to express an outcome-oriented "consensus" position of the scientific community on an issue of policy-relevant science. Finally, private sector entities can package science by commissioning books and review articles that purport to summarize the state of science, but do so in biased and incomplete ways.

Federal Science Advisory Panels

The most troublesome packaging efforts are initiated by federal agencies through the formal processes established by law for the creation and use of official federal advisory panels. These bodies enjoy the privileged position of providing direct guidance to regulators while at the same time appearing to the outside world to be respected and credible repositories of science advice. Because of the de facto power these and other formal advisory committees can wield, Congress enacted the Federal Advisory Committee Act to prescribe procedures for the agencies to use in convening and using them. The statute directs the relevant agency to assemble panels that are "fairly balanced in terms of the points of view represented" and to ensure that they will "not be inappropriately influenced by the appointing authority or by any special interest."[1] In practice, however, the actual operational rules leave the agencies with sufficient discretion to assemble panels that might satisfy legal requirements but are nevertheless badly imbalanced and therefore likely to reach a predetermined outcome.[2] This is one strategy for bending science that does not easily escape the attention of the scientific community, perhaps because it has an obvious interest in the composition of prestigious federal advisory panels.

"GAMBLING" WITH SAFE LEAD LEVELS IN CHILDREN

The CDC's Advisory Committee on Childhood Lead Poisoning Prevention enjoyed a long history of providing highly respected, scientific advice to HHS. In 2002, however, HHS Secretary Tommy Thompson shuffled the membership in ways that suggested he hoped to skew the panel's recommendation to be more in keeping with the views of the Bush Administration. Just weeks before the panel was scheduled to meet to consider whether to lower the 1991 standard for blood lead levels of 10 µg per deciliter (µg/dl) to even more protective levels, Secretary Thompson intervened to reconfigure the panel. One of the longstanding members, a respected pediatrician whom an agency staff member had already contacted and asked whether he would be willing to serve as the next chair of the committee, was dismissed. Two other nominees for open slots, both respected health scientists who had been checked and approved by the CDC staff, were rejected by Secretary Thompson. Dr. Susan Cummins, a former chair of the CDC's lead advisory committee, stated that this was the first time an HHS secretary had ever rejected nominations by CDC staff or members of the committee.[3]

In their place, Thompson selected four new members, all of whom had direct or indirect (through their employer) ties to industries that would be economically harmed by a more stringent lead standard.[4] Moreover, at least two of the nominees had done effectively no work on the effects of lead on children, the very purpose of the panel's deliberations. The most suspicious nominee had served as an expert witness for the lead industry in lead-poisoning litigation and had testified that he didn't "believe that any of the epidemiological studies have established any relationship between lead ingestion and adverse cognitive, behavioral, or emotional status"[5] and that the current CDC level was already too protective by seven- to tenfold.[6] Another nominee admitted that he had been contacted first by the lead industry asking if he would be willing to serve on the advisory panel before receiving the invitation from CDC.[7]

The reconfigured panel seemed to achieve Secretary Thompson's goals. In spite of a widely held view by scientists within and outside of the CDC that the "safe" level of lead should be lower, the panel ultimately recommended that the lead standard should remain at the existing level of 10 µg/dl, a result that was no doubt beneficial for the lead industry.[8] In defending this result against vigorous criticism, the HHS spokesperson argued that it was "disingenuous to criticize the Bush Administration for installing like-thinking individuals 'when every Administration does that. . . . That's like saying, Gosh, there's gambling going on in this casino.'"[9] ❐

In theory, federal science advisory boards provide an essential service by offering agencies expert guidance on technical issues or scientific information that might otherwise become politicized or misunderstood. Indeed, when an agency ignores the guidance of such a panel, reviewing courts tend to create a soft presumption that the agency is behaving arbitrarily.[10] Not surprisingly, as science has played an increasingly important role in health and environmental regulatory decisions, these advisory panels have grown both in number and influence. The NIH alone has 144 of them, and the FDA and EPA both rely heavily on them for advice.[11] Data from 1989 suggest that roughly half of the EPA's major activities in one form or another were debated, reviewed, or influenced by its umbrella expert panel, the Science Advisory Board.[12]

Many observers of science-based decision-making in regulatory agencies view the science advisory panel as a rare island of objectivity in a swirling sea of politics and interest group competition.[13] Meddling with this sacred ground is therefore a serious business. It not only has the potential for changing lay understandings of policy-relevant science but also can distort policy outcomes. While the political appointees who head the agencies inevitably express "preferences" in forming panels, particularly on highly salient issues, most administrations over time have appreciated that too much intervention will lead to unsupportable decisions and even to unpleasant political repercussions.[14] Because the science advisory process is generally viewed as vital to the legitimacy and soundness of regulatory decisions, those who are caught meddling in the process tend to receive harsh condemnation from a broad array of actors, but most especially the scientific community.

Although accounts of significant stacking of federal science advisory boards are rare, some administrations have been unable to resist the temptation to try. The Nixon administration raised a few eyebrows in this regard;[15] but the Reagan administration's rather overt efforts to skew science advice set off serious alarms in the scientific community. Most notorious was the internal circulation of a hit list of existing and potential science advisors who should *not* be included in any advisory processes.[16] These scientists were perceived as holding scientific positions that conflicted with the administration's deregulatory policy perspective or were economically threatening to favored administration constituencies. Once the hit list appeared in the national media, however, the administration was so roundly criticized that it backed off considerably from its initial efforts to stack the panels.

A second epidemic of stacking occurred twenty years later during the George W. Bush administration. In addition to the CDC lead panel, specific examples of stacked panels include panels that were assembled to provide advice on "environmental health, genetic testing, reproductive health, the protection of research subjects, and workplace safety."[17] Each of these incidents provoked harsh criticism from the scientific community. A highly publicized letter signed by a large number of scientists;[18] several critical General Accounting Office reports;[19] resolutions formulated by two mainstream scientific organizations—the AAAS and the American Public Health Association;[20] and unflattering news stories in *Science,* the *Wall Street Journal,* the *New York Times,* and the *Los Angeles Times*[21] all took the administration to task for its activities.

This harsh criticism did not, however, deter the administration, and, if anything, it grew even more aggressive and creative in its packaging efforts. In 2003, for example, the White House's OMB created a new program to allow agencies to assemble "expert panels" on any significant scientific issue of concern without any added requirements to guard against stacking, and it even mandated this type of review for particularly "influential scientific information that the agency intends to disseminate."[22] Even more alarming, OMB initially proposed to structure the program in a way that tended to favor private sector experts over academics and government scientists.[23] While strong objections to the proposal from scientists and public interest groups ultimately led the OMB to retract its controversial approach to membership selection, much of the remainder of the program is still in place.

As the foregoing examples suggest, the legal constraints on assembling representative scientific panels are rather easily avoided or even ignored. A determined agency can stack a scientific advisory committee despite the Federal Advisory Committee Act's limited and generally ambiguous requirements that panels be "balanced" and not subject to "undue influence." One way for an agency to accomplish this without fear of detection is to employ a litmus test to exclude or remove outspoken scientists whose views are likely to clash with those of the administration or its favored constituencies. This technique rids the committee of problem votes without casting doubt on the remaining panel's overall "balance."[24] For example, the Office of the Secretary of HHS rejected three out of six of the scientists recommended for scientific advisory committees on occupational safety and health grants, and in at least one instance, the rejection was apparently based on the nominee's position on a scientific issue.[25]

Federal agencies can also attempt to shift the consensus that the larger scientific community might otherwise reach by sprinkling in a few scientists who are likely to be dependable advocates for the administration's position on controversial issues of policy-relevant science. For example, the FDA leadership in 2002 appointed a new member to its Advisory Committee on Reproductive Health Drugs who had very few scientific credentials but considerable ties to the Christian Medical Association lobby.[26] The same year, fifteen of the eighteen members on the advisory committee to the National Center for Environmental Health were replaced, in quite a few cases with scientists with strong industry connections.[27] Moreover, the ambiguity of existing conflict of interest disclosure requirements allows agencies to add conflicted members to advisory committees, often without overstepping any legal bounds.[28] Finally, since many of the advisory panels are completely discretionary, they can simply be disbanded if manipulating their membership proves too difficult. As the old adage goes, "Don't ask for advice that you don't want." The Bush administration, for example, has disbanded the National Human Research Protections Advisory Committee and HHS's Advisory Committee on Genetic Testing, both of which attracted ideological opposition from religious groups.[29]

The overt attempts by the Reagan and the second Bush administrations to stack federal advisory committees are probably not as common as upper- and midlevel agency officials' more subtle attempts to manipulate the advice that they receive from advisory panels by gradually populating them with scientists who over time are more likely to ratify the agency's positions on the relevant scientific issues than to reject or question them. The extent to which this more subtle manipulation of panel membership occurs cannot be determined due to the limited and incomplete records supporting agency panel selection decisions. Tracing the allegiances of each member of a federal science advisory panel is a time-consuming and often impossible task because so much of the information surrounding each member is private. Unless an agency requires rigorous disclosure of employment, financial, and consulting relationships, as well as ideological positions, many of these conflicts will not be on record.[30] Moreover, even when disproportionate numbers of committee members appear to be suffering from financial or ideological conflicts of interest, as has been the case with some FDA and

EPA panels,[31] the point at which the panels themselves become imbalanced is never clear. The extent to which this imbalance, when identified, actually affects the outcomes of committee deliberations is also unclear, although the pressure toward consensus that such committees typically feel may ensure that minority views are well represented in advisory committee reports.

Even without an ally in the executive branch, an advocate with a continuing economic or ideological stake in a regulatory program may be able to infiltrate science advisory committees with their consultants or sympathetic scientists in an effort to tilt the outcome away from what might otherwise be the consensus view. The limited case studies in the literature suggest that a healthy percentage of scientists serving on EPA science advisory committees consulted, worked for, represented, or owned stock in affected industries at the time of their service.[32] Yet some of these conflicts were neither recorded nor made available to the public. For example, in selecting a panel to review the EPA's cancer risk guidelines, the EPA did not seek more information when a panelist disclosed that he or she served in a consulting arrangement with a tobacco company and a tobacco-funded research organization; instead the agency said it "subjectively" determined that the panelist fit within the "middle spectrum" of scientific views. The agency also failed to investigate the nature of panelists' industry ties in a panel it assembled to review its risk assessment for the chemical 1, 3-butadiene. Follow-up research by the General Accounting Office revealed that two of these panel members owned stock in companies that manufactured or distributed the chemical.[33]

Even if the advocate cannot insinuate enough members onto a committee to constitute a majority, the presence of even a single mole can be quite fruitful. Consultants can effectively gather intelligence for their sponsors from their privileged positions on federal advisory committees. More important, the same pressure toward consensus that ensures that minority views are rarely squelched can effectively be manipulated by a persistent panel member to steer the committee's conclusions away from strong positions that are antithetical to the interests of their sponsors.

Conflicts of interest have afflicted FDA advisory panels even more extensively than EPA panels, in part because the FDA often assembles separate panels for individual drugs that require specialized expertise. After the FDA promulgated more rigorous financial conflict of interest guidance in 2002 to implement the FDA Modernization Act of 1997, the resulting information allowed more robust statistical studies of the extent to which committee members had such conflicts and the degree to which they affected committee deliberations.[34] A study of 221 meetings held by sixteen FDA advisory committees during 2001–4 revealed that in 73 percent of the meetings, at least one member or voting consultant disclosed a conflict of interest, but only 1 percent recused themselves. Out of a total of 1,957 members and voting consultants serving on the sixteen committees over the four-year period, 28 percent disclosed a conflict. The study also concluded, however, that excluding the conflicted members would not have had a statistically significant impact on the outcome of the committee votes.[35] Some committees have reflected stronger industry representation than others. On one committee, ten of thirty-two panelists investigating the controversial painkillers known as COX-2 inhibitors, including Vioxx, had ties to the makers of these drugs.[36] In March 2007 the agency proposed to increase the stringency of the selection process to exclude members for whom the agency's decision could have a direct and predictable effect in their financial interests valued at $50,000 or more and to limit those with financial interests of less than $50,000 at stake to nonvoting status.[37]

The lax efforts of many agencies to ensure or at least document the balance in their federal advisory committees combine with limited publicly available information and oversight to produce a perfect storm for conflicted science advice.[38] The extent to which this loose oversight of science advisory panels allows some panelists to steer panel deliberations in the direction of their employers and sponsors remains unclear. The ability of a determined minority of conflicted scientists to affect the advice provided by scientific advisory committees is nevertheless suggested by the recommendation of the recent CDC childhood lead poisoning prevention panel discussed earlier, which seemed to diverge from the views of the mainstream scientific community. While the conclusions reached by a science advisory panel will invariably depend on a range of factors, including the personalities and assertiveness of the individual members, an expert should not be included on a panel when there is a real risk that her own agenda, whether driven by ideological or economic ends, may directly conflict with the panel's mission of providing representative scientific advice. The science advisory process is simply too important an institutional tool to subject to this type of preventable, biasing error.

If You Build It, They Will Come

An advocate that is not fortunate enough to have an agency stack a federal advisory committee in its favor can always fashion an expert body of its own and populate it with scientists who are likely to share its views on the relevant scientific issues. Although they cannot match the influence of stacked federal advisory panels, these nongovernmental scientific groups that advocates assemble can still exert a surprisingly strong influence on lay persons' understanding of the emerging science. References to carefully assembled "independent" expert panels may give physicians, consumers, and even policymakers the mistaken impression that the panels were appointed by a governmental entity, when instead they operate exclusively in the private sector without even the marginal constraints that the Federal Advisory Committee Act provides. Moreover, while invitational conferences and informal assemblages of scientists are very much the accepted norm in science, these privately funded panels are strikingly different in one critical regard: because they are created for the express purpose of packaging science to advance predetermined policy positions, their pontifications are not merely *ex cathedra*—they are presumptively biased.

Creating or Coopting Professional Groups

Several "scientific" societies that appear to be scientifically grounded and representative of the larger scientific community in fact owe their existence and continued funding in large part to corporate donors who have a strong economic interest in the positions they take on policy-relevant scientific issues.[39] The International Life Sciences Institute (ILSI), for example, was established in 1978 by the Heinz Foundation and several prominent food companies, including Coca-Cola, Pepsi-Cola, General Foods, Kraft, and Procter & Gamble.[40] Its official mission is "to improve the well-being of the general public through the advancement of science,"[41] but a better sense of its real mission may be gleaned from its actions. It played a prominent role in defending sugar against nutritionists' concerns about its adverse effects in 1998 when the United Nations Food and Agriculture Organization (FAO) convened an expert consultation on the nutritional value of carbohydrates (including sugar) to provide guidance to governments around the world on appropriate dietary intake of carbohydrates. Unbeknownst to most of the participants and to the governments receiving the panel's report, the consultation was funded by the ILSI and another industry-funded group, and the ILSI was allowed to nominate scientists of its choosing to sit on the panel.[42] The FAO's assistant director general later admitted that although the arrangement did not specifically violate any FAO rules, it "did contravene commonsense norms of transparency and the avoidance of perceived conflict of interest."[43] One of the FAO participants who was unaware of the ILSI's covert role later observed that it helped "explain why there was a very strong attempt to dilute everything we said about sugar."[44] The skewed panel report was a boon to the sugar industry, which issued a press release on it entitled: "Experts see no harm in sugar—good news for kids."[45] Several years later, the ILSI managed to situate itself as a consultant to the WHO; its role may help explain why all statements regarding the adverse role of sugar in diets were eventually dropped from the first worldwide Nutrition Plan of Action.[46] After an environmental group challenged ILSI's consultancy status, the WHO executive board in 2006 excluded it from its standard-setting activities, but allowed it to continue its status as a nonvoting participant in WHO governing bodies.[47]

Much like the ILSI, the American Council on Science and Health (ACSH) was founded in 1978 as a "consumer education and public health organization"[48] that purports to represent "mainstream science";[49] in fact, it receives much of its funding from regulated corporations.[50] Although the sponsors play a less prominent role in its day-to-day activities, statements by its founder and most prominent spokesperson, Elizabeth Whelan, suggest a close relationship with industry funders. For example, when Shell Oil Company decided to quit funding the ACSH in 1992, Whelan complained to her staff that "[w]hen one of the largest international petrochemical companies will not support ACSH, the great defender of petrochemical companies, one wonders who will."[51] The ACSH has defended the industry position in a number of public health controversies involving such products as alar, saccharine, formaldehyde, ethylene dibromide, and genetically engineered bovine growth hormone.[52] Arguing in 1994 that the government should not waste its time with the mandatory nutrition labeling on food products that has now become quite accepted within the industry, Whelan told the media that fat and cholesterol labeling requirements "conferred no benefit that I'm aware of."[53] In January 2005, the ACSH published a book titled *America's War on Carcinogens: Reassessing the Use of Animal Tests to Predict Human Cancer Risk*.[54] As the title suggests, this was not an animal rights tract. It was instead the culmination of the ACSH's twenty-five-year crusade to prevent regulatory agencies from continuing the generally accepted practice of relying on animal testing for the purpose of assessing human health risks.

Although the ILSI and ACSH are extremely effective at advancing their funders' interests through their scientific positions, advocates have found that they need not always construct nonprofit scientific groups from the ground up. Coopting existing reputable professional associations can be at least as effective.[55] The American College of Occupational and Environmental Medicine (ACOEM), an organization consisting of more than five thousand physicians and other health care professionals, publishes position papers on the causes of and ways to prevent and treat various occupational diseases.[56] Defendants in toxic mold litigation have frequently relied on the ACOEM's 2002 position paper on the adverse effects associated with molds in indoor environments, which refers to "[t]he present alarm" over human exposure to "so-called" toxic mold and concludes that "[c]urrent scientific evidence does not support the proposition that human health has been adversely affected by inhaled [mold toxins] in the home, school, or office environment."[57] However, the document itself—as well as the ACOEM website from which it may be downloaded—neglects to disclose that two of the three authors of the paper were consultants for GlobalTox, Inc., a firm that specializes in providing expert testimony for defendants, and were consulting for the defendant in a toxic mold case at the time they were working on the document to the tune of $375–500 per hour. The third author, a professor at the David Geffen School of Medicine at the University of California–Los Angeles (UCLA), has also consulted for defendants in toxic mold cases and is paid $500–720 per hour for his efforts.[58] Other examples of advocates' influence on the output of professional associations include the following:

- The American Psychiatric Association publishes the highly influential *Diagnostic and Statistical Manual,* which the FDA relies on to identify the conditions that warrant treatment with drugs. Of the 170 members of the association who have contributed to the manual, 95 had one or more financial associations with the pharmaceutical industry, and 100 percent of the members of the panels on "mood disorders" and "schizophrenia and other psychotic disorders" had financial links to the industry.[59]
- The Society for Women's Health Research (SWHR) grew out of the conviction that federally sponsored research was not devoting sufficient attention to women's health issues, but over time it received much of its financial support from pharmaceutical companies. When the NIH halted a major clinical study of hormone replacement therapy in 2002 because one of the drugs had significantly increased the risk of heart attacks, stroke, blood clots, and breast cancer, SWHR defended the drug's manufacturer. It also testified in Congress in favor of an industry-supported products liability reform bill.[60]

Not every professional organization is subject to cooptation by advocates, and not every group that advocates support takes positions that the advocates prefer. The evidence that the cooptation strategy works even occasionally, however, should serve as a red flag to courts and agencies that routinely rely on their statements and position papers.

Assembling an Army of Experts

Packaging science does not necessarily require a formal organizational structure: transient armies of experts and ad hoc expert panels can be just as effective in advocating specific, ends-oriented positions on issues of policy-relevant science. For example, Philip Morris International initiated "Project Whitecoat" in 1987 in response to the threat of increased governmental regulation following the publication of influential reports by the surgeon general and the NAS on the health risks of ETS. The project involved an extensive effort to recruit academic scientists to undertake a number of critical public relations functions for the company.[61] At roughly the same time, the Tobacco Institute established an "ETS Academic Scientist Team" to meet the same needs.[62] An "A Team" of experts was primarily engaged in research and the preparation of written testimony, while the "B Team" of "foot soldiers" was available "to testify at state and local hearings" and "to meet one-on-one with legislators" on "behalf of TI [Tobacco Institute] or a third party sponsor."[63] Assignments for both teams included "attendance at national and international conferences, participation in and submission of papers to conferences on ETS, preparing responses to articles on ETS and scientific publications, original research, and critiquing EPA's documents."[64] By 1992, the Tobacco Institute had assembled a stable of fourteen "academic scientists," all of whom were "faculty members of prestigious universities and medical schools," to "influence the scientific community's view of ETS science."[65]

Energy companies adopted a similar strategy in the late 1990s to counteract the growing scientific consensus on climate change. An American Petroleum Institute internal memo presented a strategy for investing millions of dollars to "maximize the impact of scientific views consistent with ours with Congress, the media, and other key audiences" by "recruit[ing] and train[ing]" scientists "who do not have a long history of visibility and/or participation in the climate change debate" to support the industry's position that the science on global warming was too uncertain to support regulatory action.[66] The scientists who were ultimately recruited into the energy industry's stable matched this ideal almost perfectly. Most were highly accomplished in fields other than global warming, but lacked a "long history of visibility" or, for that matter, any significant experience at all in the field of climatology. As the industry hoped, they proved to be remarkably effective at casting doubt on climate change science in the public's mind.[67]

The expert armies can also infiltrate and exert a disproportionate influence—in both numbers and participatory vigor—at scientific conferences and workshops sponsored by scientific societies and government agencies. During the 1980s and 1990s, individual tobacco companies and the Tobacco Institute sent dozens of consultants to scientific meetings to monitor presentations and to present the tobacco industry's position on issues involving indoor air quality and environmental tobacco smoke. Almost every issue of the Tobacco Institute's monthly "Public Affairs Management Plan Progress Reports" during the early 1990s reported that institute consultants had attended one or more scientific meetings where issues of relevance to the tobacco industry were discussed and debated.[68] Likewise, periodic progress reports in the Philip Morris files from those years contain a separate heading for reporting progress toward the plan's objective of "[m]aintain[ing] an expert group to participate in scientific conferences and standing committees."[69]

An even more powerful strategy, when successfully employed, is to assemble a "blue ribbon" panel to produce a consensus statement about the state of the science. For example, as several manufacturers of pet foods undertook a nationwide recall of many of their products that had become contaminated and were killing and sickening thousands of pets, one of them quickly assembled a seven-member "Veterinary Quality Assurance Task Force" to examine its production plants. The company then featured in full-page advertisements in fifty-nine newspapers the panel's commendation of the company's forthrightness and its conclusion that "veterinarians and pet owners should feel safe recommending and feeding" the company's unrecalled products. The task force report did not mention that three of its seven members had endorsed the company's products in previous ads and another was a former employee with a five-year research grant from the company.[70] Adopting a more long-range approach, the cosmetics industry in 1976 created a seven-member panel of experts called the Expert Panel for Cosmetic Ingredient Review that, among other pronouncements over the years, unanimously concluded in 2002 that phthalates were "safe for use in cosmetic products in present practices of use and concentration."[71] This consensus position was, perhaps not coincidentally, announced just twelve days after the European Commission ordered companies to remove two phthalates from cosmetics in its member countries.[72]

In other cases, advocates convene expert panels to produce statements not about what the relevant science says but about what is wrong with existing research. For example, in attempting to stave off stringent regulation of beryllium, Brush Wellman convened a panel of eight prominent scientists, six of whom had been present or past company consultants,[73] at the Cosmos Club in Washington, D.C., to critique the cancer studies on which NIOSH and OSHA were relying.[74] In a letter to two cabinet secretaries, the members of the Cosmos Club group, not identifying their industry affiliations, characterized the studies as "shocking examples of the shoddy scholarship and questionable objectivity utilized in making important national regulatory decisions."[75] The letter put OSHA on the defensive, and it never has increased the stringency of the severely outdated 1949 workplace standard for beryllium.[76] Similarly, when an article published in *NEJM* concluded that the diet drug fen-phen caused a very serious heart valve disease, the manufacturer created an expert panel to reevaluate and opine on the case studies highlighted in the publication.[77] Not surprisingly, the panel's "consensus" conclusions differed from those of the authors of the published study and played a prominent role in the manufacturer's defense strategy in subsequent regulatory battles and lawsuits.[78]

To provide a greater distance between a sponsor and the sponsored expert panels, the panels are sometimes convened by nonprofit groups whose industry ties are known only to insiders. A few examples follow:

- The plastics industry arranged for the Harvard Center for Risk Analysis to assemble a scientific panel to address the reproductive toxicity of low-level exposures to the ubiquitous plasticizer bisphenol A (BPA) in laboratory animals.[79] The panel found no "consistent affirmative evidence of low-dose BPA effects,"[80] but critics later pointed out that the panel reviewed seven of the nine industry-funded studies (all of which found no adverse effects) and only twelve of the thirty-eight available government-funded studies (nearly all of which identified some adverse effect at low levels).[81]
- At the same industry's request, the ACSH (described above) convened a "blue ribbon panel," headed by the former surgeon general C. Everett Koop, to opine on the toxicity of the plasticizer di(2-ethylhexyl)phthalate (DEHP). The panel concluded that DEHP was not harmful to human health and that banning it from medical

devices would pose a significant health risk to people relying on those devices.[82] Another panel convened by the federal government's National Toxicology Program later reached a very different conclusion, finding among other things that male infants undergoing medical treatment could be at high risk from exposure to vinyl equipment containing DEHP.[83]

- While the EPA's Science Advisory Board was reviewing a draft assessment of the health risks posed by exposure to the Teflon precursor PFOA,[84] a different panel of scientists assembled by the ACSH quickly produced a booklet and accompanying press release, based on a position paper that was "peer reviewed" by scientists likewise chosen by ACSH, concluding that PFOA posed "no likely risk" to humans in the "trace amounts" found in human blood.[85]

The close congruence in these examples between the industry position and the "scientific" conclusions that the panels reached is probably not entirely coincidental.

The blue ribbon panel strategy can backfire. A panel of eminent cardiologists selected by Johnson & Johnson to assess the safety of its heart-failure drug Natrecor for outpatient use issued a report concluding that the drug might interfere with kidney function and strongly recommended that the company notify doctors that they should cease using it for outpatient treatment. It further urged the company to continue with its ongoing clinical trial on outpatient use and to initiate a larger clinical trial specifically focused on adverse kidney effects.[86] When the company did not move rapidly enough to implement the committee's recommendations, the chair wrote several unsolicited letters expressing the committee's concern that its recommendations were not being implemented.[87] The Natrecor experience demonstrates that not every company-appointed blue ribbon panel will adhere to the company line.

Manufacturing Consensus through Symposia

Still another creative, though by no means inexpensive, strategy for packaging science is to sponsor a "symposium" headed up by a prestigious industry consultant or grantee and to limit invitees primarily to scientists who are likely to be sympathetic to the sponsor's position on the scientific issues the symposium addresses.[88] As one tobacco industry budget document related, this technique allows the sponsoring entity "to structure the location, occasions and subjects we want, rather than awaiting the accident that someone else might do it."[89] The proceedings of the symposium can be featured in press releases (or, better still, webcast), and they can then be published in a book or scientific journal. With any luck, the participants can be persuaded to arrive at a "consensus statement" at the end of the conference that adopts the position of the sponsoring entity on one or more of the controversial issues the participants have addressed.

Because groundwater at many locations around the country had become contaminated with the chemical perchlorate, a constituent of the fuel used in many weapons, the Department of Defense and several of its munitions contractors faced huge financial liabilities. They formed a consortium called the Perchlorate Study Group (PSG) to conduct toxicity testing on that substance, since the extent of the cleanup at these sites would depend on the dose at which perchlorate was toxic.[90] After the EPA proposed stringent guidance for perchlorate cleanups based on a number of studies, including some funded by the PSG, the Department of Defense sponsored a three-day "Perchlorate State-of-the-Science Symposium" at the University of Nebraska to "provide a review of five fundamental science issues related to the potential health risks from low-level exposure to perchlorate."[91] The speakers at the symposium were drawn primarily from scientists working for the Department of Defense or as grantees of or consultants to the PSG, but included no scientists from the EPA or any environmental group. The authors of the two epidemiological studies finding toxic perchlorate effects on children were not invited. A week after the conference concluded, the University of Nebraska published a press release relating the "consensus" of the attendees that the data on rats that the EPA had relied on were "unreliable" and should be set aside until new studies could be undertaken.[92] They further recommended that "more human research related to sensitive populations should be considered,"[93] but did not suggest how scientists would go about ethically conducting human research on developing fetuses and small infants.

As with blue ribbon panels, sponsors often find it beneficial to channel their financial support for a symposium through an "independent" entity with no apparent ax to grind.[94] One of the helpful functions that the ILSI performs for its food and beverage industry sponsors is to host conferences and symposia that deliver more carefully controlled results than an open conference. In 1985, Dr. Peter Greenwald, the director of the National Cancer Institute's Division of Cancer Prevention, wrote a memo to complain that an upcoming ILSI conference "appear[ed] to have assigned key roles to a few persons predisposed to denying ill effects of dietary fat and postponing as long as possible any specific recommendations in this area."[95] To bolster the industry's case for the proposition that sugar was not per se bad for consumers, the industry-funded Oldways Preservation and Exchange Trust (OPET), a self-described "food-issues think tank," convened a conference on "Sweetness, Sugar, and Health" in Mexico City. Invited scientists and selected journalists traveled at OPET's expense to the conference, but it was in fact paid for by Coca-Cola and other companies with an interest in preserving demand for sugar and sugar-based products. The conference produced a "consensus document" featuring the participants' conclusions that "[t]he biological preference for sweetness is universal" and that "sugars and all approved low-caloric and non-caloric sweeteners are safe, and they offer useful options that can help consumers manage sweetness."[96]

Occasionally, the sponsoring industry will press so hard to present the appearance of a consensus that it perverts the outcome of the conference. In the spring of 1990, as the EPA was preparing a major assessment of the health and environmental risks posed by the chlorine industry byproduct dioxin, an industry trade association called the Chlorine Institute and the EPA jointly convened a "scientific" conference—the "Biological Basis for Risk Assessment of Dioxins and Related Compounds"—at the famous Banbury Conference Center in Cold Spring Harbor, New York.[97] Although the participants debated the issues vigorously, many were later surprised to read in a press release prepared by a public relations firm hired by the Chlorine Institute that the group had reached a "consensus" that the EPA's risk assessment should be revised to lower the predicted risks.[98] An outraged participant wrote a letter to the conveners and other attendees expressing her "dismay that a prestigious institution like Banbury would tolerate such conduct" and complaining that she "did not expect to be manipulated by industry or government spokespeople."[99] Despite these objections, the Chlorine Institute launched a remarkably successful "public outreach program" to "capitalise on the outcome of a recent Banbury conference on dioxin."[100] *Science* reported that the scientists had "reached an agreement that surprised almost everyone."[101] *Time* wondered whether "[a]fter years of warnings about its ability to cause cancer, is it really true that [dioxin] is not so dangerous after all?"[102] The government official who had years earlier ordered the evacuation of Times Beach, Missouri, because of dioxin contamination told the *New York Times* that it now "looks as though the evacuation was unnecessary."[103] Two researchers at the National Institute for Environmental Health Sciences, however, found no support for the apparent "consensus" in studies considering very low levels of dioxin exposure, and a decade later there was still nothing approaching an actual "scientific consensus" on the question of a threshold "safe" level of exposure to dioxin.[104]

If You Pay Them, They Will Publish

Resourceful parties can also pour money into commissioning books, review articles, and related work that gives the false impression of what mainstream scientists might view as settled and unsettled issues in their specialties. Books and review articles are a staple in the scientific literature, and scientists frequently rely on them to learn about areas with which they lack intimate familiarity. In addition, regulatory agencies constantly refer to books and review articles in drafting regulatory support documents, and they are frequently introduced as evidence in tort litigation. While this technique overlaps heavily with the attempts to shape sponsored research that we explored in Chapter 4, these efforts to summarize the state of the science at a given time are somewhat more ambitious than the attempts to change the outcome or trajectory of a single study that we discussed in that chapter.

A "SOLUTION" TO THE RISKS OF BERYLLIUM

In early 1987, the country's primary producer of the industrial metal beryllium, Brush Wellman, was concerned that the scientific literature published during the previous twenty years had been "very damaging" and was "constantly being cited" by "doctors at medical meetings" and, perhaps more distressingly, by "potential customers as the cause of their unwillingness to use our products."[105] In addition, federal agencies like the EPA and OSHA were citing the literature "to support regulatory activities that are harmful and expensive to Brush."[106] According to an internal Brush Wellman memorandum, the "solution" was to "combat this situation" with "a complete, accurate and well written textbook on beryllium health and safety" and "a number of medical papers published in prestigious medical books."[107]

The book would be financed by Brush Wellman, and the bulk of the work would be done by two Brush Wellman employees. It would, however, "have to be published under the auspices of some not-for-profit organization such as a university or medical groups," and it would cost about $50,000 and take two years to complete. The review articles would have to be written and published more rapidly so that they could be "used as up-to-date" references for the book. They would have to be written "in co-authorship with renowned secondary authors."[108] The memorandum went on to suggest a "sequence" of industry-commissioned articles to be either coauthored or ghostwritten by Brush employees. In fact, two journals had already agreed to publish two of the six proposed but still unwritten papers.[109]

According to the Five-Year Strategic Plan for Brush Wellman's Environmental Affairs Office, that office would "retain consultants," "allocate funds," "[r]etain editor and authors," and "develop [a] first draft" of a "comprehensive book on the environmental health aspects of beryllium" by the end of 1987, and the book would be out by 1988.[110] The plan was brought to completion when *Beryllium: Biomedical and Environmental Aspects* was published somewhat later than expected in 1990, and Brush Wellman distributed it to hundreds of medical schools and libraries throughout the country.[111] Referred to in Brush Wellman correspondence as "the first comprehensive update on beryllium health issues in over 25 years,"[112] the book set out the "company line" on beryllium, including the highly questionable conclusion that workers exposed to beryllium at levels under 2.0 μg/m^3 had not contracted berylliosis.[113] The book was an important step in Brush's larger project of "educat[ing] the university professors on our materials and health issues in order to train the next generation of engineers on the truths and myths about beryllium-containing materials."[114] ❒

The strategy of commissioning books and review articles is more likely to have a significant impact on the scientific community, as opposed to the political world, when the scientific debates are still relatively unsettled. During the early 1990s when the research on the adverse effects of ETS was just beginning to appear in the scientific literature, the tobacco industry maintained an active program for commissioning experts to prepare review articles and books taking the position that ETS was at worst a harmless nuisance. One of the strategies offered in a lengthy British American Tobacco Corporation document of November 1990 consisted of "assigning to scientific consultants the task of preparing review articles, for submission to scientific/professional journals, on appropriate topics."[115] Scientists hired by tobacco companies were paid up to $25,000 apiece for full-fledged review articles.[116] A Tobacco Institute document describing various "EPA Projects" describes several recent publications the institute had recently financed, including a book on "ETS epidemiologic studies" written by the consultant Peter Lee for $20,000, along with the following review articles:

- Article on the EPA's ETS meta-analysis for submission to the *Journal of Clinical Epidemiology* (Gross), $19,000.
- Analysis of the EPA's ETS meta-analysis for submission to the *American Journal of Epidemiology* (Lee), $8,000.
- Extended analysis of the EPA's ETS meta-analysis for submission to the *Journal of Risk Analysis* (LeVois), $7,000.
- Article comparing the EPA's approach to risk assessment in the case of ETS with their approach in the case of electromagnetic radiation, for submission to *Science* (LeVois), $21,000.
- Article on the EPA's ETS risk assessment for submission to the *International Journal of Epidemiology* (Switzer and Layard), $50,000.
- Response to Repace and Wells articles in *Environment International* (Lee), $2,500.[117]

Not surprisingly, review articles prepared by consultants to the tobacco industry tended to conclude that ETS posed minimal risks. In one survey of such articles, 94 percent of review articles on ETS prepared by industry-affiliated authors concluded that passive smoking is not harmful, while only 13 percent of review articles prepared by authors without tobacco industry affiliations reached that conclusion.[118]

Other industries have followed in the footprints of Brush and big tobacco. The manufacturer of neurontin, an epilepsy drug, secretly spent $303,000 on a book on epilepsy in order to ensure that its product was portrayed favorably.[119] The manufacturer of Xanax commissioned an introductory review article on the efficacy of its drug that was published in a prominent journal supplement.[120] A group of eighteen sympathetic scientists were convened at a commodious location to prepare a rebuttal to EPA's nine-volume draft dioxin risk assessment that they published as a joint paper in *Science*.[121] Even when these works are scientifically condemned,[122] they may still enjoy positive play in policy debates. In the case of climate change, for example, Senator James Inhofe devoted half of a Senate hearing to a review article commissioned by the American Petroleum Institute that he promised would "shiver the timbers of the adrift Chicken Little crowd."[123]

More worrisome still is evidence that some journals are themselves sponsored by affected parties. The *Journal of Regulatory Toxicology and Pharmacology*, for example, is the official journal for an association called the International Society for Regulatory Toxicology and Pharmacology.[124]

This organization's sponsors include many of the major tobacco, chemical, and drug manufacturing companies, and a longtime industry consultant sits on its board.[125] One prominent epidemiologist complained that some of the articles published in the journal contained such "elementary mistakes" that they would receive "an F" if a doctoral student had submitted them in an academic setting.[126] Perhaps not surprisingly, that journal does not require submitting authors to disclose conflicts of interest.[127]

Journal editors can also take advantage of their privileged roles by using their editorial power as yet another asset in marketing their services to private sector clients. In writing to a potential client, the Phosphine Group, the editor-in-chief of the *Journal of Risk Analysis* did precisely that when she offered them her candid assessment of the best strategy for obtaining a more favorable EPA risk assessment for the pesticide phosphine. She concluded that "the approach with the greatest likelihood of affecting EPA's position is to prepare and publish in a peer-reviewed journal a scientific paper or article that describes the current science on the toxicity of phosphine." She went on to offer: "Since I am currently Editor-in-Chief of the international journal *Risk Analysis,* perhaps the peer-review process could be expedited, if we decide that it is the journal of choice."[128] In the wake of the incident, that journal has now instituted a conflict disclosure policy.[129]

Conclusion

Packaging science is a strategy that takes place at the interface between the level at which scientists design studies, gather data and analyze results on a day-to-day basis and the level at which public sector decision-makers use scientific information to decide cases and make regulatory policy. Dominated by experts, often at the pinnacle of their profession, this intermediate level is where scientists collect and assimilate the completed studies that make up a body of research, synthesize large amounts of information into a coherent whole, and communicate the overall message to policy-makers and the public. At this level, advocates cannot influence study design, manipulate the way the data are collected, or skew the interpretations and analysis of individual studies. They cannot hide studies from the scientific community and the public, though they can belittle or ignore them in providing their advice, preparing their reports, and writing their review articles and books. Although advocates have less discretion at this level to affect what the science says, their efforts to package science are in many ways more important from the public's perspective, because the work product at this level is more comprehensible to agencies, courts, and the general public. Since much of the output at the packaging level is consciously framed for consumption by agencies, courts, and other policy-makers, advocates' subtle efforts to influence that framing exercise can have a larger impact on real-world decisions than efforts to shape and hide science at the lower level occupied by the bench scientists.

The examples of subtle attempts to pack government scientific advisory committees, create and coopt professional societies, convene private sector blue ribbon panels, assemble traveling teams of experts, and determine the content of books and review articles detailed in this chapter strongly suggest that advocates understand the importance of packaging science. These examples also suggest that many advocates are willing to pay considerable sums in order to play an influential role in that process. Advocates in the private sector and the government can effectively employ surreptitious strategies for packaging science—strategies that public interest groups rarely take, because of their expense—to shape science where it counts. As with the strategies for bending science discussed in Chapters 4–7, this is not a desirable development from the standpoint of sound public policy. These practices, if allowed to continue and expand, will erode judges' and regulators' confidence in the objectivity of these otherwise extremely valuable tools for converting raw science to comprehensible and usable information for public sector decision-making, and they will ultimately reduce public regard for the overall scientific enterprise. Unfortunately, as we shall see in the next chapter, fighting against this powerful tactic, particularly when it occurs wholly outside the reach of regulatory control, is neither easy nor straightforward.

CHAPTER NINE

Spinning Science

The Art of Manipulating Public Perceptions about Credible Science

For as long as controversies have erupted over scientific issues in litigation, regulation, and policy-making, advocates have devoted large sums of money to persuading government officials and a skeptical public that the positions of their adversaries are based on bad science and emotion. When Rachel Carson published *Silent Spring* in 1963, for example, the chemical industry raised $135,000 for a "crash project" to educate the public about alleged errors in the book.[1] Educating the public about policy issues is called "public relations," and that function has become so critical to public policy-making since the early twentieth century that it has evolved into a profession that is practiced by thousands of consultants working in corporations, small private consulting firms, the government, and nonprofit public interest groups.[2]

The primary goal of the public relations campaigns discussed in this chapter is to manipulate public perceptions about the scientific underpinnings of a client's or an opponent's position in an environmental controversy.[3] Thus, the strategy of spinning science plays out almost exclusively in the realm of policy and not in the realm of science, where knowledgeable experts are generally able to detect and dismiss these attempts to manipulate perceptions. The public relations specialists who launch these campaigns usually focus them instead on educated members of the general public who read the papers and watch the evening news.[4] They carefully frame the issues, often on the basis of polling and focus group sessions, and they employ advertising and other media strategies to repeat the message over and over again until, like the brand name of a major detergent, it is indelibly etched on the minds of ordinary citizens. As one might expect, these sophisticated campaigns are invariably expensive. The spinning strategy is therefore not as routinely available to ordinary citizens and small entities that lack the resources to employ it effectively.[5]

SPINNING VARELA

In February 1990, a Philip Morris scientist, Thomas Borelli, discovered an abstract for a dissertation written by a Yale University graduate student named Lewis R. Varela reporting the results of his analysis of data from 439 lung cancer cases in nonsmokers gathered by his dissertation advisor, Dr. Dwight Janerich, under a grant from the National Cancer Institute.[6] At that time, the Varela study was the largest case-control study of ETS ever undertaken in the United States. Varela found no increased risk associated with spousal exposure to the three lowest levels of exposure, and his ultimate conclusion was that "the previous studies may have overestimated the effect attributable to passive smoking."[7]

This conclusion was a godsend for the tobacco industry. Borelli later enthused: "We couldn't believe it when we saw it."[8] Sensing that Philip Morris could make good use of the Varela dissertation in its new public relations offensive against a forthcoming risk assessment of ETS being prepared by the EPA, Borelli turned the matter over to the Newman Partnership, a South Carolina public relations firm specializing in scientific controversies.[9] Newman in turn shipped the dissertation for peer review to the Boston epidemiologist Dr. Alan Leviton, who concluded that it was "a high-quality study with some intriguing findings"[10] that should be "condensed to one or more journal articles so that it could enter the more widely-available epidemiologic literature."[11] When Borelli then asked Janerich why it had not already been published, Janerich explained that Varela (who had died not long after completing his degree) had merely performed "the first cut" through a huge amount of data. Janerich and other coauthors were in the process of following up and "refining the data."[12]

Newman then drafted a letter for Borelli's signature to William Farland, the EPA official responsible for the ETS risk assessment, suggesting that the Varela dissertation be included in the EPA's draft risk assessment.[13] In fact, EPA scientists had discovered the dissertation almost a year earlier and had concluded that a further breakdown of the data was necessary. Seeking additional details, EPA's consultant unsuccessfully attempted to contact both Varela, who was by then deceased, and Janerich, who did not respond.[14] Without the necessary breakdown, the staff decided to include the dissertation in one of the risk assessment's two qualitative analyses, but not in its quantitative analysis.[15]

Newman urged Philip Morris to use the Varela study as the impetus for an aggressive preemptive strike against the still-pending EPA risk assessment by distributing it to various scientists and giving a sympathetic reporter a scoop in advance to "inoculate" the story against negative press.[16] With the assistance of a much larger public relations firm, the Edelman Group, Newman drafted a strategy memorandum setting out various options for capitalizing on the Varela windfall.[17] The "long-term" goal of the proposed public relations effort was "to establish that the ETS health question has not been resolved and is, in fact, still open to controversy."[18] The two options for "intermediate goals" were "to generate favorable ETS news prior to the release of EPA risk assessment" or "to use Varela to generate controversy as an example of the fact that opponents have not included the full range of existing data in their conclusions."[19]

Having been forewarned that the publication of the EPA risk assessment was imminent, Philip Morris, on April 30, 1990, mailed the Varela dissertation (now christened the "Yale study") to more than three hundred scientists, along with a cover letter alluding to "publication bias" and explaining that "critical evaluation of *all* major studies" was important.[20] On the same day, the company sent an almost identical letter, over the signature of Philip Morris chair emeritus Joseph Cullman, to his longtime friend the EPA Administrator William Reilly.[21] Newman phoned journalists throughout the country to alert them of Philip Morris's action.[22]

Pursuant to the Newman strategy, a copy of the Varela dissertation and Philip Morris's take on it was given to a reporter for the *Saint Petersburg (FL) Times,* which published the story the day before Philip Morris's mass mailing.[23] According to the article, "a company considered as canny as they come in public relations matters [was] coming out of the closet on the issue of health and smoking" to contend that the government was "setting policy on smoking in public by throwing scientific integrity out the window."[24] The story quoted Tom Borelli to the effect that Philip Morris had discovered the Varela study in February and immediately sent it to the EPA but the "EPA said it would not consider the study, that it didn't have time to review it."[25] The "inoculation," however, was not a complete success. A story in the *Richmond (VA) Times-Dispatch* the next day mentioned the mailing to scientists and Philip Morris's criticisms of the EPA, but also quoted the epidemiologist Lawrence Garfinkel to the effect that the "salvo is nothing more than a replay of the tobacco company's former tactic of saying there is some doubt about the danger of first hand smoke."[26]

The public relations action did have the intended effect of putting the EPA on the defensive. Within a week, it issued a press release defending itself against Philip Morris's charge that it had ignored the Varela study. Calling the allegations simply "incorrect," the release told of the EPA's discovery of the study and its consultant's attempts to contact Janerich.[27] At that point, however, the story took on a life of its own, when someone at the EPA leaked the entire draft risk assessment to a reporter for the *Los Angeles Times*. This paper's story announced that the EPA would "soon declare second-hand cigarette smoke to be a known carcinogen" and that it was responsible for more than three thousand cases of lung cancer among nonsmokers per year.[28]

In the final analysis, the bureaucrats in the EPA may have outmaneuvered the industry's expensive public relations experts. The real news in the minds of the reporters covering the ETS story was a risk assessment (albeit one in draft form) classifying ETS as a Class A carcinogen and estimating that it caused three thousand deaths per year. To make matters worse, some papers paid as much attention to the company's public relations tactics as they did to the Varela controversy itself.[29] If the mass media did not turn out to be an effective ally for the tobacco industry in its battle with the EPA, however, the industry could still turn to other (potentially more powerful) allies in Congress, as we shall discover, where the alleged "omission" of the Varela study retained a great deal of traction. ❐

At the outset of the wave of regulatory interventions enacted during the consumer-environmental movement of the early 1970s, the executive vice-president for public relations for the Tobacco Institute, William Kloepfer, delivered a speech entitled "Public Relations in the Nation's Capital," in which he distinguished between the role of the corporate lobbyist and the role of the corporate public relations operation.[30] He stressed that public officials "who take seriously their responsibility for the public's welfare *are* responsive to what they *think* the public thinks—or what the public will think with a little guidance from authoritative sources."[31] Those authoritative sources must be put before the public by the relevant industry's public relations people. The public relations representative "must be in a position to sense, to foresee, to know the concerns of [industry] regulators."[32] To do that, "you have to be steadily oriented toward their sources of input, and to be in a position to deal directly and steadily with those sources."[33] In many cases, the sources of input are scientists and their research.

This chapter dissects the public relations campaigns of industries, plaintiffs' attorneys, and public interest groups and examines the scientific component of their larger strategic operations. Spinning often involves both reframing issues and communicating them in ways that are most likely to produce favorable public responses. Teams of scientists who serve as "truth squads" can be carefully assembled to convey the message from a credible source. In this outreach process, litigation can also serve as a useful device for attracting the media's attention. The chapter closes with an exploration of how each of these techniques exploits the media's own weaknesses in determining what issues to cover and their limited resources in pursuing and researching stories.

Framing the Message

The immensely successful insinuation of the term "junk science" into the public conversation about litigation and regulation, discussed in Chapter 6, is a perfect example of a public relations technique known in the business as *framing*. Originally popularized by the Manhattan Institute fellow Peter Huber,[34] the term "junk science" took on a life of its own after companies and trade associations recognized its value as a universal pejorative for scientific data and analysis that did not comport with their view of the world. They quickly popularized the term with campaigns conducted by public relations firms that often used front groups like the Advancement of Sound Science Coalition (discussed later).[35] In a 1999 publication, the University of Pennsylvania professor Edward H. Herman surveyed 258 articles in mainstream newspapers from 1996 to 1998 that employed the term "junk science." While only 8 percent of the articles used the term in reference to corporate-sponsored science, 62 percent used it in reference to arguments advanced by environmentalists or personal injury lawyers.[36] It seems unlikely, in light of what we have shown in the previous pages, that such a lopsided characterization reflects an objective assessment of the arguments.

When officials discovered the first U.S. mad cow in Washington state in December 2003, consumer groups urged the USDA to expand its program for testing cattle for mad cow disease to determine the extent of the problem.[37] The USDA Secretary, Ann Veneman, announced that the department would double the testing program from twenty thousand to forty thousand suspect cattle, but it would not test any of the rest of the 35 million cattle slaughtered annually. Testing cattle for mad cow disease is time-consuming and expensive, but Secretary Veneman's explanation for declining to expand the testing program did not even mention cost. She dismissed consumer group demands for more testing on the ground that additional testing would not reflect "sound science."[38]

From a purely scientific perspective, this was a curious explanation. Why would science frown on a decision to test more cattle? Science is generally greedy for information. Assuming that tests are accurate—and no one suggested that these were not—more data points will yield a more robust understanding of the phenomena being studied. While Veneman's explanation would no doubt seem odd to a scientist unfamiliar with the political terrain, her invocation of "sound science" to justify a decision that must have been driven by other considerations was entirely predictable. Appeals to science are more politically salable than appeals to economics. The secretary's use of the term "sound science" was no doubt intended to create the impression that the USDA's decision was driven by objective criteria, solid empirical data, and rational analysis. Nearly everyone believes that society is better off when governmental interventions in private market arrangements to protect public health and the environment are driven by solid evidence and analysis. Appeals to economics, by contrast, are nearly always motivated by somebody's self-interest, even if over the long haul society as a whole may be better off when governmental decision-makers attempt to achieve efficient results.

Because appeals to economics are inherently suspect in an increasingly cynical society, it makes sense to frame self-interested claims as appeals to science. This fundamental political reality was captured very nicely in a memorandum from the political consultant Frank Luntz to Republican leaders. In discussing the global warming debate, Luntz observed:

> The economic argument should be secondary. Many of you will want to focus on the higher prices and lost jobs that would result from complying with [the] Kyoto [Climate Change Treaty], but you can do better. . . .
>
> The most important principle in any discussion of global warming is your commitment to sound science. Americans unanimously believe all environmental rules and regulations should be based on sound science and common sense. Similarly, our confidence in the ability of science and technology to solve our nation's ills is second to none. Both perceptions will work in your favor if properly cultivated.[39]

Thus by shifting the focus from economics to science, Veneman was employing a tried-and-true public relations framing strategy for defending politically unpalatable decisions. Other groups, including public interest groups, have employed this science framing strategy in their efforts to capture the public's attention and loyalty.[40]

IMPLICATING ALAR

The Natural Resources Defense Council (NRDC), a prominent nonprofit environmental group, became concerned in 1989 about the risks of a widely used plant growth regulator called Alar that orchardists applied to apples to make them redder and larger. The NRDC was concerned that Alar was carcinogenic and became concentrated in the process of making applesauce, apple juice, and similar products. Since children consume these products, often in high quantities, the carcinogenic risks were likely to be particularly high for this already susceptible population.[41] The available research on Alar was badly incomplete, however, and the NRDC faced a challenge in communicating these possibly significant but still unproven risks to the public.

In what at that time was a rare move, the public interest group turned to a public relations firm for help. In the resulting media campaign, the group framed its position as compelled by science rather than driven by a policy of risk aversion. The NRDC presented its worst-case risk assessments with misleading quantitative precision, implying that the scientific research was decisive. For example, NRDC staff scientists predicted that Alar caused an "estimated 240 cancer cases per 1,000,000 population among children who are average consumers of Alar-treated food, and a whopping 910 per 1,000,000 . . . for heavy consumers," without identifying or quantifying the considerable uncertainties surrounding those numbers. An NRDC spokesperson charged in a *60 Minutes* interview that "Alar is 'a cancer-causing agent that is used on food' and 'the EPA knows [it] will cause cancer in thousands of children over their lifetime.'"[42]

The industry's science-based counter-framing of the debate was equally misleading. It argued: "There is no scientific evidence that residues in food from regulated and approved use of pesticides have ever been the cause of illness or death in either adults or children." The NRDC's allegations that Alar caused cancer in children were without "scientific merit" because they were "based exclusively on studies that expose[d] rats and mice to enormous doses of chemicals."[43] Even the EPA did not attempt to dispel the mischaracterization of the Alar debate as one exclusively over science. Instead, the EPA stuck to its own quantitative risk assessment, which adopted more middle-of-the-road assumptions,[44] and, under pressure from the Apple Institute, issued a joint statement with the USDA and the FDA that apples were safe to eat.[45]

As a result of the NRDC's powerful science-framed public relations campaign, a confused public panicked, and many parents simply stopped serving apples and apple products that might have contained Alar. A few months later, Uniroyal, the manufacturer of Alar, voluntarily withdrew the product from the market.[46] The scientific community continues to debate the risks of Alar, and economists disagree about the extent of the economic damage done to the apple industry.[47] It seems clear in retrospect, however, that misrepresenting the science as the determining factor in deciding what to do about its presence in apples was not an honest way to communicate health risks, even if it did prove to be an effective public relations strategy.[48] ❐

Regulated industries have used the mirror image of the NRDC's science-based framing to summon the need for "sound science" before "rushing to regulate." The implicit suggestion is that science—and lots of it—must accumulate before protective regulation is justified. Since the mid-1990s, this familiar science-oriented framing device has become the leading deregulatory strategy, as evidenced by a number of "regulatory reform" bills and executive policy statements that insist on a large body of scientific research as a prerequisite for regulatory action.[49] Relying on the need for "sound science," companies and trade associations insist that "the human data are not representative, the animal data are not relevant, or the exposure data are incomplete or not reliable" and "more research is needed before protective action is justified."[50] Yet as many of the proponents of "sound science" know all too well, it is impossible to meet these ambitious scientific demands in their entirety, even with infinite research resources. The physicist Alvin Weinberg's path-breaking 1972 article on "trans-science" made it clear that many of the environmental and health questions that perplex regulators cannot be resolved definitively by science because of either ethical or scientific limitations on experimentation.[51] In fact, regulatory programs that require protective standards to be based only on best technology or other information that is not in short supply are promulgated about three to six times faster than those that require standards to be based on scientific proof that they are needed to protect the public from a "significant risk."[52] For those eager to slow or even halt health and environmental regulation, the "sound science" frame provides a magic bullet.

According to the head of one of the country's largest public relations firms, "[i]n this era of exploding media technology, there is no truth except the truth you create for yourself."[53] Frank Luntz likewise believes that in the public policy arena, perception trumps truth and that truth is, in any event, "a malleable thing" that can be shaped by "changing perceptions, usually through the use of emotion and fear."[54] The beauty of the "science" frame is that it feeds into the public's great respect for but general unfamiliarity with science. Opinion polls indicate that the American public vastly prefers scientists over politicians for resolving environmental disputes[55] and that scientists are held in highest esteem relative to all other professional groups, with the possible exception of physicians.[56] Surveys also show that the public has a limited understanding of the scientific enterprise and, as a result, may tend to underestimate the multiple constraints on science and the limitations on the kinds of answers it can provide to socially relevant questions.[57] As a result, American cultural preferences make science and scientists the ideal "good fairy" for diverting attention from the moral and economic imponderables that lurk just beneath the surface. In the world of public relations, where a media consultant strives to create a reality in the mind of its audience that comes as close as possible to the reality that its client wants to project, science provides the perfect delivery vehicle.

Communicating the Message

Most public relations campaigns are, like the *Silent Spring* initiative mentioned earlier, defensive in nature. Typically, an industry calls in the public relations professionals to help it weather a political crisis or ward off an agency's regulatory proposal. For example, soon after twenty-five hundred climatologists from around the world issued the 1995 report of the IPCC predicting "widespread economic, social, and environmental dislocation over the next century" if the world community did not take action to restrict emissions of greenhouse gasses,[58] a public relations specialist for the American Petroleum Institute circulated an eight-page "Global Climate Science Communications Action Plan" recommending a $5 million public relations campaign that would, among other things, recruit scientists to question and undercut the report, provide media training, arrange for meetings with science writers, and generate a "steady stream of columns and letters to the editor" on climate change topics.[59] The strategy was apparently successful. A study of coverage of the climate change issue in the four leading American newspapers between 1988 and 2002 concluded that the reporting did not reflect the consensus expressed in the IPCC report.[60]

Many public relations campaigns, however, seek to influence public opinion over the long haul beyond the immediate crisis or election season. Politicians consult the polls in deciding how to vote on important issues, and public relations initiatives can influence the public opinion that the polls attempt to measure. High-level regulatory agency officials, who depend on politicians for their jobs and agency funding, also know how to read the polls. Scientists and other professionals read the newspapers and watch the evening news and are therefore possible public relations targets as well. Successful public relations efforts can subtly affect the climate of public opinion as it slowly evolves over the years. Then, when ordinary citizens sit as jurors in individual cases, opinions previously formed with the help of public relations firms may affect the way they perceive the facts and arguments of the parties. Thus, the U.S. Chamber of Commerce, as part of its long-term strategy to change public thinking on the impact of common law tort actions on the economy, purchased the *Southeast Texas Legal Record,* a small newspaper in a region that had produced a large number of hefty jury verdicts against corporate defendants, and began to run frequent antiplaintiff editorials.[61]

Whatever the objective, the appointed scientific spokespersons must "stay on message." In the game of manipulating public opinion through the media, the observer Jim Naureckas notes that the message "needs to be repeated over and over for a long period of time."[62] In the case of controversies over policy-relevant science, that message is almost always some variant of the message that the Newman and Edelman firms urged for the Varela study: "The science is still very uncertain, and we don't know for sure. We should not act precipitously until we have more information." The FDA head David Kessler discovered that the tobacco industry's strategy for avoiding regulation by the FDA and liability in lawsuits was "[d]evised in the 1950's and '60's" and "embodied in a script written by the lawyers" from which no deviation was permitted. According to Dr. Kessler: "The basic premise was simple—smoking had not been proved to cause cancer. Not proven, not proven, not proven—this would be stated insistently and repeatedly. Inject a thin wedge of doubt, create controversy, never deviate from the prepared lines. It was a simple plan, and it worked."[63] The plan worked for the tobacco industry for many decades, and it is currently working for other advocates as well.

A LONG-TERM PUBLIC RELATIONS PROPOSAL FOR BERYLLIUM

When OSHA was threatening to promulgate a more stringent standard for beryllium in the workplace, the beryllium industry asked Hill & Knowlton, a public relations firm (the firm that devised the tobacco industry's 1950s strategy) to submit a proposal. The firm responded with a proposed campaign to "educate various audiences with the facts about beryllium and reinforce these facts consistently over time to dispel myths and misinformation about the metal."[64] Audiences would include the media, scientists, engineers, legislators, government executives, government regulators, medical personnel, and the general public.[65] Hill & Knowlton would first prepare a comprehensive white paper on beryllium's health and environmental impacts, and it would include a public opinion survey, fact sheets and brochures, media training, legislative monitoring, emergency planning, and paid advertorials. Hill & Knowlton would then assemble "[t]hird party scientific and engineering experts from academia and industry" to provide "testimonials" that would be featured in the white paper and "liberally referenced in all support materials."[66] The experts would also "be used to provide objective information that is supportive of [the client] in a variety of forums, such as at congressional hearings or to media representatives."[67] Finally, Hill & Knowlton would identify and monitor "groups hostile to beryllium" and develop "face-to-face meetings in order to exchange information, communicate the facts about beryllium, and dilute opposition group's effects."[68]

A brochure accompanying Hill & Knowlton's pitch letter detailed the services that its Division of Scientific, Technical, and Environmental Affairs was prepared to provide to its clients. Among other things, Hill & Knowlton was capable of "[p]roviding industry information to environmental, science, and consumer writers and other media contacts," "[h]elping clients present industry positions before local, state, and federal regulatory bodies," "[a]iding in establishing and operating professional seminars dealing with health issues," and "[o]rganizing and conducting a nationwide media tour for eminent scientists to present a balanced view of the role that chemicals play in the earth's atmosphere."[69] ❐

The Hill & Knowlton proposal reveals that public relations firms have a variety of tools at their disposal. An attachment to the firm's memorandum described a number of successful public relations campaigns it had conducted in several historic public health controversies, many of which are described elsewhere in this book:

Asbestos—For the Johns-Manville corporation, Hill & Knowlton over several years prepared testimony for hearings and meetings, helped organize and conduct symposia, helped members of the scientific community explain "in an understandable fashion to various audiences" their research on the "hazards (or lack thereof) of small amounts" of asbestos. As a result, "OSHA and EPA standards on asbestos developed in the early '70s were sufficiently lenient so as to give the industry time to develop and install appropriate technological controls and to develop substitute products for some asbestos applications."[70]

Dioxin—Working for "a major chemical manufacturer," Hill & Knowlton trained research scientists "to serve as spokespersons to explain the scientific issues," arranged "broadcast and print media interviews with science, medical, environmental, and editorial writers," and arranged "for the scientists to [be] participants in radio talk and call-in shows." To bring "greater scientific credibility to the company's position," the firm "suggested and helped organize a scientific symposium on dioxin so that many researchers in the field who had important positive data to report could do so in a situation designed for maximum media exposure." The firm designed and publicized a research program for its client, and as a result of over four hundred media interviews in five months, the client "established itself as an authority on dioxin research."[71]

Saccharin—After a study showed that saccharin caused cancer in laboratory rats, Hill & Knowlton recommended a strategy of "[c]oncentrating publicity efforts on the studies and statements of scientists, medical authorities, and other credentialed third-party sources (rather than industry or association spokespersons)." The firm then coordinated and publicized "third party" scientific testimony during public hearings held by the FDA and established "news banks" of "lay and scientific spokespersons" to address issues as they arose during the hearings. Perhaps in partial response to the firm's effort, Congress wrote legislation ensuring that saccharin could remain on the market with a label warning of its effects on laboratory animals.[72]

Vinyl Chloride—When the polyvinyl chloride industry was "under attack from the press and regulatory agencies" during the early 1970s, Hill & Knowlton prepared and updated "general background papers" on occupational health, air pollution, birth defects, and risk/benefit analysis. The firm also "prepared testimony for congressional hearings and various state regulatory and legislative initiatives concerning vinyl chloride" and organized an "effort to bring congressional pressure to bear on OSHA on the industry's behalf." On the media front, the firm carried out a "nationwide media briefing program" and wrote "numerous articles and letters to the editor on vinyl chloride" for industry

> officials. The "resultant standards were significantly less onerous" than proposed.[73]

In addition to the services listed in the Hill & Knowlton brochure, public relations firms can produce syndicated columns for their clients' signatures and deliver press releases to the journalists via their own wire services.[74] Given sufficient resources, public relations firms can even produce sophisticated movies for distribution to policy-makers, schools, and local television stations. The Western Fuels Association, for example, spent $250,000 to produce a video on global warming that enthusiastically conveyed the encouraging message that global warming would enhance agricultural productivity, turn deserts into grassland, and generally result in a much better world.[75]

Other public relations firms specialize in "offensive" work aimed at neutralizing potential opponents. The public relations firm DuPont retained to deal with the PFOA controversy detailed in Chapter 5 wrote in its prospectus to DuPont: "DUPONT MUST SHAPE THE DEBATE AT ALL LEVELS" by using "a strategy at the outset which discourages governmental agencies, the plaintiff's bar, and misguided environmental groups from pursuing this matter any further."[76] The firm boasted that for more than twenty-three years it had "successfully guided clients through myriad regulatory, litigation, and public relations challenges posed by those whose agenda is to grossly over regulate, extract settlements from, or otherwise damage the chemical manufacturing industry."[77] Another firm's website boasts that it "stay[s] on the offensive" by "[g]athering intelligence, recruiting allies, gaining public support—and creating risk for your opponents."[78] Using ex-CIA and ex-FBI agents, the firm specializes in "psychological profiling" of company adversaries in the public interest community and "following the money of our clients' adversaries."[79]

When all else fails, advertising in the mainstream media offers an effective, if expensive, way to reach target audiences. The Tobacco Institute spent $500,000 on an advertising campaign to discredit Dr. Oscar Auerbach's "smoking dogs" studies and to embarrass the American Cancer Society for supporting Auerbach's research.[80] The lead industry conducted a fifty-year advertising campaign touting the safety of leaded paint, featuring the "Little Dutch Boy" and specifically targeting children.[81] Advertisements run by the petroleum industry featured cowering chickens and suggestions that people who supported global warming hypotheses also thought the world was flat.[82] A public relations firm's plan for reacting to the EPA's anticipated risk assessment for ETS noted that "[w]hile advertising has many downsides and is generally considered to be only a last resort strategy, it may be that it is the only way for tobacco to get a message out in a pure and untrammeled form."[83]

Scientific "Truth Squads"

One favorite public relations strategy of companies that want to sway public opinion on an issue of policy-relevant science, especially one that might be addressed at the state or local level, is to take a selected group of media-trained scientific consultants on a road trip to carry a carefully crafted message to the people. For example, one of the public relations strategies outlined in the Tobacco Institute's Public Affairs Operating Plan for 1991 was to "[c]ontinue efforts to focus greater attention on the inconclusive nature of the scientific data regarding the alleged health effects of ETS" and "[i]ncrease awareness of the need for more and better research on the relationship between ETS and health claims."[84] To accomplish this, the institute planned to "continue to maintain a well briefed and up-to-date scientific consultant team, capable of conducting briefings on ETS research with state and local public policy makers, of testifying before state and local legislative bodies, and of conducting media tours on the scientific issues."[85] The institute later referred to the team as its ETS "truth squad."[86]

The technique is not limited to the tobacco industry. In 1994 the Global Climate Coalition, a consortium of industry groups opposed to greenhouse gas controls, sponsored a media tour by Dr. Sally Baliunas, an astrophysicist at the Harvard-Smithsonian Astrophysical Institute, to popularize her theory that recent increases in atmospheric temperature were attributable to sunspots.[87] Four years later, the American Petroleum Institute budgeted $600,000 to establish a cadre of twenty "respected climate scientists" and to "identify, recruit, and train a team of five independent scientists to participate in media outreach."[88] The outreach consisted of traveling around the country to meet with science writers, newspaper editors, columnists, and television network correspondents, "thereby raising questions about and undercutting the 'prevailing scientific wisdom.'"[89] The Chlorine Chemical Council in 1994 hired a public relations firm to launch a thirty-city tour of sympathetic scientists to meet with news representatives and community leaders to challenge the scientific basis for an upcoming EPA reassessment of the health risks of dioxin.[90] After the manufacturer of one of the most heavily used herbicides in the country began to fund much of Dr. David B. Baker's research, he became a consultant to the pesticide industry-funded Council for Agricultural Science and Technology and began to travel the country in damage-control operations as reports of pesticide contamination of drinking water began to make news in the 1990s.[91]

Think Tanks and Front Groups

As we showed in Chapter 6, affected parties can attack scientific studies indirectly by relying on allied attacks from think tanks that they support. According to the journalist Trudy Lieberman, "think tanks . . . have become idea peddlers extraordinaire, every bit as skillful as the sellers of toothpaste and detergent."[92] Some think tanks have gone to great lengths to project an academic patina by establishing endowed "chairs," housing retired politicians as "senior fellows," and employing "scholars in residence."[93] The Oregon Institute of Science and Medicine provides one of the more blatant examples of how front groups can be employed in attempts to bend public perceptions.

THE OREGON PETITION

In 1998, an obscure think tank called the Oregon Institute of Science and Medicine prepared and circulated a document called the Oregon Petition. Financed by the energy industry, the petition challenged the growing consensus among climatologists that global warming was likely to result in catastrophic changes in the ecology of the planet. The petition was spearheaded by a former NAS president, who was at that time the chair of another industry-funded think tank, the George C. Marshall Institute. An accompanying article supporting the petition was formatted to resemble an article in a prominent NAS publication.[94] The list of "scientists" signing the petition contained very few climate experts, but was instead dominated by local doctors, weather reporters, and high school teachers, whose credentials in climate science were at best questionable. It also contained a few highly dubious "scientists" like B. J. Hunnicutt and Benjamin Pierce (the fictional doctors on the popular television show *M*A*S*H*).[95] Although the NAS immediately disavowed the petition, it was nevertheless cited by politicians and editorial writers in prominent mainstream newspapers to support their opposition to greenhouse gas controls.[96] ❐

As with public relations campaigns, advocates on both sides of science policy debates can establish and rely on think tanks, but the side with the most resources winds up with the most think tanks. Thus, while a small number of think tanks like the Economic Policy Institute, the Urban Institute, and the Center for National Policy have traditionally supported governmental intervention to solve social problems, their efforts have been overshadowed during the last thirty years by institutions like the Heritage Foundation, the American Enterprise Institute, and the Cato Institute that advocate free market approaches to social problems.[97] In the context of health and environmental regulation, the message of the industry-funded think tanks is "invariably the same: government regulation—most especially environmental protection—is bad, and any science that justifies it is 'junk.'"[98]

The think tanks have become ubiquitous in Washington, D.C. politics and in media coverage of health and environmental issues. In addition to attacking studies published by scientists, government agencies, and public interest groups, they write reports, develop attractive websites and media information centers, provide testimony and briefings on Capitol Hill, meet with executive branch officials at the White House and in the agencies, appear as "talking heads" on television news shows, write op-ed columns and letters to the editor, circulate petitions, and pitch news stories to reporters.[99]

It is sometimes equally effective, and much less expensive, to establish a less formal front group with an attractive-sounding name and a sophisticated media presence that has as its primary objective conveying the advocate's scientifically packaged message to the press. For example, Philip Morris in 1991 created a front group called the Advancement of Sound Science Coalition and housed it in a public relations firm.[100] In the words of one tobacco industry executive, this front group soon became "part of a larger mosaic that concentrate[d] all of the EPA's industries against it at one time."[101] The Global Climate Coalition was a similar multiindustry front group run by a public relations firm out of the offices of the National Association of Manufacturers to oppose the growing scientific consensus on global warming.[102] During the congressional debates over "regulatory reform" measures introduced in Congress in the mid-1990s, public interest groups and labor unions established the group Citizens for Sensible Safeguards to convey a uniform and coordinated message about the need for health, safety, and environmental regulation.[103] The organization quickly developed a presence in Washington, D.C. as a place for reporters to go to get a good quotation from the public interest group perspective on the relevant issues.

The best front group of all is a membership organization that advocates the position of a group of citizens who have no obvious connection to the advocates that sponsor them. For example, the National Sleep

Foundation is an "independent" nonprofit research and advocacy organization founded in 1990 and "dedicated to improving public health and safety by achieving understanding of sleep and sleep disorders, and by supporting sleep-related education, research, and advocacy."[104] When it declared National Sleep Awareness Week on March 29, 2005, it released a poll showing that 50 percent of adult Americans have difficulty going to sleep at least once a week and 10 percent of the population rarely gets a good night's sleep at all. It complained in a press release that "[i]nsomnia is a condition that is under-recognized, under-diagnosed, and as a result, under-treated."[105] None of the materials it distributed, however, disclosed that sleeping pill manufacturers provided more than half the organization's budget, paid for the poll, and hired the public relations firm that prepared and distributed the materials.[106] Similarly, the American Diabetes Association, founded in 1940 "to prevent and cure diabetes and improve the lives of all people affected by diabetes," received a large grant from Cadbury Schweppes American Beverages, the world's third largest soft drink manufacturer, to launch "a three-year, multimillion-dollar coordinated effort to fight obesity and diabetes in America."[107] Soon thereafter, the association's chief scientific and medical officer defended Cadbury's sponsorship on the ground that "[t]here is not a shred of evidence that sugar per se has anything to do with getting diabetes."[108] With organizations like these on hand, an advocate's message can reach a large audience and appear very persuasive.

Litigation as Public Relations

The most casual observer of the American news media knows that litigation is often newsworthy, especially when it features a hint of scandal. Good litigators carefully craft their cases to tell a story, and those stories are often fascinating to the public, even when the subject matter is not Michael Jackson's Disneyland Ranch but groundwater contamination in Woburn, Massachusetts. Perhaps less well-known is the fact that advocates sometimes use litigation to achieve public relations goals that are otherwise unattainable through other means.

For example, when the EPA's risk assessment for ETS was published in final form, the public relations firm Burson-Marsteller wrote a lengthy memorandum to Philip Morris outlining a public relations strategy for minimizing its impact on the tobacco industry. First and foremost (as we mentioned in Chapter 6), it was necessary to "Sue the bastards!"[109] Whatever its chances for success on the merits, a lawsuit against the EPA challenging the risk assessment would convey "the substance and symbolism of our principal message: We are right! We shall fight!"[110] A lawsuit "would establish both focus and direction" and "[i]noculate all audiences for follow-up activity."[111] Supported by "appropriate communication overlays," litigation "could regain some percentage of industry credibility and it could stimulate others, who have heretofore been too timid to fight back against the EPA, to summon up their own courage for their own battles."[112] The industry did in fact launch what can only be characterized as a legal jihad against the EPA's ETS risk assessment that cost the industry and the government millions of dollars and was not fully resolved for nearly a decade.[113]

Since litigation is always very expensive, it is certainly not the preferred technique for an advocate to get its point across, but it is usually available to those who have the resources to use it. It is, of course, unethical for an attorney to bring a frivolous lawsuit solely for public relations purposes, but challenges to administrative agencies are easy enough to justify, because one or more aspects of an agency's technical explanation and analysis can nearly always credibly be characterized as "arbitrary and capricious." The challenger may know full well that reviewing courts are supposed to be especially deferential to agencies that address issues "on the frontiers of scientific knowledge," but winning is not necessarily the point of litigation that is designed primarily to send a message to the agency, other advocates, and the general public.

The Mechanism: Exploiting the Media

The tendency of the rich and powerful to maintain their wealth and power at the expense of the diffuse public interest is an inequity the mainstream media have often exposed and, through the indirect power of enlightenment, helped to correct. Press coverage of wars, social travesties, environmental disasters, and related crises have penetrated public consciousness and often precipitated swift and strong political reactions. In keeping with their impressive contributions in raising public awareness of serious social problems, one might expect the mainstream media to be hard at work exposing the manipulations of science, manufactured disputes, and biased, self-serving research studies. Unfortunately, when advocates employ the strategies detailed in this book to bend science to predetermined ends, the media seem to reinforce the carefully crafted message as often as they expose the manipulation. Apparently reporters and their editors rarely have the capacity or the will to challenge the bona fides of the scientific arguments in these technical debates. The tobacco industry's successful insinuation of its public relations message into the pages of *Forbes* is a good example.

PURSUING FORBES

In late November 1991, Burson-Marsteller reported that it was "pursuing the possibility of a major *Forbes* article that would attack the long sordid history of the EPA."[114] The article would concentrate "on a wide variety of issues as per the outline" the firm had "previously provided" to Philip Morris. Burson-Marsteller "hoped that ETS would receive some attention" in the article.[115] The next day, the firm reported that it had already made contact with *Forbes,* and there was "stated interest, but approaches like this can go any way."[116] A set of Burson-Marsteller talking points would "be used with *Forbes* in an attempt to convince the publication that the EPA, in general, deserves a hard journalistic look from a mainstream prospective."[117] The "premise" of the article should be that the "EPA is today an agency out of control. It is a bureaucratic nightmare where ideologues push their own special agendas, where the taxpayers' money is squandered in attempts to rationalize poor decision making and where costs to business resulting from agency actions are irrelevant and dismissed out of hand." The talking points listed several areas of "professional incompetence," including asbestos, dioxin, radon, battery recycling, Grand Canyon pollution, electromagnetic fields, chlorinated organics discharges into the Great Lakes, ETS, and the cost of environmental regulations in general.[118] The outline also laid out several detailed charges of "mismanagement" with respect to the agency's Superfund program for cleaning up hazardous waste disposal facilities.[119]

On July 6, 1992, *Forbes* ran a major cover story entitled "You Can't Get There from Here" and subtitled "How Politics Is Strangling Environmentalism."[120] Highly critical of EPA Administrator William Reilly, the article conveyed the strong impression that the EPA was an agency out of control, where bureaucratic fiefdoms were advancing their own ideological agendas and the taxpayer's money was being wasted on poor decisions. The article focused especially on the EPA's allegedly cavalier treatment of costs to industry, quoting a former EPA general counsel's claim that EPA rules were costing $100 million to $30 billion for every life saved.[121] A separate box on Superfund mismanagement, entitled "Much ado about very little," took up almost a full page of the six-page article.[122] The article briefly mentioned "mainstream scientific opinion" that "agreed that the danger from toxic waste was vastly exaggerated," but the article did not specifically mention ETS.[123]

It is impossible to know for certain whether the *Forbes* article resulted from the Burson-Marsteller initiative for Philip Morris or would have run in substantially the same form without that overture. The theme of the article was virtually identical to the one Burson-Marsteller was hoping a *Forbes* article would convey. Philip Morris officials might have been disappointed by the article's failure to mention ETS specifically, but they must have been pleased by the strong attack on the EPA, which must have had the effect of discrediting it in the minds of readers. ❐

Much of the media's complicity in attempts by advocates to bend science is no doubt unintentional. Fair-minded editors and journalists are committed to telling both sides of the story in public controversies. In practice, this translates into a need to quote someone from both sides of the debate. However, sometimes the search for "balance" leads the media to seek quotations from scientists or even pseudoscientists who take positions far removed from mainstream science.[124] The journalist Chris Mooney observes that the journalistic norm of balance "has no corollary in the world of science."[125] Scientific theories and interpretations must pass the test of peer review and publication in reputable scientific journals, replication by other scientists, and ultimately general acceptance in the scientific community.[126] In science, unlike most political journalism, there are not necessarily two opposing, yet credible sides to every question.

Limited time, resources, and print space also prevent most reporters from deeply exploring any one story.[127] Reporters generally lack the technical expertise to probe beneath the contours of the scientific debates to learn whether the disagreements are scientifically credible, badly biased, or not scientific at all.[128] That type of careful probing generally requires the assistance of scientists who are willing to disengage from the dispute, and such credible contacts are often difficult to find.

Even for those reporters who are inclined or able to dig deeper, the reward is unlikely to justify the extra effort. Editors want exciting stories that sell papers and advertising space. If a story is attractive in its current garb—even as a concocted controversy—so much the better. When a disagreement appears to rest on complex, technical facts, it is hard to justify a deeper investigation in light of the weak or nonexistent sales benefits it is likely to yield. Probing beneath the surface of the dispute will also require a large investment in scarce reporting time and resources. This investment may reveal that a dispute is manufactured, or it may not turn up anything of interest. And when it does hit pay dirt, one or more of the advocates may well accuse the reporters of officiously going beyond their proper professional role of reporting the news as it unfolds.

Still, following up on charges of fraud and bias can be quite rewarding for a journalist if a publicly despised culprit is the source of the problem. This may explain both the more frequent revelations of sponsor-dominated science discussed in Chapters 4 and 5 and the general absence of exposés on nuanced attempts to deconstruct science outlined in Chapter 6. Even easily comprehended stories about fraud and bias, however, seem to run only when reporters have a very sound basis for suspecting such problems, perhaps because investigative journalism into fraud can be a risky project. Sponsors, as we have seen, can be quite creative in devising techniques for laundering or concealing the sources of bias in ways that fool even their scientific publishers. The sources who are in a good position to identify fraud, like disgruntled former employees and whistle-blowers, often have their own axes to grind and may deceive or mislead even cautious reporters.

To the extent that the press fails to probe beneath the surface of campaigns to bend science, its complacency serves only to reinforce the apparent legitimacy of spun science. And this message is not lost on the public relations firms. Sheldon Rampton and John Stauber note that "[t]he PR industry has mastered the art of putting its words into journalists' mouths, relying on the fact that most reporters are spread too thin to engage in time-consuming investigative journalism and therefore rely heavily on information from corporate- and government-sourced news releases."[129] Advocates with the resources to educate the reporter may fare better in the ultimate coverage than busy scientists who return phone calls reluctantly and provide measured responses to a reporter's questions. For a reporter working under a deadline, any assistance in covering a story and identifying additional sources of information is usually received enthusiastically. Well-funded interest groups thus find it quite profitable to invest resources in informing the press regularly and developing stables of experts to be available to answer reporters' inquiries. The result, however, is an imbalance in coverage that parallels the imbalance in resources that lead to disproportionate control over legal processes.

The public relations strategies for manipulating the media are, of course, available to advocates on all sides of the relevant issue, but they are not employed equally on both sides of most disputes over policy-relevant science because of their obvious expense. Public relations firms like Fenton Communications, the firm that launched the Alar initiative for the NRDC, work primarily for plaintiffs' attorneys and public interest groups like the Campaign for Tobacco-Free Kids.[130] When a public interest group decides to launch a major public relations initiative, it can therefore call on high-quality public relations expertise. As a general matter, however, companies and trade associations with an economic stake in the outcome of policy disputes can devote resources to public relations campaigns that are orders of magnitude larger than what public interest groups can afford. During 1990–1991, for example, Philip Morris alone paid Burson-Marsteller a total of $7,731,000, far more than twice the entire annual budgets of many public interest groups, for help in attacking a single EPA risk assessment.[131] In the mid-1990s the coal and oil industries likewise spent millions of dollars on a public relations campaign to downplay the threat of global warming, much of which was devoted, in the words of Ross Gelbspan, to "amplifying the views of about a half-dozen dissenting researchers, giving them a platform and a level of credibility in the public arena that is grossly out of proportion to their influence in the scientific community."[132]

As the previously described Burson-Marsteller overture to *Forbes* suggests, advocates can sometimes even persuade the mainstream media to become willing participants in advocates' attempts to bend science. News networks like the Fox News Channel are not especially discreet about their eagerness to purvey critiques of regulatory agencies and courts that hew closely to the positions of advocates on one side of scientific debates. One frequent Fox News commentator on science issues in litigation and regulatory policy is Steve Milloy, who is known as "the junkman" because of his frequent and unfailingly strident critiques of the approaches public interest groups and some regulatory agencies take to science issues in health and environmental policy-making.[133] Now a fellow at the Cato Institute, Milloy began his career not as a scientist but as a lawyer working for Jim Tozzi, a Washington, D.C. insider introduced in Chapter 6, who ran a business consulting firm and a stable of industry-supported think tanks and was responsible for the appropriations rider that became the Information Quality Act.[134] Milloy moved from there to the EOP Group, a public relations firm that specializes in environmental regulatory issues for a host of clients, including the American Petroleum Institute, the Chlorine Chemistry Council, and the National Mining Congress.[135] At that time, EOP was in charge of running the front group Advancement of Sound Science Coalition (discussed earlier), and Milloy was appointed executive director of that group. With his background as a public relations specialist for the tobacco, chemical, and mining industries, it should come as no surprise that his Fox News appearances consist primarily of diatribes against regulators and public interest groups.[136]

Other media outlets can be persuaded on a case-by-case basis to portray the world as it is viewed by an advocate. For example, some public relations firms specialize in video news releases that simulate local news broadcasts so carefully that local television stations can run them as if they were produced by their own reporters, and some local stations are willing to broadcast them virtually unedited.[137] During the 1990s, the Chemical Manufacturers Association produced videotaped news releases on a semimonthly basis for distribution to fifteen hundred stations around the country. One of the videos reported that a "new study says that more than 10,000 Americans may be dying prematurely each year because of government regulations."[138]

Conclusion

Advocates have a strong incentive to frame the issues that arise in health and environmental disputes as scientific issues rather than as economic or policy questions, even when such disputes have little to do with science and everything to do with economics or ideology. Sophisticated public relations advisors know very well that the public is far more receptive to claims that an agency action is not based on "sound science" than to claims that government action to protect health and the environment costs too much. Regulatory agencies and courts also know that a controversial decision is much easier to justify if is framed as the outcome required by science, rather than merely the policy choice of an unelected judge or bureaucrat, however well grounded the decision might be in the underlying statute. As a result, many important policy disputes to which science is relevant are framed by virtually all the participants as disputes for which science is determinative.

For a number of reasons, framing science policy issues as scientific issues is not an approach that is conducive to sound regulatory or judicial decision-making. First, suggesting that current laws or policies rest on "junk science" and can be magically improved merely by the application of "sound science" distorts reality and misleads the public. When the message is filtered through expensive public relations experts who believe that "there is no truth except the truth you create for yourself,"[139] the quality of the public debate over issues critical to the public health and welfare is likely to be the worse for it. The University of Colorado climatologist Roger Pielke warns that "the use of science by scientists as a means of negotiating for desired political outcomes . . . threatens the development of effective policies in contested issues."[140]

Second, framing policy as science elevates the power of those claiming scientific expertise over other equally legitimate participants in the decision-making process. Science can be relevant to many critical health and environmental issues, and scientists must therefore play a role in resolving such disputes. Many of these disputes, however, are transscientific in nature: the critical questions can be posed of science (and therefore asked of scientists) but cannot be answered by science. Since the scientist is no more qualified to opine on policy questions than any other educated member of the public, framing transscientific issues as questions that must be answered by "sound science" cedes too much power to the scientists and takes power away from the lay people who are more directly affected by the decisions, in the case of agencies, or who are the institutionally appropriate fact-finders, in the case of juries.

Third, when advocates who are expert at framing policy issues as scientific questions then demand an unachievable degree of scientific certainty as a prerequisite to governmental action or judicial awards of damages, the government can easily become paralyzed, and justice can be denied. In the regulatory context, this works for or against the regulated industry, depending on how the regulatory process itself is structured. In a few regulatory programs, like the FDA's regulation of new drugs, a private actor may not sell a product or engage in a proscribed activity without the permission of a government agency. If public interest groups are allowed to frame every issue in the drug approval process as a scientific issue and if they also demand a high degree of scientific certainty as a precondition to drug approval, fewer new life-saving drugs will appear on the market. Conversely, the EPA has the burden of justifying its decision to promulgate most environmental regulations under statutes like the Clean Air Act and Clean Water Act. If polluting industries are allowed to frame all of the issues that arise in the rule-making process as "scientific" and demand a high degree of certainty prior to promulgating those rules, then the EPA will be able to promulgate fewer protective rules. In both cases, large sums of money and the public health are at stake. In the judicial context, by contrast, the strategy virtually always works to the advantage of defendants, because the plaintiff normally bears the burden of proof.

The media have been far too susceptible to manipulation by advocates who have the resources to deal with them in sophisticated ways.

Reporters may believe that this is an inevitable consequence of their inability to distinguish good science from bad science. But that is no excuse for failing to ask probing questions about where the source got his or her information and who provided the resources to make it easily available to the press. As the media and those who follow it disclose more and more instances of media manipulation, the claim of scientific naiveté becomes disingenuous. As one longtime journalist observes, "[b]y embracing the . . . spin, giving it independent credibility, and spreading its messages uncritically, the press has become a silent partner, and the public is none the wiser."[141]

CHAPTER TEN

Restoring Science

Forcing Bent Science Out into the Open

Bent science, at first blush, seems like just another depressing and intractable social problem. Much like the influence of money on political campaigns, the emergence of science bending during the late twentieth century and its apparent growth since then underscore the extraordinary power of interest groups and advocates in our political system. We have all known for some time that special interests exert a heavy hand in determining who will represent the public in state and national government. It is now time to recognize that they have also infiltrated the scientific pipeline that provides most of the information that governmental institutions rely on to implement and execute the laws that govern many critical aspects of our day-to-day lives.

One aspect of science bending, however, clearly sets it apart from other social challenges, like campaign finance reform. In bending science, the casualties are not limited to the powerless members of a diffuse public and the fragile democratic processes highlighted in the political science literature and the editorial pages. Some of the largest casualties of bent science include the bedrock institutions of science itself. Scientists on the margins of professional respectability thrive as their well-supported, outcome-driven research dominates some fields of policy-relevant science, while other scientists honestly practicing their craft find themselves sorted into camps by special interests who either praise their work and support it or attack it and attempt to stifle it. The editors of our best international scientific and medical journals are chagrined by their inability to weed out unreliable research emerging from a funding regime that is increasingly driven by the expectation of future economic gain. Having implemented stringent conflict of interest policies, they find that they lack sufficient resources and influence to ensure meaningful compliance. Even our great research universities are threatened, as evidence mounts that they frequently operate more like profit-driven corporations than institutions dedicated first and foremost to the pursuit and transmission of knowledge. The public's faith in science, while never easy to measure, may finally be eroding under the steady flow of reports of manipulated and distorted research.

Fortunately, this unique threat that science bending poses to organized, powerful, and respected institutions in the realm of science offers a key to successful reform that may be lacking in other areas, like campaign finance. In his path-breaking work *Imperfect Alternatives,* Neil Komesar outlines the basic ingredients for large-scale reform of social problems. Reform can occur, even in the worst of circumstances, either when the stakes for at least some of the affected interests become so high that they are willing to invest time and resources in addressing a problem or when the information that such groups need to take effective action becomes so easily available that action becomes less resource-intensive.[1] Classic tipping points, where social reform follows, thus arise either when intensely affected subgroups of larger affected groups are stimulated to take action and catalyze their dormant members, or when the problem becomes quite salient and accessible to all members of the larger group, as in the case of a highly publicized scandal or major catastrophe.

Both of Komesar's preconditions for effective action—a group of highly affected parties and more accessible information about the problem—are now present in the current controversies over bent science, and both are slowly moving the relevant institutions in the direction of reform. Most dramatic is the increased engagement of individual scientists and the scientific community as a whole in addressing the problem of science bending. Highly respected mainstream scientific organizations like the NAS and the AAAS have assumed a leadership role in speaking out against ideological, political, and market-driven distortions of science. Their unambiguously negative response to the Barton letter, to early OMB proposals for peer review, to the stacking of advisory committees, and to climate change advocacy are just a few examples of an incipient activism on the part of important subgroups of the larger scientific community. The scientific journal editors, an extremely influential subpopulation, have been even more sharp in their criticisms and concerns about the changing world of privately funded science. These responses are not surprising. Loss of public respect for science and threats to scientific freedom are very serious matters for these groups and their members. And to the extent that these important subgroups become engaged in pursuing reforms, their efforts can stimulate action on behalf of not only the scientific community but also the characteristically more passive general public.

At the same time, these scientific organizations and their members are dedicating some of their energies to spreading the word about the pervasiveness of science bending and are generating discussions on how best to reform these problems. The number of panels, reports, university-sponsored conferences, and other events that highlight some piece of the bent science problem seems to rise each year. These efforts should reduce the information costs to nonscientists, thereby enticing more significantly or even mildly affected citizens to join the scientific community in advocating reform.

Assuming for the moment that the stage for reform is now set, we turn to the substance of the necessary reforms and the institutions that should implement them. To return to legal theory: Komesar observes that in identifying reform options, the reformers should not limit their options to minor tweaks in existing judicial or regulatory rules. Reformers should instead step back and examine the entire toolbox of market, regulatory, and liability possibilities to determine how best to offset the participatory imbalances introduced by dominant interest groups. The best institutional approach will emerge from a comparative institutional analysis that considers the ability of each of the relevant institutions to minimize the opportunities for dominance by high-stakes interest groups.[2]

To address a problem, like science bending, that represents an almost total failure of all the relevant institutions, however, we may conclude that multiple and redundant reforms across the market, political, and common law systems are needed to offset the enormous pressures that high-stakes special interests directly and indirectly bring to bear on esoteric science. Although the ultimate goal in each institutional setting is to provide some counterbalance to the strong pressures special interests exert on science, we may decide that no single institutional fix will suffice. The proposals we set out below therefore include recommendations for multiple institutions simultaneously to enhance advocates' accountability for bending science by making their efforts more transparent, by relying more heavily on formal science advisory groups for ballast against advocates' dominance of decision-making processes, and by providing incentives for uniquely situated opponents or skeptics to root out bent science.

In this first of our two chapters on reform, we focus on how information disclosures—which exert pressure indirectly through increased transparency—can begin to increase the accountability of advocates who create bent science and manufacture scientific critiques. Although most of these reforms will require some changes to existing regulatory or legislative rules, their common goal is not to regulate bad behavior directly, but instead to force undesirable practices out into the open, where other scientists, members of the public, and even insurance companies and industry competitors can judge them. In most cases, the reforms advocated here do little more than codify into law the practices many scientific journal editors have already begun to initiate with respect to their own publication policies. For example, disclosures of conflicts of interest and data sharing are now relatively well established in most of the major scientific journals. The next step is for agencies and courts to implement similar disclosure policies for all policy-relevant science that informs governmental decisions involving public health and the environment.

While we do not imagine that greater transparency will always lead to better quality science, these information disclosures can counteract stakeholder manipulations of policy-relevant science in at least two different ways. First, by making both underlying data and information on the provenance of a study available to other scientists in an enforceable manner, the suggested reforms should enable interested scientists in both the private sector and the government to assess the reliability of the reported results. In some cases, this transparency will also assist journalists, public interest groups, competitors, and the general public in making their own evaluations of the reliability of a scientific study or related critique. Second, both regulatory theory and limited empirical observations suggest that positive behavioral adjustments can result from legally required disclosure requirements imposed on regulated parties, even when the information is not routinely used by other stakeholders and the government. According to this theory, mandatory disclosure should shift the internal cost-benefit calculus of a potential sponsor of an outcome-oriented study or critique, because disclosure reduces the overall benefit of the manipulation. Empirical research in areas as wide-ranging as environmental law, medical care, and securities regulation reveals this effect of transparency in changing internal practices, even in the absence of a vigorous audience engaged in overseeing or judging the disclosures.[3]

Finally, we suggest how two institutions that are uniquely situated to serve as gatekeepers over bent science, the universities and the media, can amplify this greater transparency in ways that should increase its impact on judicial, administrative, and legislative decision-making. While neither the universities nor the news media have taken much of a leadership role in combating bent science in the past, the greater accessibility of information on the provenance and technical bona fides of policy-relevant research should allow and even encourage them to do just that.

Disclosures of Conflicts of Interest

Philosophers of science can legitimately debate the attainability of "objectivity" and "disinterestedness" in the abstract realm of "pure" academic science, but in the earthy world of policy-relevant research, the statistical correlations between sponsored research and sponsor-friendly results are strong and unambiguous. Even if the vast majority of scientists would never consciously allow the potential for financial gain to influence the outcome of their research, literally dozens of studies have found a positive correlation between financial sponsorship and favorable outcomes for the sponsors. Ideological conflicts of interest, though less well studied, can also threaten the integrity of scientific research.[4] The resulting need to eliminate, or at the very least demand disclosure of, significant conflicts of interest has not been lost on the scientific community.

Most scientists seem to agree on the desirability of at least some degree of disclosure when significant financial conflicts are present,[5] and their efforts have played out in scientific conferences, the professional journals, and the media.[6] Beginning as early as 1984, some of the more prominent medical journals, such as *NEJM* and *JAMA,* began to require authors of original research to declare their financial associations, including "consulting fees, service on advisory boards, ownership of equity, patent royalties, honorariums for lectures, fees for expert testimony, and research grants."[7] The Council of Science Editors takes the position that science journal editors are responsible for "establishing and defining policies on conflicts of interest."[8] Some journals now refuse to publish literature reviews or editorials by authors who have a conflict of interest in the outcome, because the extent and effect of the bias in such contexts is difficult to detect through the usual methods of replication and validation familiar to science.[9] After its strict disclosure policy limited the range and diversity of the review articles and editorials it published, however, *NEJM* changed its policy in 2002 to require disclosure of only "significant" financial interests.[10] Other journals' disclosure requirements have become so lengthy that journals post them on their websites, rather than devote valuable journal space to them.[11]

Despite this strong trend of many journals toward requiring conflict disclosures, editors have nevertheless struggled with ensuring compliance with those requirements.[12] For example, *JAMA,* which has one of the country's most stringent disclosure policies, published a paper by scientists at the Massachusetts General Hospital, Emory University, and UCLA concluding that hormonal changes during pregnancy do not protect women from depression to the extent that previous studies had suggested, without disclosing that seven of the thirteen authors had been paid consultants or lecturers for manufacturers of antidepressant drugs.[13] It also published a study finding that women suffering from migraine headaches had an elevated incidence of heart disease by authors who failed to disclose that they had received money from manufacturers of painkillers.[14] Noting that the journal received more than six thousand submissions per year, its editor cautioned: "I do not know what is in the hearts, minds and souls of authors" and "I'm not the F.B.I."[15] Other journal editors share this sentiment.[16]

Empirical studies reinforce these concerns. In one study, Professors Sheldon Krimsky and L. S. Rothenberg surveyed more than sixty-one thousand articles published in the top English-language scientific journals during 1997 for conflict of interest statements and discovered to their surprise that although almost one-half of those journals had official policies requiring disclosure of conflicts of interest, only 327 disclosures (0.5 percent of the total number of articles) were published. Worse, sixty-six of the journals with explicit conflict of interest policies reported no disclosures at all during the year. Krimsky and Rothenberg concluded that one likely explanation for this very low rate of disclosures is poor compliance, not a high percentage of researchers free of conflicts.[17] A more detailed study conducted by the Center for Science in the Public Interest of conflict of interest disclosures in four prominent medical journals in 2004 found that in 8 percent of the 163 articles, one or more authors failed to comply with the journal's disclosure policy, narrowly interpreted.[18] Another 11 percent violated a broader interpretation of the disclosure policy that requires authors to identify when they are "employed by or consulted for industry, owned stock or stock options, received research support or honoraria from industry, held or applied for patents or had similar financial arrangements with any private firm, no matter how distant its relationship from the subject of the article."[19]

One reason for poor compliance may be some scientists' strong resistance to the notion that conflicts of interest are an important issue for the scientific community. A prominent epidemiologist at the Harvard School of Public Health who is a former editor of *Epidemiology* views disclosure requirements as equivalent to "McCarthyism," and he argues that they violate the principle that a work of science must be judged solely on its own merits without regard to authorship or source of funding.[20] A prominent Harvard hematologist who regularly engages in sponsored research accuses "academic socialists and the conflict-of-interest vigilantes" of stifling critical progress in biotechnology.[21] It may take more than a vague threat of losing publication privileges in major medical journals to persuade scientists with these convictions to report significant conflicts to their institutions and the public.

Some of the less prestigious journals are reluctant to implement and enforce conflicts disclosure policies because they believe that authors will simply take the articles to competing journals.[22] Even editors of the most prominent journals worry that papers they reject for conflict of interest reasons will wind up in the citable peer-reviewed literature in journals with less restrictive policies.[23] Even if all of the relevant scientific journals achieved 100 percent compliance with robust disclosure requirements, however, serious conflicts would still plague the generation and use of policy-relevant science because many of the studies and critiques that regulators and the courts rely on are never published in any journal. Indeed, vigorous enforcement by journals of strong conflict of interest disclosure policies might simply induce sponsors to rely even more heavily on unpublished studies in dealing with regulators and courts.

The obvious remedy for this "race to the bottom" is for regulators and courts to reinforce journal-imposed requirements with a legally required disclosure requirement that identifies possible sources of sponsor influence. If sponsors and scientists understand that their control over research will ultimately be made public and that the value of the resulting research might be discounted because it lacks critical indicia of objectivity and reliability, then these requirements might cause them to adjust their arrangements in ways that promote greater research independence from sponsors. Furthermore, the prospect of public criticism for basing important policy decisions on research studies produced under the heavy hand of advocates might induce agencies and courts to rely on such research less frequently in the future. If the perceived benefits of control over sponsored research do not outweigh the costs, sponsors may change their practices to ensure that future research products will not be encumbered with embarrassing qualifications. Thus, while these legal disclosure requirements may not resolve all types of disturbing, undisclosed conflicts that can plague research, they should curb at least a significant portion of the problems that arise with sponsored research destined for regulation or litigation.

We propose a short, but mandatory, conflict disclosure requirement that applies to all scientific information or critiques submitted to courts and regulators. The requirement should also extend to postmarket research conducted on the safety and efficacy of federally regulated products and activities, even when formal regulatory submissions are not required.

Although mandatory conflict disclosures for policy-relevant research is an obvious first step toward more objective decision-making, existing legal institutions rarely, if ever, demand such disclosures. Agencies like the EPA, OSHA, the Consumer Product Safety Commission, and the National Highway Traffic Safety Administration all lack any formal mechanisms for identifying potential conflicts of interest or for determining the extent of sponsor influence over sponsored research.[24] The FDA is one of the few agencies that has instituted a conflict policy that requires financial disclosures for safety research conducted by private parties in support of a license to market a drug or food additive.[25] The required FDA disclosures do not, however, discriminate between sponsored studies in which the sponsor retains control over the research design and those in which the sponsor relinquishes control over any aspect of the research process.[26] They also do not extend to postmarketing research that is not intended for FDA review but may have a dramatic influence on doctor or consumer understanding of the safety or efficacy of drugs. Thus, most real-world agencies have failed to take advantage of a readily available and easily implemented tool for encouraging greater freedom for researchers, enhancing the objectivity of the resulting research, and increasing public respect for the resulting governmental decisions.

Similarly, while the relevant judicial opinions are filled with expressions of concern about the "testability" of expert opinion in determining whether it is sufficiently reliable for jury consideration, they pay surprisingly little attention to ensuring that juries receive information about debilitating conflicts of interest that may compromise the underlying research. Juries learn of such conflicts only if the parties bring it to their attention, and the obvious fact that both sides are relying on paid experts to present scientific information may mask the subtler influence sponsors have over the underlying research that forms the basis for the expert's opinions. When a jury relies on bent science, the consequences for society may not be as severe as those resulting from bad regulatory decisions. But they can have devastating consequences for the individual parties to the litigation.

Requiring full conflict disclosure in both the judicial and regulatory spheres will advance a number of important policy objectives simultaneously. First, such disclosure makes information available to courts, regulators, and the public that can be very useful in assessing the credibility of research, especially when scientific controversy erupts over the implications of individual research results. Standardized disclosure policies would also assist scientists who serve on scientific advisory boards, sit on judicially appointed panels of neutral experts, or are otherwise involved in reviewing policy-relevant science for courts and agencies. Second, full disclosure of the details of privately funded research arrangements should turn the current incentive system around by attaching long-overdue rewards—in terms of perceived reliability—to research that is funded by an interested party but is not controlled by it. In a full disclosure regime, more prejudicial forms of influence—like contractual restrictions on the conduct or publication of research or sponsor collaboration in study design or data interpretation—should become more visible and, consequently, less popular. Third, disclosed conflicts may expose potentially influential studies that require further investigation or even replication to ensure their reliability. Finally, full disclosure should have a positive impact on the quality of judicial and agency decision-making by allowing decision-makers to take into account the origin of the information before them in deciding how much weight to allow it in balancing all of the relevant considerations.

Assuming that existing legal rules should be reformed to require full disclosure of the level of sponsor influence affecting policy-relevant research, several implementation issues require further attention. First, the new rules must err on the side of disclosure, much as the *JAMA* and *Science* guidelines do, to prevent advocates from finding ways to circumvent disclosure requirements. *Such a federal requirement could demand, for example, that scientists "disclose all affiliations, funding sources, and financial or management relationships that could be perceived as potential sources of bias" in their submission and certify that "all authors have agreed to be so listed and have seen and approved the manuscript, its content, and its submission."*[27] *In addition, scientists should describe the role any sponsors played "in study design; in the collection, analysis and interpretation of data; in the writing of the report;*

and in the decision to submit the report for publication."[28] *Violations of these disclosure requirements should be subject to penalties and even criminal sanctions under applicable laws, including the Federal False Claims Act.*[29] With regard to scope, the disclosure requirement should clearly demand disclosure when a researcher is in doubt. So, for example, if a think tank sponsored or collaborated with a researcher, a disclosure of that relationship would be necessary even when the source of funding for the think tank is unknown to the scientist. Similarly, scientists filing extensive or lengthy technical comments should be required to provide disclosures of the nature of sponsor influence over their critiques. In order to encourage disclosures, moreover, the relevant legal institutions may have to assume the worst regarding the provenance of research if a researcher refuses or is no longer available to provide this information. Thus, when a submitter presents no basic information regarding provenance, regulators and courts might assume as a matter of law that the sponsor designed the study, reserved the right to suppress adverse findings, and made all major decisions regarding data collection and interpretation. Since sponsor involvement is likely to be much less serious than the "worst case," a "worst case" assumption should help draw out information about sponsor influence, but it will also serve as a protective device to err against relying on research when the independence of the researcher is unknown.

Disclosures of sponsor influence do not directly address the parallel problem of ghostwriting and redundant publications, however. Accordingly, legal decision-makers must also anticipate and redress misleading authorships through additional disclosure requirements. Several journals, for example, require primary authors to sign a statement accepting full responsibility for the conduct of the study and to certify that he or she had full access to the data and controlled the decision to publish.[30] The purpose of the requirement is to demand that "an academician put his or her reputation on the line, and that of the institution."[31] Regulatory disclosure requirements should include similar certifications for submitted studies. To deter the tendency to publish or submit redundant studies, submitters of information to agencies and courts should also be required to state at the outset of the submission whether the reported results are part of a larger study and to disclose any related paper involving that data of which they are aware.[32] Together, these added certifications will also alert regulators to issues regarding authorship and redundancy and presumably begin to deter these practices.

Although conflicts disclosure is valuable in its own right because of the additional information it makes available to decision-makers and the public and the positive incentives it provides to researchers and sponsors, it should also have an impact on how agencies and courts actually use scientific information in making real-world decisions. Some research might be so seriously conflicted as to warrant its exclusion altogether from the decision-making process. For most studies, however, regulators and courts will probably need to examine conflicted research on a case-by-case basis, explaining in individual cases how they considered conflicts and the weight that they afforded such studies in their decision-making. Forcing courts and regulators to explain how they treat various conflicts in weighing the scientific evidence is a sensible compromise for ensuring that conflicts are factored into the relevant decisions without losing the value of the underlying scientific information altogether.

The Role of Agencies and Courts in the Interim

It is, of course, always possible that our recommendation for mandatory disclosures of conflicts of interest will not come to pass. In the interim, the courts and agencies should at the very least ensure that the provenance of all research that they rely on to support regulatory and judicial determinations is, as far as is possible, divulged and considered in the decision-making process.

The provenance of information is in some ways a less difficult issue for the courts. Since the scientific information they rely on comes almost exclusively from experts testifying for advocates, both judges and juries know that it does not necessarily comply with the scientific norm of disinterestedness and presumably adjust their assessments accordingly.[33] The provenance of the scientific studies the experts rely on is somewhat less accessible to the fact-finders, but often obtainable. The Supreme Court in the famous *Daubert* case assigned to trial judges the role of gatekeeper in civil litigation with the responsibility of ensuring that the scientific information that experts make available to the fact-finders is scientifically reliable.[34] When the research underlying an expert's testimony consists entirely of studies commissioned for the litigation, the courts are aware that the likelihood of bias is high, and they should subject it to careful scrutiny.[35] Thus the court in the silicosis litigation we discussed in Chapter 2 subjected the diagnoses of the plaintiffs' experts to very careful scrutiny and issued a 127-page opinion, replete with 147 pages of appendices, in which she excoriated some of them for the bias she detected.[36] Since the *Daubert* Court suggested that peer review is a factor in assessing the reliability of research, some testifying experts have gone to the extra effort of submitting commissioned research to peer-reviewed scientific journals.[37] That fact standing alone, however, should not cause the courts to let up their guard if the research was commissioned solely for the purpose of litigation.

Some courts and commentators have taken the tenuous position that if the commissioned research was completed prior to the onset of litigation, "the expert's opinion is not biased and is thus reliable."[38] We have seen, however, that advocates are perfectly capable of bending science far in advance of a lawsuit, either to meet regulatory requirements or in anticipation of future litigation. Thus, the mere fact that the study was published in a peer-reviewed journal prior to the onset of litigation should by no means carry a presumption of validity without further inquiry into its provenance. *The courts should allow the parties to conduct discovery and present evidence on the extent to which some or all of the research underlying an expert's testimony was in fact commissioned by an entity with a direct or indirect interest in the litigation and on the degree to which that entity exercised control over the outcome of the research.*

The problem is worse for the agencies, because they cannot always depend on advocates on the other side to uncover veiled provenance. Unable to subpoena documents from the parties to their proceedings or to take sworn depositions of their experts, agency staff may be less adept at ascertaining the sponsorship of the studies that parties cite or the books and reports on which they rely. As we have seen, paid consultants, some even working for other government agencies, make presentations at agency advisory committee meetings without identifying that fact or the sources of the information that they present. The problem may also be worse for society as a whole, because the health and environmental consequences of a bad decision by an agency charged with protecting the general public in a major rule-making effort or product approval proceeding can be far more costly than the consequences of a single erroneous judicial decision.

Since agencies cannot depend on the parties to the proceedings to reveal hidden provenance, *agency staff should devote resources to probing the provenance of each study that the agency proposes to rely on in a significant way in the technical support documents, preambles, and statements of reasons for its nonroutine regulations and licensing decisions.* For example, when an agency proposes to rely on a published scientific study by one or more authors who are affiliated with or are funded by a party with an interest in the outcome of the proceeding, the staff should contact the authors and attempt to find out how extensively the sponsor of the study controlled the research. Likewise, when a scientist making a presentation before an advisory committee relies on a particular scientific study, the agency staff should make similar inquiries of the scientist and the authors of the study.

Access to the Data Underlying Completed Studies

Despite a general commitment to scientific honesty and open communication that includes sharing data and methods,[39] researchers in hotly contested areas of policy-relevant science are sometimes reluctant to share the data underlying published studies with affected parties. This reluctance may reflect an understandable concern that their generosity will lead to harassment and illegitimate challenges to their conclusions, but it may also stem from an inappropriate desire to escape peer review. From an external vantage point, however, these motives can be indistinguishable. Either way, a researcher's resistance to data sharing impedes the ability of other scientists to evaluate controversial research that may play an influential role in informing policy.[40] Furthermore, to the extent that the data are derived from clinical trials involving human subjects, the researcher may have a moral obligation to share the redacted data, so as to fulfill the expectations of the volunteers that their participation would be used to advance scientific knowledge.[41]

CIRCUMVENTING RESEARCHER RELUCTANCE WITH A RIDER

By the early 1990s, a massive epidemiological study of the health effects of ETS conducted by Dr. Elizabeth Fontham and her colleagues had surpassed the Hirayama study (described in Chapter 6) as the bellwether study on ETS risks.[42] The tobacco industry believed that once the EPA had incorporated the Fontham study into the comprehensive risk assessment it was in the process of finalizing, the risk assessment would have "a profound effect, direct and indirect, on public opinion, business restrictions, and government policy, influencing a large number of federal, state and local laws that regulate indoor smoking."[43] The industry was eager to ensure that Fontham's study was reliable, but also seemed willing to attempt to undermine its credibility. In apparent pursuit of the latter goal, Philip Morris and its consultant, Jim Tozzi, wrote to Fontham and her colleagues demanding access to the "raw data" used in the study and promising to take extensive steps to protect the participants' privacy and any proprietary information.[44] Fontham and her coauthors refused, explaining that while they had "always been willing to provide data to the publishing journals under agreed-on safeguards for protecting confidential and proprietary information" and were "exploring several options for an independent reanalysis" of the data, they were "concerned that these data be used solely for furthering science and that they not be distorted by the economic interests of other parties who analyze them."[45]

Having failed at the nice-guy approach to obtaining the data, Philip Morris elected to subpoena it in ongoing litigation in which plaintiffs were claiming that ETS had caused their diseases. On the basis of prior experience with the litigation subpoena strategy, a company executive predicted that this would be "a very difficult task,"[46] and that prediction proved accurate.[47] RJR then sued Fontham's home institution, Louisiana State University, to obtain access to the data, but the Louisiana court dismissed the lawsuit.[48] A federal district court in California ultimately granted Philip Morris access to the raw data, but with the stipulation that it could not be used or disclosed for any purpose other than defending itself in a Florida lawsuit filed by a plaintiff allegedly damaged by ETS, and all copies had to be returned at the conclusion of the case.[49] This effectively precluded Philip Morris from publicizing any reanalyses of the data.[50]

The industry next decided on a legislative solution to the data access problem. As it happened, six hundred companies from several industries were having similar difficulties gaining access to the data underlying another major epidemiological study, the Six Cities Study, which found premature mortality correlated with increased concentrations of particulate emissions in six major U.S. cities. According to an internal Philip Morris memo, this unhappy coincidence for a diverse group of regulated parties gave the tobacco industry a "unique and unprecedented opportunity" to "direct the established political and business coalitions."[51] A "best case" scenario for this approach would be federal legislation that required data sharing. Using this law, the tobacco industry could "get the data from the Fontham study and prove it does not show any association between ETS and disease." This, in turn, "prevents OSHA from acting and stops/repeals smoking bans."[52]

The industry strategy to access the data underlying controversial policy-relevant research succeeded in 1999, when Congress passed the appropriations rider now known as the Shelby amendment or the Data Access Act, providing any person with full access to data underlying any federally funded research that informs regulation.[53] As it is for most appropriations riders, the legislative history contains no evidence that Congress debated the bona fides of the Data Access Act or knew of its existence.

Indeed, debates on similar legislation several years earlier indicated that had Congress considered the legislation as a stand-alone bill, it would have precipitated much debate, attracted many amendments, and might not have successfully negotiated the legislative gauntlet.[54] ❒

As might be expected from its questionable origins, the original text of the Data Access Act not only gives private entities access to federally funded data but also provides them with tools to harass researchers with onerous and unjustified data access requests. Drafted broadly enough to include all federally funded studies, even those that were ongoing, the legislative language placed no limits on how much data private parties could demand from federal researchers. Instead, "[f]ederal awarding agencies [were required] to ensure that *all* data produced under an award will be made available to the public through the procedures established under the Freedom of Information Act."[55] The Act further failed to specify how far back in a researcher's research program the requirements extended, thus potentially empowering any private entity to demand data from studies published thirty years earlier.

Despite its worrisome beginnings from the perspective of policy-relevant science, the Data Access Act may ultimately have a happy ending. Concerned about the potentially broad reach of the provisions, the scientific community reacted vigorously to the legislation, and scientists inundated the OMB, the agency charged with its implementation, with thousands of critical comments.[56] As a result of this critical outpouring, the agency rules implementing the law now roughly resemble what most scientists seem to accept as reasonable requirements for data sharing. Under the regulations, data from federally funded research must be shared only for "published studies" and not ongoing research; data disclosure is limited to those data needed to "validate the study"; and requestors must pay researchers for the reasonable costs incurred in responding to data requests.[57] While the scientific community has not put aside all concerns about the law, particularly with respect to its potential as a tool for harassment, scientists are apparently much more comfortable with the rules promulgated by the OMB.[58]

The one glaring exception to the scientific community's general satisfaction with the Data Access Act is its limitation to federally funded research.[59] Congress could easily correct this by amending the existing federal requirements for sharing data to include both privately and publicly funded policy-relevant research, while keeping in place the protections OMB has added. Since privately commissioned research is more likely to be manipulated through one or more of the techniques described in this book, it is in far greater need of mandatory data sharing requirements. One longtime expert on pharmaceutical research has observed that "[m]aking the raw data available would do more to keep researchers and universities honest than any of the research safeguards or ethical frameworks that universities are busily trying to put in place to manage links between industry and academia."[60]

Our second recommendation is thus that data sharing requirements, along the lines developed by OMB in Circular A-110, be extended to all research that informs regulation and litigation, even when it is financed exclusively by a private party. Ideally, the data should be accessible through an open electronic database available over the internet. If this method of access does not prove realistic, other means should be taken to ensure that the data are not only technically available, but practically available to those who are interested in accessing them.

Although some of the private data underlying regulation are already available in theory without this suggested amendment, in practice, public access to these data can be quite limited. In licensing programs for drugs, pesticides, and toxic substances, for example, the FDA and the EPA do demand access to data in advance as a condition to accepting studies for regulatory purposes, but the data often remain buried in agency files, or in some cases in even more inaccessible, "trade secret" files, until a party specifically requests them under the FOIA.[61] On occasion, the implementing agency has even conspired with regulated industry to leave relevant data in company files rather than store it in agency files, where it might be subject to FOIA requests.[62] Accordingly, in amending the Data Access Act, it may be appropriate to take the additional step of requiring the privately produced data submitted in at least an application for a license or permit to be posted on the internet to ensure ready public access.

Any new requirement that facilitates public access to private data would have to include enforcement provisions with sufficient bite to ensure compliance. In 1997, Congress moved a small step in the direction of data access when it enacted a little-known law requiring private sponsors of clinical trials on drugs that treat serious life-threatening diseases to make the underlying data available to doctors and the public in a centralized registry.[63] That law, however, has proven singularly ineffective in inducing companies to make the data underlying company-sponsored clinical trials available to the public.[64] For example, as of mid-2004, fewer than half of the ongoing clinical trials of cancer-treating drugs were on the registry.[65] The FDA official in charge of the site concluded that "[m]any pharmaceutical trials are not participating" in the program or "are not fully participating."[66]

Widespread noncompliance with these less rigorous data disclosure laws suggests that any new statute should include provisions requiring vigorous agency surveillance of compliance and strict penalties for noncompliance. For example, one way to get the attention of companies needing premarket approval would be a provision placing a moratorium on marketing prior to full compliance with all data sharing requirements.[67] An additional way to ensure compliance, suggested by Dr. David Michaels, would be to require certification by a specified official within a company that all data have been submitted in compliance with applicable laws.[68] The designated corporate official would also be responsible for justifying any trade secrecy or related claims of confidentiality. This central certification, modeled after the Sarbanes-Oxley Act, would help ensure that companies paid serious attention to the disclosure requirements. Like the Sarbanes-Oxley Act, the statute should also include criminal penalties.

The primary complication likely to arise from an expanded approach to data sharing involves information the submitter can legitimately characterize as a commercially valuable trade secret. For example, companies submitting health and safety testing data in connection with premanufacture notifications to the EPA under the Toxic Substances Control Act often argue that the precise chemical identity of a tested product constitutes trade secret information that, if revealed to potential competitors, would destroy the product's economic value to the company, thereby reducing private sector incentives to develop potentially useful products in the first place.[69] Epidemiologists have expressed a similar concern that large databases, which can take years to assemble and are often the source of data for multiple publications, are the equivalent of intellectual property and should also be protected from disclosure to potential competitors. Forcing epidemiologists to share their full databases at the time of the first publication will encourage them to delay publishing any individual study until they have completely mined all of the data, thereby depriving agencies and the public of valuable public health information in a more timely fashion.[70]

The solution to intellectual property concerns could lie in a rebuttable presumption that the public may have access to all privately sponsored data unless the private entity is able to rebut that presumption by establishing that the requested data are protected by legitimate trade secrecy or compelling privacy concerns. This presumption would effectively flip the burden of proof to those claiming protections, an approach that the EPA does not currently take with respect to trade secret protections in a number of its regulatory programs. Moreover, the submitting entity would have to demonstrate that the protected information could not be redacted, while still providing public access to the remaining data. In the case of intellectual property, as opposed to privacy, the protections should also be limited in time (for example, three years) to the period when the cost to a researcher or sponsor from disclosure is highest.[71]

A less easily resolved concern with this proposal is its potential use for harassment. Indeed, this was the basis for the scientific community's strong resistance to the original statute.[72] Ultimately, more sophisticated data sharing requirements may be needed. Trial and error may point the way to a better approach. Until then, disclosure requirements that track the scientific community's own requirements, as evidenced in the journals, seem the most sensible and prudent first step toward combating bent science.

Discouraging Suppression of Adverse Effects Information

Providing access to the data underlying completed studies will not necessarily prevent sponsors from suppressing or prematurely terminating studies that are yielding undesirable results. Indeed, a sponsor's awareness of the fact that the data underlying completed studies are subject to data access requests may increase incentives to suppress unwelcome research. Because suppression is both an effective and largely undetectable vehicle for bending science, it is particularly difficult for the legal system to correct. The difficulties are compounded in that scientific norms discourage scientists from disseminating research results until they have been checked, double-checked, and vetted by peers, if possible. These ethical guidelines can create a special dilemma for health scientists. If they publicize preliminary findings before they have been fully vetted in the scientific community, they risk scientific opprobrium and public embarrassment if subsequent analysis requires the findings to be qualified or withdrawn. Failure to make preliminary findings public, however, may delay action needed to reduce human exposures to dangerous workplaces, products, or pollutants. This was the dilemma Bayer faced in deciding what to do with its early research on Baycol.

BAYER'S RESEARCH LAID BARE

When the FDA approved Bayer Corporation's Baycol in June 1997, it was one of a promising family of drugs called statins routinely prescribed to persons susceptible to heart disease to reduce levels of bad cholesterol.[73] Both the company and the FDA knew that statin drugs can cause rhabdomyolysis (rhabdo), a rare disease in which a chemical in the blood breaks down muscle tissue into toxic proteins that in turn can cause kidney failure and death.[74] Bayer's clinical trials, which were conducted to support its new drug application for Baycol at doses of 0.2 and 0.3 milligrams, reported no cases of rhabdo.[75] Because Baycol at these doses did not outperform existing statins, however, Bayer soon applied for approval at doses of 0.4 and 0.8 milligrams.[76] One of the studies the company commissioned to support the new application showed that a dose of 1.6 milligrams caused an increased incidence of elevated creatine kinase (CK) blood levels in patients, a clear warning sign for rhabdo. The minutes of the company's Cerivastatin Communication Committee reported that "[t]he large percentage of patients experiencing CK elevations led to a consensus by the committee not to publish the results of this study."[77]

Not long after receiving these clinical results, a company scientist reported that suspected adverse drug reaction (SADR) reports from doctors in the field (where Baycol was being used at the lower dose of 0.3 milligrams) "indicate[ed]" that Baycol "substantially elevat[ed] risk for rhabdomyolysis compared with other statins."[78] Bayer did not disseminate this information, however, because Bayer's scientists concluded that its internal analysis of SADR reports "was not substantive data."[79] On July 21, 2000, the FDA approved the company's application for a higher dosage of Baycol of 0.8 mg.[80] By November, another internal company analysis of SADR reports concluded that patients taking Baycol were five to ten times more likely to contract rhabdo than patients on the other statin medications.[81] Again the company declined to publish or publicize this analysis, and the drug remained on the market. When the company finally withdrew Baycol from the market on August 8, 2001, the FDA had received postmarketing reports linking the drug to thirty-one deaths from rhabdo.

Bayer decided not to share its preliminary adverse findings with the FDA in part because they were incomplete. Rather than set off false alarms, Bayer opted for collecting more information and conducting more analyses to determine whether Baycol was in fact the cause of health problems in some users. As a scientific matter, Bayer's decision was acceptable: preliminary findings are generally not appropriate for publication or dissemination. But when public health is at risk, scientific reticence may have to yield to precaution, and information on adverse effects should be shared with public officials who are in a position to take protective action. This is the point at which Bayer's conduct and, in the view of many observers, the entire postmarket surveillance system established by the FDA fell short.[82] ❐

A number of environmental and health statutes require regulated parties, under threat of both civil and criminal sanctions, to report adverse information, including preliminary findings discovered in the course of ongoing research.[83] The laws therefore resolve the dilemma for the scientists and sponsors who are subject to them by overriding their natural conservatism in favor of an early warning system to maximize health and environmental protection. If Bayer had immediately shared the preliminary discoveries of Baycol's potential adverse effects with regulators, some of the premature deaths might have been avoided. Since the agency professionals who are assigned to assessing adverse effects reports are generally doctors or scientists, they understand the preliminary nature of the results and can take that into account in deciding the next steps. Instead of imposing stringent regulatory restrictions on the basis of preliminary research results, the agency might, for example, require the manufacturer to undertake specific follow-up research under a tight deadline.

Even though Congress has demanded early reporting in some regulatory programs, the agencies have been slow to implement the requirements. Moreover, the vagueness of the reporting requirements, which typically employ terms like "substantial risk," gives violators a great deal of leeway to justify noncompliance.[84] *We recommend that agencies amend their adverse effects reporting rules or, if necessary, Congress amends the underlying statutes to establish a more rigorous reporting system that errs on the side of overreporting. In particular, the agencies should broadly define "adverse effect," prepare lists of adverse effects of relevance to their regulatory programs, and limit researcher discretion in deciding whether a finding is too preliminary to warrant reporting.* The EPA has in fact promulgated requirements that satisfy these more ambitious standards under the pesticide statute.[85] Agencies, including the EPA, should follow this approach in implementing similar reporting regimes for toxic substances, pharmaceuticals, risks to workers, and disclosures of spills or sudden releases of large amounts of chemicals or pollutants. Furthermore, the agencies receiving this information should ensure its immediate dissemination to environmental and health professionals by posting all adverse effects reports on the internet in a form that is searchable through a variety of queries, including chemical name and manufacturer.[86] The postings, by highlighting the most salient information, could counteract submitters' natural inclination to dilute damaging reports with "data dumps" of all in-house toxicity studies. Finally, Congress should extend adverse effects reporting and posting requirements to all regulatory programs in which agencies routinely rely on regulatee-submitted health, safety, and environmental studies.

Because suppression is easily disguised, agencies should implement rigorous enforcement programs capable of detecting and vigorously punishing reporting violations. As we explained in Chapter 5, very little meaningful enforcement of adverse effects reporting requirements occurs under existing adverse effects reporting regimes because the agency enforcers typically devote their limited resources to ensuring adequate compliance with substantive requirements. *We also recommend that in order to send a clear message that data suppression is likely to be a futile strategy, Congress should expand agency enforcement resources and require the agencies to conduct a specific number of unannounced inspections of research facilities and research contractors per year.* Since the information most likely to be suppressed may not be documented and may be difficult to locate, agencies should initiate educational programs on the internet and at research facilities used by regulatees to highlight the types of information that must be reported and the penalties for failing to report. Strong protections for whistle-blowers and generous bounties should be made available to employees who bring unlawful suppression to an agency's attention. Unlike the disclosure requirements of the vigorously enforced securities laws, violations of adverse effects reporting requirements under the health and environmental laws can result in death or serious injury. The Justice Department should therefore prosecute under the criminal laws, companies and high-level corporate officials who are responsible for reporting violations.[87] And finally, Congress should impose more significant civil penalties for failure to report adverse information, perhaps by including explicit causes of action for any victims who have suffered from the suppression of health-related information.

In promulgating reporting rules under the Federal Insecticide, Fungicide and Rodenticide Act, the EPA initially attempted to include pesticide manufacturers' agents, including research scientists, in the class of persons responsible for reporting.[88] The EPA ultimately concluded that it lacked legislative authority to broaden the category to that extent;[89] but Congress could amend this Act and other reporting laws to include these agents explicitly.[90] *We recommend that Congress amend the adverse effects reporting requirements to broaden the category of persons responsible for reporting.* Scientists working under contracts that bar them from reporting adverse effects would then have an overriding legal obligation to report adverse effects, or risk civil and criminal sanctions. Indeed, Congress should further broaden the requirements to include attorneys who learn of adverse information in the course of litigation. This would discourage plaintiffs' attorneys from collaborating with defense counsel in sealed settlements to suppress information that agencies might use to protect future victims, while at the same time avoiding the need to adopt broader reforms of settlement practices merely to address undesirable suppression of this limited body of health and environmental research.

It may, in fact, be desirable to clarify more explicitly, through environmental regulation or legislation, that these statutory adverse effects reporting requirements override contract law. Gag contracts that prevent scientists from conveying important health-related information to the appropriate regulatory authorities are probably void as a matter of contract law on public policy grounds. Unwitting signatories to the contracts, however, must risk multimillion-dollar lawsuits to test this proposition.[91] A statutory or regulatory clarification or even an appellate court opinion clarifying that federal adverse effects reporting requirements override contractual provisions between private parties should help to discourage these contracts and enable informants to speak out without fear of breach-of-contract claims.

Congress could further supplement adverse effects reporting requirements by enacting legislation prohibiting companies sponsoring research intended for use in federal licensing programs from entering into and enforcing gag orders on researchers who conduct the required testing, even when adverse effects reporting requirements are not triggered. A bill introduced in the 109th Congress called the Fair Access to Clinical Trials Act would have accomplished this in the limited case of clinical trials conducted on pharmaceutical products.[92]

Congress could also require sponsors of studies intended for use in premarket approval programs to log individual projects into a public, internet-based registry at the time data are first collected. After several years of experimentation, manufacturers of drugs and medical devices are now required by Congress to register certain clinical research studies at their onset to allow regulators, scientists, and the public to track their fates from start to completion.[93] The resulting permanent registry should discourage sponsors from suppressing or prematurely terminating the applicable clinical studies when results are adverse to their interests. The registry mechanism may, however, be difficult to transfer to research on other potentially harmful products and polluting activities. Since most studies in these areas do not involve human subjects and therefore do not have to pass through a human subjects review process, companies tend to initiate them in a more ad hoc and informal fashion. The variable nature of the research project would also complicate any efforts to specify and enforce clear registry requirements. Finally, given the varied nature of the studies that presumably could be logged into the database, it is not clear that regulators or the public would be willing or able to use such a large registry to identify incidents of premature termination or suppression.

A very different and we believe more promising way to approach adverse effects information gathering is to task regulatory agencies with greater responsibility for learning about adverse effects, thereby avoiding their dependency on sponsors that are often strongly inclined to conceal bad news. The Food and Drug Administration Amendments Act of 2007 (FDAAA) requires FDA to compile and analyze a number of sources of information on adverse effects and events, including data available from Medicare, the Department of Veterans Affairs and other public databases in addition to screening the adverse effects reports in its own regulatory data base on a regular basis.[94] FDA can also require manufacturers to conduct post-market surveillance research.[95] While the Act continues to impose demands on manufacturers to report adverse effects information, Congress evidently concluded that an equally important approach to uncovering information about adverse effects is to make it a well-funded agency priority. The FDAAA's "cadillac" version of adverse effects reporting and analysis may be beyond the current means of agencies responsible for other areas of public health and environmental regulation. It seems clear, however, that empowering agencies to take a more comprehensive and aggressive approach to ferreting out information on the risks posed by the products and activities they regulate, including more rigorous analysis of the adverse effects and related information already in their possession, may be the best way for them to combat the strong incentives that private sector sponsors have to hide relevant science.

The Special Role of Universities in Disclosing Researcher Conflicts

Universities are in many ways better positioned than scientific journals to ensure that the pipeline of scientific information does not become contaminated by advocates' efforts to bend science. In many ways, universities possess the keys to the kingdom of science. Tenure at a good research university is the holy grail for research scientists, most of whom spend many years as graduate students and postdoctoral researchers at universities at the outset of their careers. Even though industrial laboratories often have large pools of resources at their command, the cream of the crop in most fields can be found in the often considerably less commodious confines of university labs. At the same time, companies that find themselves in the midst of political controversy are willing to pay considerable sums for the prestige associated with prominent universities. Thus, for example, soon after several energy companies entered into a $225 million arrangement with Stanford University to fund their Global Climate and Energy Project in 2002, the Exxon Corporation ran an advertisement announcing its alliance with the "best minds" at Stanford to address the critical energy issues, including the "lively debate" over the role of greenhouse gases in climate change. The ad was signed by a Stanford professor and bore the official seal of Stanford University.[96]

After the scientists themselves, the universities bear the primary responsibility for the integrity of the output of researchers that labor within their walls. The grants from federal agencies and foundations that support researchers and their graduate students and postdocs go to the universities, and not to the individual scientists.[97] Most large-scale contracts with private sector entities to support research in universities are signed by officers of the university, not individual scientists. The universities have the responsibility of installing institutional review boards to oversee research with human subjects and implementing procedures for identifying and punishing scientific misconduct.[98] And the universities are just as likely as their errant researchers to receive a black eye when scientific fraud comes to light. Thus it behooves university administrators to pay serious attention to the integrity of what goes on within their institutional bailiwicks and to ferret out and prevent inappropriate intrusions by outsiders.

For several reasons, discussed at more length in Chapter 4, the universities have not played an especially prominent role in ensuring the integrity of policy-relevant science. They have no control, of course, over the large volume of policy-relevant science that does not go on within their laboratories, and it appears that at least some of the policy-relevant research that has in the past been undertaken within the universities may be migrating into the private sector. As noted, much policy-relevant research is rather standardized and uninteresting to scientists operating at the cutting edges of their fields. There is little universities can do to push or cajole them into undertaking or overseeing science that simply does not appeal to them. Nevertheless, a great deal of policy-relevant research does go on within universities, and they are still in a good position to ensure that the research for which they are ultimately responsible meets the highest standards of scientific integrity. The performance of the universities in this regard, however, leaves a great deal to be desired.

The universities are well aware of the desirability of disclosures of potential conflicts of interest. In fact, the NIH and the NSF promulgated rules in 1995 for university grant recipients that require primary investigators to disclose any "significant financial interests," defined as any equity interest of greater than 5 percent and any salary, fees, honoraria, and the like exceeding $10,000 per year from a single company, to their universities. The university must then manage, mitigate, or eliminate any real or apparent conflicts of interest and report them to the funding agency. According to the regulations, a conflict of interest exists when the interest could "directly and significantly" affect research design, the conduct of the study, or the reporting of the study.[99] Other federal granting agencies, however, do not impose similar conflict of interest requirements on their grantees, and the NIH and NSF regulations do not require that the conflicts be disclosed to the public.[100] As we showed in Chapter 4, only a slight majority of universities do in fact make this information publicly available.[101]

To ensure that the public can confidently rely on universities to identify and manage conflicts of interest, universities should require that all researchers report significant financial interests on an annual basis and make these reports available to the public on a website or other convenient location. This is a natural extension of our earlier recommendation that individual studies submitted to the agencies include a full conflict-of-interest disclosure.

The universities can also address, in part, the problem of sponsor suppression. Although the vast majority of research universities have adopted policies forbidding scientists to sign contracts that give private sector sponsors veto power over the publication of project results or the power to delay publication for more than thirty to ninety days, the extent to which researchers actually comply with these policies is not at all clear.[102] One easy way for universities to gain better control over the contents of contracts between sponsored researchers and their sponsors is to review the contracts prior to execution. *To ensure that policies against constraints on publication are enforced, universities should require that all contracts with outside sponsors be cleared with the university's general counsel for compliance with university disclosure policies prior to execution.*

Sponsors that cannot control the publication of research output may still bend science by exerting control over research design and data collection and analysis and interpretation, either as granted explicitly in contracts or as assured by tacit understandings between the researcher and the private sector entity. Professor Lisa Bero, a longtime student of university-industry relationships, notes that even if a scientist is free to publish the results of sponsored research, "you can still get hassled, still get pressure put on you for fear that you won't get any future funding."[103] Although the university's general counsel should be able to police contracts to ensure that sponsors are not explicitly given undue control, it would be very difficult for university administrators to police the day-to-day interactions between university researchers and sponsors, especially when the researcher may in fact have an equity interest in the corporate sponsor. In a 2006 article in *Science and Engineering Ethics,* the head of the federal Office of Research Integrity stressed that when a financial conflict is "very impactful, such as a large stake in the research project," the university has an obligation to "assume direct responsibility for ensuring the quality and reliability of the data," and if that "cannot be accomplished, the institution may need to completely bar the investigator from participation in the study."[104]

The NIH and NSF regulations leave the management of conflicts of interest entirely up to the universities, with little guidance as to how to go about that task. Consequently, the substantive rules are usually quite vague and tend to vary widely among institutions.[105] The universities still face a great deal of market pressure to refrain from placing substantive limits on conflicts of interest arising from a general fear that the best scientists with the most outside funding will simply seek out the schools with the least restrictive policies.[106] The journalist Jennifer Washburn's comprehensive examination of privately sponsored university research concludes that "efforts at self-regulation have been distinctly unimpressive."[107] She quotes the Washington University molecular biologist Garland Marshall, who has founded two outside companies that engage in pharmaceutical research: "There's no way to avoid conflict; there is no way to serve two masters. You can be the best-intentioned person in the world, but there is going to be a conflict."[108]

The difficulty in limiting sponsor control over research is considerably exacerbated by the fact that many major research universities have become collaborators with private sector entities in financing and profiting from the research undertaken within university laboratories.[109] For example, Clemson University in 2002 entered into an agreement with BMW in which BMW contributed $10 million to the university (and a fine car for the president's use) toward the university's planned $1.5 billion automotive research and educational center, in exchange for the right to submit a list of potential professors to be hired, input into curricular decisions, and the right to review student publications to prevent publication of "proprietary information." The university's vice-president for research and economic development acknowledged that the agreement gave the company significant control over research results.[110] British Petroleum Company entered into an even bigger joint agreement with the University of California–Berkeley and the University of Illinois at Champaign-Urbana in early 2007 to give the two universities a total of $500 million over ten years for a new Energy Biosciences Institute that will conduct research on biofuels. Under the arrangement, fifty scientists employed by the company will work on campus with university researchers, but in separate laboratories; the company will have the right to appoint the second-in-command and other high-level officials; and it will have a major say in setting the institute's research agenda.[111]

Material transfer agreements imposed by at least one university on recipients of reagents, gene sequences, and other research tools developed by their own scientists (often with federal funds) have even required the prepublication submission and review of manuscripts to protect the university's intellectual property.[112] Under these conditions, scientists sometimes get into extended disputes with their own universities over their efforts to share scientific information with other scientists and the public.[113] Indeed, many universities have even set aside venture capital funds to finance start-up companies to bring faculty discoveries into the private sector.[114]

The public may legitimately wonder whether it can trust institutions that are so heavily invested in the commercialization of scientific research to police conflicts of interests. If such conflicts persist, the federal government may need to play a stronger role in implementing and enforcing substantive conflicts management rules. One longtime observer, for example, has recommended that the federal government require all researchers who receive any federal monies to post all of their sources of financial support along with patents and equity interests in companies using their research on a convenient website.[115] The strong resistance that the director of NIH encountered when he imposed a similar requirement on scientists working for that institution suggests that university scientists and their institutions would strongly resist such a requirement, making it very difficult to enforce. An added, market-oriented incentive to induce universities to engage in the problem of conflicted research might be to treat private research funding that comes with some strings attached as business income for tax purposes. As long as the private sector enjoys some control over how the research is performed, the philosophy behind tax exemption for university research seems attenuated at best.[116] Before Congress intrudes that heavily on the existing freedoms of university researchers, however, the universities should perhaps be given a chance to show that they can responsibly implement and enforce effective conflict of interest disclosure requirements.

Adjusting the Spin Cycle with Information Disclosures

Since the primary target of advocates' attempts to spin science is the media, the reporters and editors who absorb and interpret the news have a responsibility to examine research carefully and to point out sources of bias. Getting to the root of research conflicts is not always easy, but many journalists are in fact undertaking this responsibility in an admirable fashion. Much of the information we have on advocates' attempts to bend science comes from investigative journalists' painstaking efforts to unravel scientific disputes over policy-relevant science and to probe the bona fides of the contentions of the various participants in those disputes. The coverage by the *Los Angeles Times* of serious conflicts of interest on the part of government scientists at the NIH that we discussed in Chapter 4 is a good example.[117]

Unfortunately, other journalists and editors have yielded to the easy temptation to accept spun science from advocates at face value.[118] Part of the problem may result from the fact that many journalists apparently believe they have an obligation to quote an expert on each side of scientific disputes and to remain agnostic on the issue of whether either expert represents a consensus of the scientific community.[119] For busy journalists who doubt their ability to evaluate the bona fides of scientific disputes, balancing an expert on one side of the dispute with an expert on the other can serve as a convenient substitute for fact checking.[120] William Serrin, a professor of journalism at New York University, worries that "[w]e may be using objectivity as a shield or device that allows us not to be better reporters."[121]

Social scientists have in fact studied this phenomenon in the context of the climate change debate. Maxwell T. Boykoff and Jules M. Boykoff examined coverage of "global warming" in prestigious U.S. newspapers from 1998 through 2002 and concluded that "adherence to the norm of balanced reporting leads to informationally biased coverage of global warming," which has allowed "the US government to shirk responsibility and delay action regarding global warming."[122] A subsequent study of one year's worth of reporting on climate change in 251 newspapers, including many local and community newspapers, concluded that "[n]ot only were there many examples of journalistic balance that led to bias, but some of the news outlets repeatedly used climate skeptics—with known fossil fuel industry ties—as primary definers" of the scientific issues.[123]

With a readily available source of information on conflicts of interest under our proposed disclosure rules, journalists should be able to identify more easily the conflicts associated with research information that comes from scientists who have a financial interest in the outcome of a policy dispute. In an age of electronic communications, where journalists can easily gain access to much of the recent scientific literature on policy-relevant science, they can already examine disclosures in the peer-reviewed literature in evaluating whether to rely on the scientists making them.[124] Our proposal would provide an even more accessible repository for this information. Indeed, if conflict disclosures were federally required, journalists reporting on a heated dispute over policy-relevant science might be seen as abrogating their journalistic responsibility by omitting mention of disclosed conflicts. Even without our proposal, journalists and editors should at least inquire into the funding sources and ideological views that might cause an expert to be biased, and inform readers of serious conflicts they uncover. As two longtime observers of the public relations industry observe, "[e]ven if money does not create bias, it is a leading *indicator* of bias."[125]

Finally, with our disclosure proposal in place, members of the public could also inquire more easily into the provenance of the scientific information that becomes available to them through advertising and the media. One does not have to be a jaded cynic to recognize that scientific information that arrives by way of paid advertising should be taken with a large grain of salt. Members of the public should make an effort to become more sophisticated readers of newspapers and watchers of television news as well. When a talking head or a quoted expert has something to say on a scientific issue, the viewer or reader should pay careful attention to the identification of the institution with which that person is associated. The internet offers a quick and convenient way for any interested citizen to find out who is financing the think tanks, public interest groups, and ersatz scientific organizations that are most heavily involved in spinning science. All it takes is an appreciation for the degree to which advocates can and often do bend science to their own ends and a bit of scientific curiosity.

Conclusion

The legal system has done little to penalize advocates who bend science, while at the same time indirectly rewarding their efforts by taking the results of bent science at face value in determining regulatory standards and liability. The journals, universities, and scientific societies can do a great deal to ensure that advocates do not bend science while scientific studies work their way through the scientific pipeline, but the ultimate solution to the problem must come from the regulators and courts that demand and use these studies after they emerge into the policy realm. The legal system should follow the recent lead of the scientific community by requiring full disclosures of the data underlying research that informs public health protection and any financial conflicts of interest on the part of the scientists who produce that research. The government should also encourage universities and the media, perhaps with some monetary inducements, to take a leadership role in monitoring contractual relationships between researchers and the private sector. The reforms suggested in this chapter are not naive: they are politically and legally straightforward. All that is needed now are a few dedicated leaders who are committed to overcoming several decades of inertia.

CHAPTER ELEVEN

Reforming Science Oversight

Instituting More Vigorous Oversight Processes

For determined advocates, the suite of information disclosure requirements we proposed in the previous chapter present little more than an annoying road bump that they can either negotiate or circumvent on their way to molding policy-relevant science to suit their needs. Indeed, some advocates may simply choose to disclose their conflicts and data and leave it to the courts, the regulators, and the public to sort out the fine distinctions needed to assess the reliability of their submissions. Information about provenance may suggest that decision-makers should view a study or critique with a grain of salt, but it will not necessarily tell them how deeply to discount that study in comparison to studies untainted by potential conflicts of interest. Likewise, access to underlying data can be very valuable to a decision-maker with sufficient time and expertise to engage in an independent analysis and interpretation of the results, but it will not ordinarily play a role in day-to-day decisions of busy governmental bodies and consumers. Therefore, a second layer of reforms is necessary to compel the government, with the help of organized scientific input, to play a more active role in weighing and assessing this information in order to weed out bent science.

Rather than a passive approach that relies exclusively on disclosure, this chapter considers proactive reforms that either circumvent, blunt, or supplement currently skewed participatory processes to yield more balanced oversight and decision outcomes. These reforms are legal in nature, and they span both the regulatory agencies and the courts. As in the case of information disclosure requirements, the goal of each of the reforms is the same—to counteract the dominant and generally unreviewed role that advocates for various special interests play in bending science. This is accomplished in three different ways—first, by instituting neutral science advisory panels that largely circumvent stakeholder influence; second, by blunting some of the tools that stakeholders have historically used to bend science; and third, by nurturing adversarial processes that provide individualized benefits to a diverse group of otherwise dormant affected parties to encourage them to engage in greater scrutiny of the science used in policy-making. We discuss each set of reforms in turn.

In the first set of reforms, we emphasize the vital role science advisory boards can play in providing a neutral, moderating influence in hotly contested debates over policy-relevant science. Science advisory panels, staffed by independent and representative experts and asked to review a specific body of policy-relevant science, can serve as ballast against stakeholder dominance and help to identify and appropriately discount bent science. Professor Sheila Jasanoff observes that these panels can also serve as a stopping mechanism to protect against advocates' seemingly endless attacks on reliable but unwelcome research.[1] While this approach is certainly subject to abuse (a concern that occupied much of Chapter 8), the current system can be reformed in relatively straightforward ways that should ensure more neutral advice in the future. Since scientific advisory panels are expensive and time-consuming, however, we also offer suggestions for improving government peer review processes to provide high-quality, balanced scientific oversight of the remaining policy-relevant research on a less formal basis.

Our second set of reforms focuses on changes to the legal tools advocates currently employ to harass researchers and bend science. The toolbox includes such worthwhile but easily abused legal strategies as scientific misconduct allegations, third-party subpoenas, state open records requests, and defamation claims. Our reforms in this category simply reduce these tools' susceptibility to abuse by placing modest evidentiary burdens on advocates who invoke them and by penalizing advocates who make accusations that turn out to be wholly without merit.

The third set of reforms attempts to achieve substantive change indirectly by infusing inherently imbalanced processes with skeptics who can tangibly benefit, usually through fees or damage awards, by rooting out bent science. In particular, we suggest that several overlapping adversarial processes, rather than one process centralized in a single court or agency, may provide subgroups of otherwise dormant affected parties with greater incentives to challenge bent science. For example, in tort cases, plaintiffs' attorneys can play a valuable (and lucrative) role in challenging sponsored science, and they typically undertake it with a vigor that far surpasses most regulatory agencies' typical investigational efforts. Criminal prosecutors and citizen suits may also encounter bent science, and when they do, these subgroups are peculiarly situated to provide much-needed oversight.

While the legal reforms we propose in this chapter may appear relatively modest, considerable political pressure is nevertheless needed to bring them about. The legal system tolerates abusive strategies for bending science because powerful interest groups benefit from the current state of affairs. Only strong political pressure from those who have an interest in reducing the role bent science plays in legal decision-making can bring about a change in these status quo arrangements. The scientific community may provide some of the necessary momentum to accomplish the first two sets of legal reforms, because it has an obvious stake in the integrity of the scientific advisory committee process and in advocates' abusive use of legal processes to harass individual scientists. The scientific community is unlikely, however, to assume a leading role in advocating the third set of reforms. Other subgroups, like consumer organizations, patients' advocates, and the plaintiffs' bar will have to generate the pressures necessary to overcome existing inertial forces. Whether sufficient political momentum exists to bring about all of the necessary changes is an open question, but that should not deter us from placing them on the legislative agenda.

Counteracting the Dominance of High-Stakes Advocates with Independent Scientific Advice

Given the highly contested nature of policy-relevant science, courts and agencies have increasingly relied on formal and informal mechanisms to recruit respected scientists for objective advice.[2] Because these science advisors have the capacity to isolate "crank" scientists and identify badly biased research, such advisory processes can be especially useful in limiting the impact of sophisticated attempts to shape and attack science. They also provide an invaluable legitimation function by blessing the final scientific conclusions of an agency and thereby immunizing them from illegitimate attack.[3] Currently, agencies can elicit outside scientific advice in several forms, including formal scientific panels, peer review of individual research, and informal assemblies of scientists who offer views on scientific or policy issues. Each of these devices, however, could profit from significant improvements.

Formal Science Advisory Committees Used by Agencies

The adversarial process, as currently structured, rewards advocates who bend original research and discredit unwelcome mainstream views. *In order to counteract this adversarial pull toward extreme positions on policy-relevant science, both courts and agencies should make more productive and vigorous use of mainstream scientific advice.* Drawing on the careful assessment of independent experts will help agencies and courts isolate bent science from other research and thereby reduce their reliance on it. At the same time, this expert oversight could indirectly reduce incentives to bend science by reducing its success.

In most cases, a decision-making process that looks to independent scientists for advice and guidance will require formal science advisory panels—convened by Congress, an agency, or a court. Existing federal rules and processes for convening these panels, however, are badly incomplete.[4] While some organizations, like the NAS, take great care to ensure that their committees are balanced, fair-minded, and do not overstep their charges (a task the NAS undertakes under a fairly intense spotlight and that generates frequent criticism),[5] other science advisory boards—particularly those empaneled by some federal agencies—lack this degree of discipline. As we detailed in Chapter 8, agencies with a political agenda have found the legal elbow room they need to stack science advisory boards to increase the probability of favored outcomes. In fact, Congress became so concerned with the bias exhibited by HHS in selecting advisory panel members that it added a requirement in HHS's appropriations bill that it "cease and desist from" employing a political litmus test in selecting its science advisors.[6] Advisory panels thus have the same potential for abuse as the science-bending techniques they are supposed to be correcting.

By far the most important step in eliciting the advice of a formal scientific advisory panel is the selection process, where the goal should be to ensure that it is made up of respected experts who are free of significant conflicts and, as a group, are broadly representative of the larger scientific community. Accomplishing this is no small feat, but clues for success can be found in organizations, like the Health Effects Institute, that have succeeded in evaluating hotly contested scientific research in a credible way that withstands special interest attack.

THE HEROICS OF THE HEALTH EFFECTS INSTITUTE

When Harvard researchers published the Six Cities Study suggesting that fine particulate pollution led to an unexpectedly high mortality rate, particulate-emitting industries were understandably concerned.[7] A number of affected industries requested the original data supporting the study, but the Harvard researchers refused, because they were concerned that even the redacted data could be used to identify original study participants who had been assured of confidentiality.[8] The refusal only made the industries that much more suspicious, and they warned the EPA against relying on the study for regulatory purposes because its conclusions were, in their view, inherently suspect. After considerable controversy erupted over the study, the EPA commissioned the Health Effects Institute (HEI) to perform a reanalysis of it.[9] Established in 1980 by the EPA and the motor vehicle industry for the purpose of providing "high-quality, impartial, and relevant science on the health effects of air pollution," this institute supports original research, writes reports, and conducts reviews of reports written by others.[10] Its board of directors consists of distinguished academics and nonprofit leaders, most of whom have no ties at all to the auto industry.[11] Its Research Committee, which guides its research projects, is likewise composed entirely of academics of somewhat less senior status.[12]

The Health Effects Institute was an excellent choice to serve as a reviewer of the Six Cities Study, since it was well regarded, both in terms of its expertise and its neutrality.[13] The institute's scientists were also able to view the original data under assurances that privacy-related information would be kept confidential. Ultimately, the institute "found the original data to be of high quality, and essentially confirmed the validity of the original findings and conclusions."[14] Because it was the agreed-on umpire, its by-and-large positive review effectively insulated the Six Cities Study from serious criticism in the ensuing regulatory and judicial proceedings over the EPA's revisions to the national ambient air quality standard for fine particulates.[15] This review, however, was not a simple or cost-free exercise. One institute analyst cautioned that if the job had come at a different time, the review process would have delayed critical regulatory decisions, and he ventured that "most folks would agree [that such an extensive reanalysis] is really rarely justified."[16] ❐

The happy ending for the Six Cities Study controversy provides guidance for future efforts to empanel respected scientists to review contested studies or science-based conclusions. The Federal Advisory Committee Act, which applies to all advisory committees assembled by federal agencies (but not to state and federal courts), requires that advisory committees be "balanced," without providing any real guidance on what that term means in the context of scientific advisory committees.[17] The agencies, moreover, have done little in the way of promulgating regulations or issuing interpretative guidelines to fill in the gaps.[18]

The first item of business should be for the General Services Administration to amend its government-wide regulations implementing the Federal Advisory Committee Act to require all agencies to obtain comprehensive disclosures of conflicts of interest for each panel member serving on a science advisory board that is subject to the Act. The current rules, which were promulgated in 2001, merely make agency heads responsible for assuring compliance with unspecified conflict of interest disclosure requirements unless the agency considers its science advisors to be "special governmental employees," in which case more extensive, but still incomplete conflict disclosures are required by the Office of Governmental Ethics.[19]

The NAS provides a workable model for the scope and specificity of mandatory disclosures. Potential appointees to NAS committees must file conflict of interest disclosure forms detailing relevant business relationships and remunerated or volunteer nonbusiness relationships, present and past government service, research support, examples of relevant articles, testimony, speeches, and any other information "that might reasonably be construed by others as affecting [their] judgment in matters within the assigned" NAS activity.[20] These potential conflicts are vetted at the first meeting of the committee and must be updated during subsequent meetings.

The Food and Drug Modernization Act of 1997 contained a provision requiring the FDA to demand comprehensive disclosures from potential members of its advisory committees.[21] After considerable delay and the threat of a lawsuit, the FDA promulgated draft conflict of interest guidelines that required detailed disclosures from science advisors, including the names of companies with whom advisors have relationships in the nature of consulting contracts, expert testimony, and stock holdings, along with ranges of remuneration for each of these categories.[22] Growing dissatisfaction with the conflicts on the FDA's advisory boards recently led to even more rigid policies governing conflict of interest disclosures and related requirements.[23] Whatever the specifics, more stringent policies governing the conflict disclosures of the scientists serving on science advisory boards are essential.

Second, the agency-convener should strive to assemble science advisory boards that serve as a roughly representative proxy for all of the credible views of the larger scientific community, rather than attempting merely to balance one bias against another in identifying individual panel members. Since there is no map of scientific expertise that one may consult on any particular issue, assembling such a representative group is by no means an easy task. Nevertheless, instead of presenting advisory panel selections as *faits accomplis,* agencies should at least "show their work," by identifying the types of disciplinary expertise they believe is needed and stating their reasons for selecting the individual members from these disciplinary areas.

To ensure better transparency in the process of empaneling science advisory boards, agencies should also publish in the *Federal Register* and on the internet a statement explaining the logic behind the composition of a particular panel and providing information on each panelist, including at the very least any potential conflicts of interest.[24] Substantively, the agency's explanatory statement should have to overcome a strong presumption against appointing any member who has a substantial conflict of interest.[25] Agencies should also provide an opportunity for the public to comment on the nominees and to propose other qualified panelists.[26] Both the NAS and the EPA have greatly profited from their attempts to follow both of these steps.[27] To ensure that special interests do not overwhelm agencies with challenges before they have had an opportunity to promulgate a rule or standard, any judicial review of alleged violations of the requirements governing advisory panels should have to await the challenge to any final agency action that relies on the advisory committee's recommendations.

Since these measures may still be inadequate to gain the public's trust or protect against political manipulation of science advisory processes, agencies could, on their own initiative, go a step further toward ensuring that these panels are fairly representative by asking independent institutions, like the NAS, the Health Effects Institute, or the AAAS to assume the responsibility for selecting the panels to preside over particularly important issues. Since these organizations' survival depends on their reputation for independence and since they are more removed from the political process, they are likely to do a better job of empaneling a representative and disinterested group of high-quality experts than the agencies.[28] In his book on the Office of Technology Assessment, an agency created in the 1970s to provide scientific advice to Congress, Bruce Bimber argues that much of that agency's success in preserving its reputation for scientific objectivity was similarly attributable to its careful effort to survive in a bipartisan congressional environment in which it could never be certain whether Democrats or Republicans would be in control.[29] When it is not already engaged in issuing its own report on the subject at issue, a highly respected, independent scientific institution modeled after the NAS or now defunct Office of Technology Assessment, could serve as an outside consultant charged with mapping the expertise of the larger scientific community and identifying whether a proposed panel would match that constellation of views. Agencies would still be held legally responsible for choosing panels and for ensuring that they were representative and met the balance requirements of the Federal Advisory Committee Act and other legislative requirements. The highly respected contractors would be employed to improve the agencies' performance, not reduce their responsibility.

A second challenge for agencies that use science advisory boards is the threshold question of whether and when to do so. Because they consume a significant amount of scarce scientific resources when properly employed, agencies should use them sparingly and only when needed to resolve an important scientific dispute. In settings where the policy-relevant science is relatively straightforward or uncontested, the laborious task of assembling a science advisory board is wholly inappropriate. Even in cases where the science is hotly contested, the agencies must ensure that differences are in fact resolvable by expert judgment and are not simply disagreements about how the available evidence should be extrapolated for policy-making purposes.

Ideally, the same highly regarded outside organization that assists the agency in assembling a panel could also help the agency decide whether or not one is needed and determine the extent of the panel's involvement. The agency could also promulgate guidelines for using advisory panels, erecting basic presumptions in favor of appointing them in specific circumstances—for example, the EPA's periodic revisions of its national ambient air quality standards or the FDA's approval of a new drug application—and against doing so in other settings that ordinarily do not raise seriously contested scientific questions. In either case, the presumption could be rebutted by a proper showing that such a panel would serve a useful function or it would waste agency and scientific resources.

Once an agency has made a considered decision to appoint a science advisory panel, the literature suggests that it should convene the panel early in the regulatory process to allow it to work collaboratively with the agency staff in reviewing and evaluating the relevant science.[30] The agency could even be required to prepare a plan of study that minimizes any regulatory delay the panel might cause. A presumption that it should add no more than six months of delay to an agency's timeline, for example, could serve as a guideline. If the nature of the scientific issues requires more time, the agency should ask the panel to meet while the agency staff is proceeding ahead with its rule-making effort. Judicial review could be available under the Administrative Procedure Act only if a panel's deliberations resulted in unreasonable delays or impeded meaningful opportunities for public comment.[31]

The agency (with the aid of its consultant) should also take great care in drafting the charge to the panel to ensure that it will not offer its guidance on nonscientific issues that are beyond its expertise and institutional competence. In fact, the panel might be asked, as a matter of course, to highlight areas where the science is badly incomplete and where more research could close the gaps. For example, panels could routinely be charged with "identifying what is not currently known or where current information is too limited for purposes of drawing conclusions" and to identify "research programs that would make meaningful progress in resolving important and relevant uncertainties in the future." Absent careful limits on their charges, these panels could easily be misused as a vehicle for delaying regulatory projects or dodging accountability for unpopular policy choices.

The final step in the process of employing a scientific advisory committee consists of incorporating its advice into the agency's decision-making process. Since the agency must have the final say and since the decision can rest as much on policy as on science, this step is in some ways the trickiest for the upper-level agency officials, especially when their conclusions fall outside the range of the panel's recommendations.

COUNTERMANDING THE CLEAN AIR SCIENTIFIC ADVISORY COMMITTEE

The Clean Air Act requires the EPA to promulgate National Ambient Air Quality Standards at a level that will protect the public health with an adequate margin of safety and to reevaluate them every five years.[32] Since changes to these standards almost always require the states to revise, and the EPA to reapprove, the states' huge, complex implementation plans to ensure that the revised standards will be attained within a decade, the process of revisiting the standards for any given air pollutant is typically a highly controversial exercise of regulatory decision-making at the critical interface of science and policy. To aid the EPA in this regard, Congress required its Administrator to appoint the permanent Clean Air Scientific Advisory Committee and charged it with reviewing critical scientific documents and providing advice and recommendations to the agency leadership.[33] When the EPA revised the standards for fine particulate matter in 1997, it located both the annual and twenty-four-hour standards well within the range of possibilities that this committee had suggested would be scientifically supportable. The agency's general acceptance of this committee's recommendations played a major role in both the EPA's explanation for revising the standard downward to make it more protective and in the judicial opinions of both the Supreme Court and the Court of Appeals upholding that standard.[34]

When the EPA revisited the standards for fine particulate matter again in 2005 (somewhat behind the required five-year schedule), the Clean Air Scientific Advisory Committee recommended that the annual standard remain the same but that the twenty-four-hour standard be revised downward from 65 $\mu g/m^3$ to somewhere in the range of 13–14 $\mu g/m^3$.[35] In late December, the agency issued a Notice of Proposed Rulemaking that proposed to leave the annual standard in place and to revise the twenty-four-hour standard downward only to 35 $\mu g/m^3$, a level that for the first time in the agency's history fell outside of the committee's recommended range of scientifically plausible alternatives.[36] It also fell outside the range suggested by the staff in the agency's Office of Air Quality Planning and Management.[37] The EPA Administrator, Stephen L. Johnson, insisted, however, that the proposal was based exclusively on his interpretation of "the best available science" that existed as of 2002 when the revision was supposed to have been completed.[38] Johnson called the committee's chair to tell her that he had based his decision on scientific advice that he had received from other sources.[39]

The unusual move precipitated an equally unprecedented reaction from the committee and its individual members. One member asked, "What is the point of having a scientific advisory committee if you don't use their judgment?"[40] The committee's chair expressed concern that the scientific advice that Johnson "heard in private trumped the scientific advice he heard in public."[41] Three months later, the full committee decided to write a letter to Johnson expressing its displeasure with the way the agency had dismissed its recommendations. In a conference call to plan the letter, committee members complained that Johnson had "twisted" and "misrepresented" the committee's recommendations.[42] One committee member who had access to versions of the decision documents that had been marked by the White House OMB complained that it had inserted language that was "very close to some of the letters written by some of the trade associations."[43] In particular, the White House had insisted on removing from the agency's public explanation the staff's conclusion that tighter standards "may have a substantial impact on the life expectancy of the U.S. population."[44] In the end, agency decision-makers were unswayed by the committee's advice, and promulgated a final twenty-four-hour standard for fine particulate matter of 35 $\mu g/m^3$.[45] Johnson continued to adhere to his "science-based" rationale for the revision, pointing out that the committee was not unanimous in its recommendations (two of its twenty-two members had supported the EPA's proposal) and that the standard was based on "complex science" about which "[r]easonable people will disagree."[46] He later amended the procedures for promulgating air quality standards to give upper level EPA policy-makers a greater role and to reduce the role that the committee plays.[47] ❒

The EPA fine particulates case suggests that the most important thing upper-level policy-makers can do to prevent public misunderstanding of their attempts to incorporate scientific advice into the decision-making process is to be candid about the extent to which their final decisions rest on science and the extent to which policy considerations fill the gaps left by scientific uncertainties. Indeed, the chair of the Clean Air Scientific Advisory Committee chided Johnson for failing to do precisely that. Rather than diminishing the committee's role, she argued that upper-level agency decision-makers should "explain why their policy choices are different than those approved by [the committee] and should not say it's based on the best available science."[48] Since this requires politically accountable policy-makers to abandon the science charade, however, it is something they are usually quite reluctant to do. Nevertheless, it is the only way the affected public can hold agency decision-makers accountable for their policy choices. Equally important, candor on the part of agency decision-makers about the nature of their decisions at the interface between science and policy may be critical to persuading competent scientists in the future to devote hundreds of hours to an exercise that otherwise appears manipulative in the extreme.

Science Advisors to Courts

Courts and juries are especially handicapped in attempting to ascertain the weight of the evidence in the typical situation in which experts rely on different studies or interpret the same studies dramatically differently. We learned in Chapter 6 that some trial courts exercising their gatekeeper roles adopt a corpuscular approach to the studies underlying expert testimony. This approach reflects the adversarial nature of the trial itself, as the advocates attempt to focus the attention of the fact-finder on the flaws in the individual studies underlying an expert's testimony and not on the validity of the overall conclusions the expert draws from the totality of the evidence. In the context of dueling experts, the corpuscular approach may seem sensible, given the fact that the experts are there because they have drawn conclusions that are consistent with the positions of the parties that hired them. From the standpoint of the fact-finder's search for the kernel of truth in what appears to be a field of flawed science, however, the situation is far from satisfactory.

The judicial system also needs a vehicle for seeking the aid of the scientific community in ascertaining the weight of the evidence on important scientific issues that arise in litigation. One vehicle for eliciting the help of the scientific community is the court's power to appoint one or more independent experts to assess the expert testimony presented in the case and the body of research underlying that testimony.[49] To the extent that courts attempt to empanel special masters or science advisors on hotly contested scientific questions, however, our analysis suggests that judges should proceed with caution. Contested scientific issues benefit from a diverse group of experts and are not easily resolved by a small group or single specialist. Ideally, courts could take judicial notice of relevant scientific advisory board recommendations and findings, rather than starting from scratch. Even when isolating mainstream experts for run-of-the-mill scientific advice in the courts, judges should ensure that an appointed expert is approved by both parties and that the expert's views are roughly representative of the larger scientific community.

Despite these qualifications, however, there are certainly occasions for which it is sensible for a court to appoint an expert or panel of experts. These occasions fall into two general categories. The first consists of cases where relatively routine but clearly ends-oriented science receives very little adversarial scrutiny and very poor information is likely to slip into the legal system as a result. The misbehavior of plaintiffs' attorneys in exaggerating numbers of victims by means of sloppy diagnostic procedures—which occurred at least in the silicosis and fen-phen litigation—provides a ready example.[50] Because the plaintiffs' attorneys enjoyed both asymmetrical information and asymmetrical resources and incentives at the diagnostic stage of enrolling victims, their methods and conclusions escaped oversight in some of the class actions that were destined to settle. To address cases in this category, courts may only need to employ a single technical expert, perhaps the equivalent of a scientific magistrate, to oversee a particular scientific practice. As long as the bent science does not implicate cutting-edge theories but instead is simply the result of adversaries taking advantage of the highly technical and expensive nature of science to outmaneuver their less well-prepared or funded opponents, the selection of a single, independent expert diagnostician should be straightforward.

The second category involves more complex cases, where the science provided by the opposing adversaries is incomplete and contested, and the court's ruling on the scientific questions has broad implications for society. A good example of this type of case is the mass litigation against breast implant manufacturers, which was based on hotly contested and very incomplete evidence of causation. In that litigation, two different district court judges wisely decided to appoint four-member science teams to review the causation evidence underlying the litigation. After both of those panels concluded that breast implants did not cause connective tissue disorders, several courts dismissed similar claims under the prevailing *Daubert* standard without appointing their own panels.[51]

Since the science is unsettled and the information is incomplete in this second type of case, more than one expert is generally needed to ensure adequate diversity of specializations and scientific perspectives. In contrast to the first type, moreover, the selection of these experts is likely to be far more contentious. As a result, the issue of how the experts should be empaneled still needs considerable work. In addition to the challenges that regulators face in developing competent and balanced panels in similar situations, courts face added constraints that further complicate their reliance on such panels. For example, courts must consider whether or by how much a panel will delay the resolution of the case, how to handle ex parte contact, how much the independent experts will cost and who should pay them,[52] and the role the litigants should play in identifying the candidates for these court-appointed experts. All of these questions are significant and deserve far more attention than we can devote to them here.[53]

Peer Review

The peer review process as ordinarily administered in the scientific community asks individual scientists acting in isolation to review and provide written critiques of studies and grant proposals in a form that the original researchers can use to fine-tune and improve their research. Peer reviewers do not communicate among themselves or attempt to reach consensus. They are, however, selected as rough representatives of the larger segment of the scientific community that has expertise in the subject matter of the research.

The peer review process is usually initiated by an entity that has a strong interest in ensuring the quality and integrity of the research, and is based on the assumption that multiple reviews of the same research by scientists with expertise will detect flaws in the design or execution of proposed and completed studies and shield against overly aggressive interpretations of the results. The peer review of scientific research that regulatory agencies routinely employ thus overlaps with, but is distinct from, the process of empaneling advisory boards to review a larger body of research and make recommendations. It is also distinct from the peer review process that the federal funding agencies use in awarding research grants and that scientific journals use to determine whether to publish submitted articles.

Over the years, several federal agencies have dedicated considerable effort to identifying appropriate peer review policies for a range of scientific products.[54] Independent organizations, like the NAS, have also recognized the importance of peer review of policy-relevant science in ensuring a scientifically rigorous underpinning for regulatory decisions.[55] Implementing peer review, however, must be context-specific, and extensive peer review is only needed for policy-relevant science when research deviates significantly from standardized methods, there are reasons to be concerned about the quality of the research, and the particular study is important to a regulatory decision.[56] An informal presumption in favor of more rigorous peer review should thus arise when these three factors are present. Unpublished, private research that does not follow a standardized protocol and that informs important regulatory decisions should receive the most rigorous forms of peer review. Basic research that is federally funded, is published in reputable scientific journals, and has few immediate regulatory consequences would ordinarily involve less regulatory peer review.

When more rigorous regulatory peer review is required, it generally should involve several fair-minded reviewers who are likely to scrutinize the research carefully. Reviewers should disclose all potential sources of conflicts of interest, as discussed earlier for science advisors. Representatives of parties that have an interest in the use the agency might make of a study should not review privately funded research. This criterion may effectively preclude heavy reliance on the internet or "open" review to provide meaningful peer review of policy-relevant science because of the significant risk that it will devolve quickly to advocate-generated critiques, without adequate attribution, oversight, or penalty. Once the reviews have been submitted to the agency, the researchers must have an opportunity to adjust and respond to any public comments, and the reviewers' comments should accompany the study if the agency ultimately uses it to inform a regulatory decision. Peer review in a published journal should act as a substitute for internal processes of regulatory peer review only in circumstances under which the journal also insists on full conflict disclosures for peer reviewers, selects fair-minded and critical reviewers, and allows comments to accompany the final study.

Unlike journal peer review, which is typically carried out at a somewhat leisurely pace, peer review in the regulatory process should be accomplished expeditiously so as not to delay the regulatory process. Judicial review should be available in limited situations to speed up agencies that tolerate unreasonable delays from peer review, just as it is in the case of delays caused by scientific advisory boards. In addition, the result of peer review should ultimately be available to all regulatory participants. In the regulatory context, it may even be appropriate to insist that reviewers be identified by name, rather than serving anonymously, in the interest of overall governmental accountability. While some of these recommendations for peer review are in place in some agencies or programs, no agency to our knowledge has instituted all of the recommendations. In our view, they are all necessary for effective and accountable regulatory peer review.

On the other hand, in settings where a government agency requires not only internal peer review but also formal approval of a study conducted by agency scientists before it can be disseminated or published,[57] alternate arrangements should be devised that allow the author to publish the paper even if it does not pass the high standards set by the agency. In this case, the agency could insist that its comments accompany the paper when sent to a journal and that its refusal to endorse the paper be included with the author's affiliation, but agencies should not be given the power to censor research.

Informal Consensus Statements

Some of the science advice that reaches agencies and the public does not come from formal science advisory boards or peer review but from informal "blue ribbon panels," joint letters, or "consensus" statements prepared by affected parties and others with an interest in particular regulatory activities. The spontaneous formation of organizations or collectives of scientists is growing rapidly as regulatory science becomes more politicized. An organization called Scientists and Engineers for America, for example, recently formed to promote the election of politicians "who respect evidence and understand the importance of using scientific and engineering advice in making public policy."[58] The British Royal Society has written a letter to Exxon demanding that the company withdraw financial support for dozens of groups that have "misrepresented the science of climate change by outright denial of the evidence."[59] Perhaps the most notable example is the June 2005 joint statement by the premier scientific academies of Britain, Canada, France, Germany, Italy, Japan, Russia, and the United States urging their nations to take action on climate change.[60]

As the Beryllium Industry Scientific Advisory Committee and other examples discussed in Chapter 8 suggest, the privately sponsored "blue ribbon panel" is an especially useful device for deconstructing scientific studies on which regulatory agencies might rely. They can provide a patina of scientific legitimacy to an industry's position that company scientists, however competent, cannot bring to the scientific debate. Agency officials may legitimately suspect that company scientists are not entirely free to express their scientific judgment when it leads to conclusions that run counter to the company's economic well-being. The views of the distinguished scientists that make up a "blue ribbon panel" are less easily dismissed, because they typically have a degree of financial independence and are presumably less likely to risk damage to their reputations in the pursuit of a single company's economic interests.

The emergence of industry-funded "blue ribbon panels" and activist scientific organizations presents the intriguing question of how courts and agencies should weigh the advice of such entities, much of which may not fairly be characterized as scientific. As a formal matter, these informal "consensus" statements and recommendations—to the extent they have bearing on a contested scientific rather than a policy issue—should be given only as much credence as their assurance of representativeness and disinterestness allows. If a covenor has taken great pains—demonstrated through an explanation or other accounting akin to the justification provided in assembling a formal science advisory board—to retain an independent and representative sample of the larger set of experts on the scientific question at hand, then their joint statement is entitled to greater weight. If the convenor instead assembles a group of scientists who all share a similar position on a contested scientific issue but does not endeavor to ensure that they are representative of the larger scientific collective, then the agency to which their advice is directed cannot presume that the group is representative. While such a group might provide valuable scientific information or evidence that could affect science policy decisions, their "consensus" view—even when it is joined by eminent Nobel Prize–winners—still represents only a selective sampling of scientists who share similar views on similar issues.

Counteracting the Dominance of High-Stakes Advocates by Penalizing Abuse of Process

As we have seen, advocates are willing to abuse available legal processes by asserting a variety of claims and privileges, not to advance justice or scientific integrity but to harass researchers and discredit credible research. Convinced that the end of advancing their economic or ideological goals justifies virtually any means, they are apparently willing to push legitimate legal tools to their outer limits to make life difficult for scientists who have the temerity to publish research containing inconvenient truths. Since they are operating outside the scientific community, the basic norms that caution against and even condemn these types of activities do not apply to them. Tragically, in many cases the law not only fails to prevent and punish such harassment, but it actually facilitates ad hominem attacks on the integrity of researchers who are engaged in policy-relevant science.

These legally backed, ad hominem attacks on scientists are not just unseemly; they discourage good scientists from conducting research in areas where the results might stir the monster into action. Since the law does little to discourage unfounded attacks on the integrity of scientists and in fact facilitates them, changes in the law are clearly warranted. In redesigning the legal tools most often employed in these attacks, however, it is not always clear where legitimate efforts to scrutinize influential scientific research end and illegitimate harassment begins. Establishing that rough borderline of legitimacy is therefore the first order of business for the reformer, and Congressman Joe Barton's encounter with Dr. Michael Mann provides an enlightening starting point for this line-drawing exercise.

BARTON'S BADGERING

In the abstract, Representative Joe Barton's letter demanding supporting information from Dr. Michael Mann, the University of Virginia scientist who developed the "hockey stick" model for global warming, seems not only within his authority as chair of an important congressional committee, but fully appropriate.[61] Alluding to "questions" that had been raised in a recent *Wall Street Journal* article about the support provided for that model by the underlying data and about the degree to which Mann had made those data available to other researchers, Barton demanded that Mann provide to the House Committee on Energy and Commerce his responses to a detailed and lengthy set of questions. In addition, Barton demanded that Mann respond to specific charges made by Stephen McIntyre and Ross McKitrick in the journal *Energy & Environment* about the accuracy of Mann's conclusions.

Since Barton's committee was deliberating over issues related to climate change, it was important for the congresspersons to feel comfortable with the quality of the applicable science. Apparent disagreements over the reliability of pivotal scientific research understandably concern congresspersons who are responsible for crafting policies that depend on that research. Moreover, Mann's refusal to share the source code of his computer model could cause a policy-maker to become suspicious. Was he trying to hide something? Finally, global warming was a matter of growing public importance, and climate change research was highly likely to inform future policy decisions in Congress and other governmental institutions. From this vantage point, Barton's effort to gain a better understanding of critical facets of Mann's climate change research appears to be a responsible course of action.

Yet mainstream scientific societies reacted in unison and quite vigorously to Barton's request. "Deep" and "considerable concern" from the ranks of the AAAS and the NAS, the two most respected scientific organizations in the United States, coupled with highly critical letters from prominent scientists, made it clear that Barton was not conducting a legitimate congressional investigation but instead was harassing Mann and a few other climate change scientists.[62] Dr. Alan Leshner, the chief executive officer of the AAAS, observed that "in their request for highly detailed information regarding not only the scientists' recent studies but also their life's work, [Barton's letters] give the impression of a search for some basis on which to discredit these particular scientists and findings, rather than a search for understanding."[63]

What exactly caused the scientific community to perceive Barton's inquiry as scientific harassment rather than justified scrutiny? The first indicia of harassment were Barton's thinly concealed insinuations that Mann's research was badly flawed. He began his letter with the observation that "[q]uestions have been raised, according to a February 14, 2005 article in the Wall Street Journal, about the significance of methodological flaws and data errors in your studies."[64] Barton continued that other researchers had "failed to replicate the findings of [your] studies, in part because of problems with the underlying data and the calculations used to reach the conclusions." Noticeably relying on the passive voice, Barton's letter noted that "[q]uestions have also been raised concerning the sharing and dissemination of the data and methods used to perform the studies."[65] It is one thing for a congressperson to ask whether a study has been replicated or whether the underlying data are accessible; it is quite another for him to allege, however obliquely, that the research is suspect before obtaining and reviewing the data and supporting studies.

Second, a great deal of the information Barton sought was irrelevant to any review of the scientific merits of the research. He asked for eight sets of information, including a list of "all financial support," "all agreements relating to [federal] grants," and the "location of all data archives relating to each published study for which you were an author and co-author" over the entire course of Mann's career. Regarding the location of data archives, Barton further asked Mann to indicate

> (a) whether this information contains all the specific data you used and calculations you performed, including such supporting documentation as computer source code, validation information, and other ancillary information, necessary for full evaluation and application of the data, particularly for another party to replicate your research results; (b) when this information was available to researchers; (c) where and when you first identified the location of this information; (d) what modifications, if any, you have made to this information since publication of the respective study; and (e) if necessary information is not fully available, provide a detailed narrative description of the steps somebody must take to acquire the necessary information to replicate your study results or assess the quality of the proxy data you used.[66]

Barton demanded all of this information "pursuant to Rules X and XI of the U.S. House of Representatives," and he graciously gave Mann a full three weeks to comply. Yet many scientists were disturbed by the fact that much of the requested information did not appear useful to any legitimate scientific effort to probe the reliability of Dr. Mann's study. Dr. Leshner of the AAAS, for example, wrote: "we are concerned that establishing a practice of aggressive Congressional inquiry into the complete professional histories of scientists whose findings may bear on policy in ways that some find unpalatable could have a chilling effect on the willingness of scientists to conduct work on policy-relevant scientific questions."[67] In another letter, twenty of the top climate change scientists in the country were even sharper in their criticism of Barton's overly broad demands, noting that "[r]equests to provide all working materials related to hundreds of publications stretching back decades can be seen as intimidation—intentional or not—and thereby risks compromising the independence of scientific opinion that is vital to the preeminence of American science as well as to the flow of objective advice to the government."[68]

Third, Barton was asking for information that was already publicly available. Dr. Leshner's letter noted that "[t]he papers in question have described the methodology as well as the findings, and additional information has been provided in on-line supplements to the papers (an increasingly common practice in the science community)."[69] In his own response to Barton's demands, Mann reiterated in detail both the articles that successfully replicated his work and the publicly available sources containing his data and methodology. Insisting that he would stand by his decision not to release the source code for his computer model, Mann underscored how this decision was based on legitimate intellectual property interests and was in keeping with NSF policies and rulings on the matter.[70] Since other scientists had already "used the methods we described and the data we archived to replicate our results," he maintained that the availability of his computer program had "no bearing whatsoever on the veracity" of the results reported in his study. The question of whether Dr. Mann had "fully satisfied established scientific standards for data-sharing" had in fact "been fully considered [and resolved in my favor] by the National Science Foundation."[71] ❐

Barton's letter illustrates several key features of the distinction between harassment and justified inquiry and highlights why existing legal tools for probing the bona fides of policy-relevant scientific research may require additional checks and balances.

Unsupported Allegations of "Fatally Flawed" Research

Chapter 7 highlights how advocates have strategically employed the Information Quality Act, scientific misconduct proceedings, third-party subpoenas, congressional inquiries, and state public records statutes to intimidate researchers and delay or halt their research. Yet, despite their clear potential for abuse, not a single one of these legal tools requires complainants or requestors to bear more than a very minimal burden of proof in support of their requests or charges. In an ideal world where scientists are the only ones drawing the battle lines and have neither the time nor the inclination to mount frivolous challenges, this type of permissive system may well be appropriate. But in an adversarial legal system, where some affected parties are willing to go to great lengths to intimidate scientists whose work has unwelcome implications for their economic or ideological interests, the trusting style of this legal structure not only tolerates illegitimate harassment, but implicitly condones such behavior.

The current rules for filing scientific misconduct charges illustrate just how naively permissive these processes can be. Despite the fact that a formal scientific misconduct complaint with a federal agency is an extremely serious matter, most universities do not require the accuser to come forward with any information whatsoever to support the complaint.[72] Wholly unsubstantiated allegations are technically sufficient to initiate a process that may go on for years, during which the scientist and (more important from the perspective of the undisclosed advocates bankrolling the challenge) his or her research remains under a cloud. As Dr. Needleman's ordeal illustrates, responding to misconduct charges necessarily becomes the scientist's highest priority, pulling him or her away from the research that motivated the attack. Thus, if moral principles or the minimal risk of sanctions administered by the scientific community do not dissuade a would-be harasser, scientific misconduct proceedings provide a legal vehicle that advocates may abuse in attempts to break down scientists and discredit their research by tarnishing their reputations.

The first step toward solving the recurring problems of harassing data demands and unfounded scientific misconduct allegations is reasonably straightforward and should not be especially controversial. *The institutions that administer the relevant procedures must place a significant evidentiary burden of justification on those who seek to invoke them.* Complainants should be required to provide at the outset objective support for any charges that research is substandard or otherwise flawed. In the case of misconduct charges, for example, the complainant should be required to prepare a report detailing the specific act or acts that constitute misconduct and explaining why the conduct identified was scientifically inappropriate. In addition, any documentary evidence should be attached along with a sworn affidavit vouching for its validity. If the claim depends on facts that are not easily ascertainable from the written record, a list of witnesses should be provided, along with information on how each witness may be contacted.[73]

In the case of demands for information or data, the law should similarly require requestors to demonstrate at the outset that the information is not otherwise accessible and is relevant and necessary to their inquiry, unless the information is already covered by the FOIA. Even if a requestor has alleged in good faith that the requested information is relevant and unobtainable, a responding scientist should be able to dispense with the inquiry by identifying locations where the information can easily be obtained or explain why the information is not relevant to replicating or validating research results. Rather than an eleven-page letter with thick attachments—as was Dr. Mann's response to Barton—a one-page response would suffice to dispense with frivolous data demands. In addition, the scientist should, of course, be allowed to invoke the traditional exceptions to FOIA requests, like the need to protect the privacy of research subjects.

Complainants who lodge misconduct charges or request underlying data (except through FOIA) should also have to disclose specific or general relationships, financial or otherwise, with affected parties who have a financial or ideological interest in the outcome of the challenged research if the proceedings have the potential to advance those interests. Even scientists who are not currently accepting compensation from a source should have to reveal any past arrangements with entities that might be adversely affected by the studies produced by the scientist under attack and any expectation of future compensation. In the case of a nonprofit organization, the disclosure should include a general list of the major foundations and corporations that support the group (or a link to a website containing that information) and a specific disclosure of any of the contributors whose interests might be affected if the agency grants its request or complaint. In the case of scientific misconduct charges, such disclosures should give the entity entertaining the complaint, the scientist who is the target of the attack, and the general public a much better understanding of the ideological or economic interests that may be motivating the attack and a better basis for evaluating its legitimacy. As important, the prospect of disclosure might discourage scientists who do not want to look like mercenaries from filing frivolous attacks. If scientific misconduct complainants fail to identify their connections to affected parties in the face of formal legal requirements requiring such disclosure, they should be subject to sanctions for defrauding the entity receiving the complaint.

No Penalties for Abuse of Process

In addition to facing no evidentiary hurdles at the front end of their data access requests and scientific misconduct allegations, those who abuse those tools to harass their fellow scientists face few penalties at the back end. Even though the tobacco industry's efforts to subpoena Paul Fischer's confidential data were frivolous on their face, neither the companies nor their lawyers received any sanctions, and they did not even have to reimburse the monies Dr. Fischer and his counsel expended in defending against the subpoena.[74] Dr. Needleman apparently defended the unfounded scientific misconduct charges aimed at discrediting his research out of his own pocket. When he ultimately prevailed, there was no day of reckoning for the challengers and the companies who financed their efforts. Despite their failure to prove their charges, the companies accomplished their ultimate goals of delaying Needleman's research for years and casting doubt on his integrity as a scientist.

To prevent the kinds of abuse highlighted in Chapter 7, penalties for frivolous challenges are essential. Creative responses under existing laws may be capable of raising the stakes for those considering abusive filings. In the case of unfounded scientific misconduct charges, for example, scientist-victims may consider filing their own scientific misconduct charges against their harassers. An NAS panel has in fact suggested this possible approach,[75] but it may require the Office of Research Integrity to amend its regulations to state explicitly that a complaint based on allegations of scientific misconduct made in bad faith or without a substantial evidentiary basis is itself scientific misconduct.[76]

Along the same lines, any reforms should include sharper deterrents for overvigorous lawyers. The plaintiffs' attorneys in both the fen-phen and silicosis litigation should be sanctioned for their behavior and even disbarred or subject to fraud charges. Serious sanctions should also accompany other abusive activities. For example, if an attorney files frivolous information quality complaints or overbroad subpoenas, this frivolous activity should be documented and sanctioned.

A truly effective remedy, however, will probably require a statute. In addition to providing a federal claim that awards scientist-victims their attorney fees and compensation for lost time and resources, Congress should also give courts the power to impose punitive sanctions of such magnitude that even economically powerful sponsors will be discouraged from engaging in such harassment in the future. The potential for punitive damages and attorney fee awards should also provide the crucial incentives needed to inspire public interest groups and plaintiffs' attorneys to come to the aid of affected scientists.[77] The American Association of University Professors has recommended that universities provide assistance to any of their scientists who come under attack without regard to the merits of the challenge.[78] The availability of attorney fees and punitive damages should make the assumption of this responsibility more palatable for universities on tight budgets.

Establishing greater penalties for abuse of process, however, comes with its own attendant risks. Because the scientific integrity procedures serve the dual functions of penalizing scientists who do engage in scientific fraud and discouraging scientists from committing fraud in the first place, the evidentiary burdens must not be set too high.[79] Indeed, the availability of sanctions could itself be subject to abuse. For example, a Good Samaritan who whistle-blows on a dishonest scientist might find himself embroiled in a frivolous lawsuit brought by the wrongdoer, and that possibility might deter future Good Samaritans from using these processes. It is therefore critical that the institutions responsible for adjudicating scientific misconduct disputes maintain a sense of balance between the need to maintain scientific integrity and the need to avoid harassment of legitimate scientists who are simply going about their work in a controversial field.[80] Some of the risks can probably not be eliminated. Again, like some of our other proposals, trial and error will likely be necessary to adjust the requirements in order to avoid unintended consequences.

Legal Tools That Do More Harm Than Good

Even with adjustments requiring evidentiary support and penalizing groundless charges, some legal tools, in the aggregate, may do more harm than good. The most obvious candidate is the Information Quality Act and its progeny, most especially OMB's peer review and draft risk assessment guidelines. The Act was passed as a rider with no hearings and no congressional discussion, and researchers have traced its origins to industries eager for a new legal tool to use in delaying agency action and obfuscating credible research.[81] Since the statute is still young and its capabilities are still evolving, it may be too early to conclude definitively that it is, on balance, a net negative addition to administrative process. The information available thus far, however, suggests that the costs of this new tool outweigh the benefits. On the benefits side of the equation, our prior chapters, particularly Chapter 6, indicate that virtually all of the challenges filed against the EPA involve disputes over interpretations, inferences, models, and similar policy issues, and not over the soundness of underlying data. Moreover, there is little evidence that the federally funded scientific information that most regulatory agencies are using and disseminating is unreliable. Since the the Act has been interpreted to exclude a great deal of privately-sponsored research, there is little reason to believe that additional procedures will bring about improvements in agency decisions.[82]

The costs, on the other hand, could be quite significant, and they are already mounting. The Information Quality Act will almost certainly damage the administrative decision-making process.[83] When agencies must implement burdensome procedures to allow challenges to scientific information, the result can only be added expense and delay in the decision-making process. Information Quality Act requests can tie up agency staff without whose expertise the regulatory process cannot move forward. The lengthy requests filed to date prove the case, and recent estimates of agency resources invested in this process confirm that these requests are often not trivial in their impact on the agencies.[84] The threat of such requests could even discourage agencies from relying on scientifically valid but imperfect studies in applying the weight-of-the-evidence approach scientists typically employ. It may also discourage agencies from conducting meaningful, publicly accountable self-evaluations of their scientific findings as new knowledge and information arise over the lifetime of a standard or rule for fear of triggering additional Information Quality Act complaints.[85]

A relatively compelling case can therefore be made for repealing the Act.[86] If Congress declines to repeal it, however, the OMB and the other agencies subject to it should put procedures into place to discourage its abuse by advocates bent on bending science. This would involve at least the adjustments recommended earlier for the other parallel legal tools—requiring evidentiary support for complaints and penalizing correction requests filed primarily in order to harass or delay. OMB should also consider narrowing the exceptions that currently insulate a considerable body of privately-sponsored research from the reach of the Act.

Ensuring a Rich Mix of Adversarial Pressures to Encourage Greater Scrutiny

In well-functioning areas of independent science, scientists are skeptical of new discoveries and examine research critically. In the area of policy-relevant science, as detailed throughout this book, the opposite is often true as bent science surreptitiously flies under the radar. Scientists, the diffuse general public, and even experts in regulatory agencies may show little interest in bent science. Consequently, advocates who distort science to produce outcome-oriented research and research critiques often receive little serious scientific scrutiny. Empaneling distinguished scientists provides one way to provide some oversight, but that is always an expensive and time-consuming endeavor that in the end is hardly fail-safe. To ensure that policy-relevant science receives a more skeptical reception in the legal system, we therefore urge Congress, the agencies, and the courts to enhance the adversarial procedures through which representatives of a diverse set of interests can expose and counteract bent science in the legal system.

In advocating increased reliance on adversarial procedures, we are suggesting, perhaps counterintuitively, that the way to reduce the impact advocates have in the realm of science is to enhance the role they play in the legal realm by increasing the range of affected interests that have access to the relevant legal procedures. The goal for this suite of legal reforms is therefore to restructure existing adversarial processes—by easing access and offering the lure of damage awards or even attorney fees—in ways that catalyze otherwise dormant subgroups of affected parties to engage in effective oversight of policy-relevant science. Simply adding adversarial processes without ensuring that they will be staffed with skeptics will not only fail to accomplish this goal, but may move the legal system in precisely the wrong direction. That is our primary concern, in fact, with the Information Quality Act, an added adversarial process that is useful primarily to the same economically powerful groups that already dominate agency decision-making. Our proposed enriched adversarial processes therefore strive to provide inducements to those entities that are likely to approach sponsored science skeptically, but might otherwise not benefit from investing time and energy in overseeing potentially bent science.

This more balanced approach to oversight of policy-relevant science also recognizes—as scientists do—the advantages of multiple opportunities for scrutiny. Within the realm of science, a scientist's research is scrutinized at multiple points, beginning with presentations at scientific conferences and peer review by journals and extending through letters to the editor and postpublication scrutiny (and sometimes replication), all the way through tenure and even posttenure reviews. Each stage involves different independent scientists and different standards of scrutiny, and each stage is important for ensuring the overall reliability of scientific research. The legal system can mimic these multiple opportunities for oversight, in part, by establishing and maintaining a number of points in the legal process at which policy-relevant science is subject to scrutiny and evaluation by a diverse group of skeptical parties. Indeed, history reveals that many of the worst abuses arise when a single interested party faces powerful incentives to bend the relevant science and both the opposing parties and the legal decision-maker lack the resources to uncover and counteract the resulting manipulation.

When the civil, criminal, and regulatory systems all engage in overseeing bent science simultaneously, the chances that bent science will be caught and rejected should increase, perhaps dramatically. Each of these institutional processes relies on different players, focuses on different issues, and uses slightly different adversarial mechanisms to root out facts. Thus, even to the extent that each of these separate legal institutions remains incomplete or even badly broken, when they all converge on the same bent science, logic and experience suggest that more bent science will be identified and addressed. *Long-term reform of the legal system therefore must employ a diversity of institutional approaches to ensure multiple points of scrutiny of policy-relevant science by a diverse group of skeptical parties.* This is important not only to provide multiple windows on bent science but also to ensure that no single interest can monopolize either the decision-making process or the scientific information used in that process.

Tort litigation can both highlight the frailties of the regulatory system and play a critical role in uncovering evidence of malfeasance. In the case of drug regulation, for example, the FDA "has neither the legal authority nor the resources to effectively identify the adverse outcomes caused by drugs already on the market."[87] Consequently, "civil lawsuits have become the primary means for protecting the public from unsafe drugs."[88] Litigation against several drugs, including the weight loss pill ephedra[89] and the sleeping pill Halcion,[90] in fact proved of critical import in focusing the FDA's attention on those products and in revealing internal company knowledge about possible harms that was not shared with the agency. The litigation in both cases triggered more vigorous regulation and ultimately led the FDA to withdraw the products.[91] In other cases, such as the litigation brought against the manufacturers of Vioxx and Prozac, documents produced during discovery revealed additional information about the manufacturers' knowledge about product harms and has led to increased public demand for more vigorous oversight of drug manufacturers generally.[92]

Litigation has also prompted regulatory action in several site-specific cases. Litigation brought by neighbors against the DuPont Corporation for damages caused by a chemical, PFOA, that one of its large facilities used in making Teflon, for example, unearthed internal company documents revealing that the company had long been aware of both the hazards posed by the chemical and its presence in nearby drinking water.[93] As we note in Chapter 5, it was only because the plaintiffs' attorney shared these documents with the EPA that the agency was able to bring an enforcement action against DuPont for its failure to report those adverse effects under the Toxic Substances Control Act. The EPA's case was settled for a fine of $16.5 million, the highest fine the agency has ever recovered for a reporting violation.[94] The litigation and media coverage surrounding the infamous Woburn, Massachusetts, hazardous waste site featured in the book *A Civil Action* was at least partially responsible for a Justice Department investigation that ultimately resulted in an indictment of the W. R. Grace Corporation for providing misleading information to the EPA.[95] Attorneys for workers suffering from serious lung ailments in a butter flavoring factory were instrumental in raising the visibility of the toxicity of the chemical diacetyl to regulators and the broader public (discussed in Chapter 5), even though this flavoring should have been regulated under several different statutes years before.[96] Although one of the attorneys began sending letters to the OSHA in 2001 demanding an investigation of the outbreak of the rare and debilitating lung disease that this chemical caused,[97] it took a $20 million verdict to generate sustained attention by the press and lawmakers to the workplace risks posed by diacetyl.[98]

In the areas of consumer and health protection, the tools available for tort litigants and even some criminal prosecutors to pry out suppressed internal documents and elicit damaging testimony from company officials and employees far surpass those available to most regulators. Plaintiffs' attorneys, for example, can compel companies to produce documents, including emails, in civil litigation that agency officials could not access even if they had the time to devote to the effort.[99] Although such documents are typically subject to protective orders, they can find their way into the public domain when cases go to trial and when public-spirited plaintiffs and their attorneys insist that documents be made public as a condition to settling litigation. The Minnesota attorney general Hubert Humphrey III did the American public an enormous service when he demanded that all discovered documents be posted on a publicly accessible website as a condition to settling the litigation his state had brought against the tobacco industry.[100] Surprisingly, most of the federal health and environmental protection agencies lack this basic power to subpoena corporate documents. When Dr. David Kessler took over as the commissioner of the FDA, he was surprised to learn that the agency "could not subpoena witnesses and documents."[101] Recent scandals involving the Cox-2 inhibitors and the SSRI antidepressants suggest that long-forgotten efforts to vest the FDA with the power possessed by any novice private attorney should be revived.[102]

The tort system can succeed where regulation fails because it presents the potential for large damage awards for a small subset of the affected public who would otherwise be unrepresented since the costs of participation in the regulatory process are so high.[103] Plaintiffs' attorneys may have an even greater stake in the litigation because of the large sums they invest in it, the possibility of large aggregated awards at the end of the day, and the ultimate prize of lottery-sized punitive damages if they uncover evidence of serious misconduct in the defendants' files.[104] These incentives in turn fuel a powerful discovery engine that has few equals in the public sector.

Victims and plaintiffs' attorneys are not the only players in this enforcement effort, however. Criminal prosecutors or state attorneys general may perceive significant legal and political gains from bringing state claims against some science bending activities. Elliott Spitzer, the attorney general of New York, for example, brought a lawsuit against GlaxoSmithKline for consumer fraud in withholding negative information about the safety of Paxil for adolescents.[105] Indeed, in some cases—where suppression leads to premature deaths or serious bodily injury and that information is withheld knowingly, for example—there is a possible homicide or manslaughter charge against corporate executives.[106] In a setting where the federal regulatory apparatus appears to be deadlocked as a political or legal matter and tort claims on behalf of deserving victims are not promising, city or state prosecutors can fill a gap in deterring bent science through criminal or state common law claims.

To counteract the natural trend toward capture of the regulatory machinery by the very industries it was designed to control,[107] then, the criminal, tort, and regulatory systems must simultaneously and perhaps somewhat redundantly address public health and environmental concerns in a fashion that is not unlike the constitutional checks and balances that the framers put in place in the U.S. Constitution. It should come as no surprise, however, to discover that heavily regulated industries are generating a great deal of pressure in the other direction. Well aware of the threat the tort system presents to their ability to dominate the regulatory process, regulated industries have persuaded several states, including Michigan, to enact legislation severely curtailing product liability actions against drug companies with FDA-approved products.[108] The Bush administration has aggressively invoked the power of the federal government to preempt state tort law claims in the case of drugs and medical devices that have received the FDA's approval.[109] These efforts not only have the potential to cut off a primary vehicle for holding companies accountable for harms caused by their dangerous products and activities but also will eliminate an important source of information about the extent to which regulated entities bend regulatory science.

A combined system in which criminal, tort, and federal regulation operate in tandem on complex public health problems also has the potential for productive institutional cross-fertilization. For example, when agencies empanel scientific advisory committees, their analyses, conclusions, and recommendations might usefully inform a court's assessment of the reliability of the scientific basis for expert testimony in tort litigation. While the courts should not necessarily take "judicial notice" of the recommendations of these formal science advisory panels in the sense of vouching for their accuracy, the panel reports could be entered as evidence to help inform the jury and judge. Likewise, conflict of interest disclosures and peer review reports might play a useful role in tort litigation. Conversely, more contemporary tort claims, such as those proposed by Professors Margaret A. Berger and Aaron D. Twerski for dignitary and emotional harms arising from a manufacturer's negligent failure to provide consumers with informed consent on the undisclosed risks of a nontherapeutic drug, could supplement regulators' efforts to encourage companies to share information about the adverse effects of their products.[110] Similar adjustments to tort law aimed at redressing unjustified uncertainty regarding product risks by reversing burdens of proof may also be warranted in limited cases. Such reforms will enhance the information base for both tort law and related regulatory programs.

Other distinct subgroups of affected parties could also act as catalysts for greater scrutiny of bent science. One very active subgroup consists of nonprofit and community groups that monitor important regulatory proceedings and can often elicit help from sympathetic academic scientists. They can also file "citizen suits" to enforce statutory and regulatory requirements and common law nuisance actions seeking abatement of risky activities. These lawsuits in particular have the potential to weed out scientific distortions that regulatory agencies might otherwise tolerate. While the resources available to public interest groups cannot match those of the companies and trade associations that often dominate regulatory proceedings, they have in the past been quite adept at deterring federal agencies' attempts to bend science.[111] Their impact on regulated entities' attempts to bend science has been less noticeable, and they have even been guilty of spinning science on their own behalf. Reforms aimed at increasing access to the regulatory process would capitalize on the positive role these groups can play, and they should encourage that role by providing, at a minimum, attorney fees and expenses to groups that successfully uncover fraud and manipulation in the use of policy-relevant science.

Harassed scientists are also among the subgroups that actively oversee bent science and, as discussed in the prior section, are arguably due compensation for the lost resources and reputational damage they suffer from advocate-sponsored attacks. Researchers at the receiving end of abusive scientific misconduct charges, subpoenas, public records requests, and good science complaints should be able to draw on an available set of legal counterclaims entitling them to damages, including punitive awards, for any delays or adverse impacts on the progress of their research.[112]

The universities find themselves in an untenable position under the current Office of Research Integrity requirements, which make them responsible for adjudicating scientific misconduct claims against the scientists they employ. Rather than abandoning scientists who become targets of such charges, as they may be inclined to do under the present system, they should create an additional office (perhaps within the general counsel's or ombudsman's office) that is separate from the investigational and adjudicatory apparatus to provide legal counsel to the target scientists. Some universities offer a great deal of support to scientists who become targets of defamation and product disparagement actions and abusive subpoena and open records requests, but others have played a passive role. In those contexts, an academic employer has a special obligation to defend academic freedom by offering legal services to scientists who become unwitting victims of harassment. Courts could further encourage universities to play a more proactive role by awarding academic scientists some form of indemnification from university employers that leave them to fend for themselves in defamation actions and harassing lawsuits brought by litigious advocates.[113]

Many of the case studies in this book highlight the vital role that whistle-blower subgroups play in exposing bent science. In contrast to other subgroups, these whistle-blowers might become more engaged if their commendable efforts entailed fewer personal costs. For example, in order to make whistle-blower reports more streamlined and less risky, Congress could empower a single agency, perhaps in one of the major research funding institutes, to receive anonymous complaints or tips about bent science. Since the complainant's identity would not necessarily be disclosed, the information contained in whistle-blower reports could not be taken at face value because of the countervailing risks of abuse by adversaries of the whistle-blower procedures. The receiving agency would therefore need to investigate and evaluate the bona fides of such complaints before relaying those that cross a threshold of plausibility to the relevant regulatory agency or court for further action.

Conclusion

Because only the most intensely affected and well-funded entities are likely to participate consistently in judicial and regulatory decision-making procedures, the legal system can become increasingly dominated by a single perspective that in turn undermines its capacity to elicit and rely on the best available scientific information and expertise. The future integration of science into environmental and health policy, in both the regulatory and common law contexts, must anticipate the areas where abuse and misuse of science are most likely and institute more proactive and productive processes to minimize attempts to bend science toward the economic or ideological ends of the affected parties who have the most to gain or lose from the legal system. When the current legal regime is prepared to penalize harassment, encourage balanced and representative scientific advisory boards, and create beneficial outlets for a diverse set of skeptics to engage in scientific oversight, it will be in a much better position to deter bent science. The legal adjustments we have recommended in this chapter, however, represent only a beginning in what should become a continuing effort to improve the way agencies and courts use policy-relevant science in health and environmental decision-making.

CHAPTER TWELVE

Final Thoughts

A Broader Perspective on the Problem and the Prospects for Change

At the outset of this book we took a step backward from the individual accounts of scientific deception in regulation and litigation to gain a broader perspective on the phenomenon the reader should by now easily recognize as "bending science." From this broader perspective it quickly became apparent that the phenomenon was not limited to isolated instances of junk science propagated by miscreant scientists and rogue advocates, but rather extended to almost every nook and cranny of health and environmental science and policy-making. Relying on the "separatist" conception of two distinct and largely unrelated worlds of science and policy, we discovered that the advocates have penetrated into that corner of the world of scientific activity where scientists create, publish, and critique policy-relevant science. Worst of all, the independent scientists who might otherwise be expected to monitor and interdict these intrusions by outcome-oriented advocates are often nowhere to be found.

The detailed descriptions in the remaining chapters of the various strategies that advocates employ to bend policy-relevant science should have inspired some degree of outrage among readers unfamiliar with the terrain. Other readers who readily accepted the separatist view of science and policy may also have accepted the correlative proposition that health and environmental regulation and litigation should be driven by "sound science." The discovery that an unidentified, but potentially significant, portion of policy-relevant science is not only unsound but also manipulated by advocates in outcome-oriented ways clashes so dramatically with the separatist formulation that some readers may be left wondering what to put in its place. More jaded readers who needed little convincing that advocates have successfully intruded into policy-relevant science may have taken away a better understanding of the breadth and depth of that penetration. But the descriptions may also have left many readers with a sense of hopeless resignation that was not significantly relieved by the relatively modest reforms suggested in the last two chapters, especially when viewed in light of the limited prospects for implementing many of the most aggressive suggestions in a world that is still dominated by the advocates.

It is therefore proper in this final chapter to take still another step backward to gain an even broader perspective on health and environmental regulation and litigation and the role that science plays in those processes. From that broader perspective, it becomes clear that as serious as the problem of bent science is for rational and effective policy-making, it is by no means insurmountable. In fact, the relevant legal institutions have already invented effective ways to bypass science altogether in environmental regulation and to a somewhat lesser extent in product regulation and litigation. The lesson offered from this broader perspective is that it is often necessary to ignore the siren calls for "sound science" in health and environmental policy-making, many of which come from the same advocates who are busily bending the science that would drive the policy-making process, and instead forge ahead by doing the best we can with available risk-reduction methods and technologies. Beyond that, there is good reason to believe that simply shining the spotlight incessantly on bent science will inspire both the well-meaning scientific community and the relevant policy-making institutions to take the necessary steps to ensure that the science these institutions do use in decision-making is both objective and of high quality.

Doing the Best We Can

An important lesson from this broader view is that it will be very difficult to keep advocates away from policy-relevant science; rather than being content with reforms that attempt to better manage the nature of this infiltration into the science pipeline, then, we may also need to develop better methods for bypassing science altogether when developing policy. Long ago, Congress recognized in the area of environmental law that if policy-makers are not able to sort the good science from the bad, the answer is not to do nothing at all. Indeed, many protective regulatory programs in the United States in fact survive and even flourish without science.

Although this is not the place for a detailed description of the regulatory regimes in which agencies like the EPA promulgate national standards governing risky products and polluting activities, it is worth noting that a large literature has documented the simple observation that "sound science" is not necessarily a magic bullet for better environmental and health policy.[1] Science, in truth, is only one of many factors that inform policy decisions about health and environmental protection. This is not because politics frequently overwhelms the regulatory decisions. Instead, it is often because there is simply not much scientific guidance available. In many settings, neither scientific theory nor a sufficiently robust set of scientific studies exists to help policy-makers assess the impacts of products or wastes on health or the environment.[2] Even when policy-relevant science does exist, it may be incapable of providing definitive answers to the highly general questions policy-makers often ask. In 1980, for example, Congress funded an exemplary panel of independent scientists to educate it about the types of emission reductions that were needed to control acid rain; ten years and $500 million of research later, the scientific group was still not ready to answer the most important of Congress's questions.[3] Science simply could not provide the kind of guidance Congress needed.

Moreover, even in settings where a body of policy-relevant science is available, that research can just as often impede as expedite efforts to achieve the consensus needed to drive science-based regulatory decisions. We saw in Chapter 6 that when the stakes are high, advocates will generate challenges to policy-relevant science, regardless of its quality, and the effect of those challenges is often to delay the resolution of important policy questions, sometimes for decades.[4] Even if regulators could isolate and remove the bulk of the badly biased science and critiques, a good deal of uncertainty would undoubtedly remain in the studies that survived for advocates to fight over. Agencies can even use science to obfuscate economic and policy factors that ultimately end up playing a decisive role in a final policy decision.[5]

Early in the evolution of the major environmental laws, Congress acknowledged the failure of its ambitious initial efforts to enact science-driven regulatory programs, and it replaced them for the most part with science-blind approaches to pollution control.[6] In some science-blind programs, the regulatory agency requires polluters to reduce their discharges by employing the equivalent of the best available pollution control technology, irrespective of the needs of the receiving environment.[7] In others, like the acid rain program, the agency requires polluters to reduce their discharges (or purchase equivalent reductions in a market) by a set percentage from a historic baseline so as to ensure progress in reducing the overall amount of pollution.[8] Indeed, some of our nation's greatest environmental successes originated from science-blind regulatory programs, including the almost miraculous transfiguration of Lake Erie, radical reductions in industrial emissions of air toxins, and the development of sophisticated computer-based emission reduction systems in automobiles.

Courts have frequently adopted a similar science-blind approach to litigation in the related field of nuisance law. A court in a common law nuisance action requires the plaintiff to establish that the defendant has engaged in activities that constitute an unreasonable interference with the plaintiff's use and enjoyment of property.[9] Once plaintiffs meet that threshold requirement, courts have frequently adopted science-blind approaches toward determining the remedy. Instead of requiring the defendant to reduce the level of the interference with the plaintiff's property to just below the point at which it no longer constitutes a nuisance, the courts typically require the defendant to modify its activities in specific ways or to install the best available pollution-control technology.[10]

Some regulatory programs could not be implemented in science-blind fashion, but might be reconfigured to be more science-limited in their approach. Currently several health protection statutes, like the Occupational Safety and Health Act, require the agency to establish that the risk posed by the product or activity passes a vaguely defined threshold of "significance."[11] Others, like the Food Quality Protection Act, require the agency to establish exposures that provide a "reasonable certainty of no harm."[12] The agency then has the burden of supporting the relevant threshold finding with scientific studies and analysis. In these programs, a more science-limited approach would task regulators only with the responsibility of setting standards or tolerances based on a reasonable, worst-case prediction of the risks using the scientific evidence available at the time. Regulated parties would then be able to rebut this worst-case starting point with their own, high quality science that reveals that pollution or products are safer than supposed. Such an approach is roughly modeled after the "penalty default" rules pioneered by Ian Ayres and Robert Gertner, which encourage actors to reveal superior information by assuming the worst about them as a default.[13]

Not all public health and environmental regulation and litigation can proceed in even this more science-limited fashion, however. The Food, Drug and Cosmetics Act requires the FDA to balance the risks posed by pharmaceutical products against their therapeutic benefits in deciding whether to approve them and determining the lawful conditions of use. In this narrow category of regulated items, the drug may pose serious risks to users, but the overall public health risks may be greater if the agency disapproves it.[14] It will also be difficult for courts to circumvent science in products liability and toxic tort cases where the plaintiff has the burden of proving that the defendant's product was a cause-in-fact of the plaintiff's damage. The courts in these cases typically require plaintiffs to present epidemiology studies showing that the relevant product or byproduct has the capacity to cause the plaintiff's disease as well as specific expert testimony concluding that the exposure caused the plaintiff's particular disease.

Thus, courts and agencies that can realistically adopt science-blind approaches should continue to avoid the serious bent science problem that afflicts science-driven approaches, and those that have the option of moving to science-blind approaches should do so until more effective strategies for identifying and discounting bent science become available. This will allow regulation and litigation to proceed in a timely fashion and at the same time avoid wasting scarce research resources on producing, critiquing, and replicating ends-oriented studies. For the remaining programs and for other situations in which the relevant legal or political considerations require the decision-maker to rely on science to support important public policy determinations, the solutions to bent science will have to lie in reforms of the sort discussed in Chapters 10 and 11.

The Power of the Spotlight

The highly publicized debates over global warming and several equally prominent expressions of outrage by the scientific community over attempts by the Bush administration to suppress scientific assessments of government scientists indicate that the scientific community appreciates the need for scientists to monitor and to focus public attention on inappropriate attempts by advocates to manipulate policy-relevant science. While ad hoc interventions of this sort will not be sufficient by themselves to purge the system of the corrupting influences of the advocates, the developments support some degree of optimism about the future integrity of policy-relevant science. The most important step toward solving a problem is often the simple recognition that there is a problem.

By focusing public attention on science bending practices and providing a strong expression of the importance of independent scientific assessments in areas of public policy that rely on science, these spontaneous actions should inspire Congress and the relevant federal agencies to put into place a legal regime that will have the effect of shining a continuous light on what has in the past been a dark corner of the scientific arena. Well-publicized congressional hearings in early 2007 on scientific integrity in public policy-making offer grounds for optimism in this regard.[15] As we discussed in more detail in Chapter 10, full implementation of legally mandated disclosure requirements should also help to expose bent science. Like the crises that led to strong majoritarian support for environmental legislation in the 1970s, strong currents of broad public outrage, perhaps combined with the powerful voices of scientists, may be capable of overcoming the monopoly special interests have enjoyed for too long over these esoteric issues.

Even if it does not precipitate additional legal reforms, legally mandated disclosure will have a salutary effect on policy-relevant science. Many effective regulatory programs in fact consist primarily of legal requirements that have the effect of shining the light on shady practices. The best known of these, perhaps, is the requirement of section 10(k) of the Federal Securities Act that companies file public reports containing audited financial statements and certain other kinds of information about the company that the Securities and Exchange Commission deems relevant to the company's economic well-being.[16] The Emergency Planning and Community Right to Know Act requires companies to file similar annual reports containing information on their releases into the environment of listed toxic substances in amounts greater than those specified by the EPA.[17] When faced with the prospect of full disclosure, sponsors may feel less inclined to insist on contractual terms allowing them to control research in outcome-oriented ways, scientists will feel peer pressure to resist sponsors' attempts to control outcomes, and universities will be more inclined to resist such contractual terms in the first place.

Greater public appreciation of the social cost of bent science may generate public support for more comprehensive solutions. One encouraging development in this regard is the enactment of the Food and Drug Administration Amendments Act of 2007, which gives the FDA considerably more power to oversee the safety of approved drugs. The greatly expanded postmarketing surveillance program will be financed by annual fees collected from the manufacturers.[18] Even though greater protection will have the predictable effect of raising drug prices and increasing public expenditures on Medicare and Medicaid, public support for providing greater power and resources for FDA to monitor drug safety is strong.[19]

A much more ambitious step would be to establish a centralized testing regime, paid for with mandatory annual assessments on the entities that benefit from the research. Centralized testing takes the discretionary testing out of the hands of the private companies and locates it in governmental agencies that have no particular interest in research outcomes. Several prominent commentators, including Sheldon Krimsky and the former *NEJM* editor Marcia Angell, have called for this type of federal control of all basic research used to determine the safety of drugs, pesticides, and other potentially dangerous products to "serve as a firewall between the drug companies and the researchers who study the safety and efficacy of their products."[20] Under this proposal, the government would collect fees from the relevant companies based on "the real costs of carrying out a clinical trial," select qualified researchers to undertake the clinical trials, and oversee the conduct of those trials to ensure adequate protection of human subjects and compliance with "high ethical standards while protecting confidential information."[21] While this is not the place for a detailed discussion of this proposal, it offers a realistic alternative to the current legal regime, which relies too heavily on the products of sponsored research.

Greater public understanding of bending science should also expose the illicit nature of some current reform proposals, including efforts by advocates to capitalize on the current, misguided separatist view. These advocates summon the separatist view to help insulate their bending activities from oversight. Onlookers who believe in the separatist view will assume that the output of science is pure and that more science and more peer review are the best solutions. But advocates also use this misguided separatist conception to persuade policy-makers that more "sound science"—more evidence and more opportunities for unrestricted peer review—should procede the rush to regulate. By contrast, not a single "sound science" reform has ever included a recommendation to fortify existing regulatory programs against common science bending practices. None of the recent "sound science" reforms would require basic disclosures of conflicts of interest, data access to privately sponsored policy-relevant research, balanced and transparent processes for appointing scientists to federal advisory boards, stronger correctives to deter suppression of relevant studies, or reforms of legal tools that advocates have abused to harass scientists.[22] These omissions from the "sound science" arsenal highlight the conceptual deficiencies in this suite of reforms and appear to betray the ulterior motivations of their proponents.

Taking the Next Step

Although the legal system is in many instances capable of adapting in ways that allow it to protect public health and the environment without resorting to science, bent or otherwise, avoidance is not always an ideal solution. Most of us would agree that when the weight of the independent scientific evidence points in the direction of a particular regulatory or litigation outcome, legal decision-makers should give very serious consideration to that result. Decisions based on solid scientific evidence and rational analysis are usually superior to those based on hunch and supposition. Furthermore, the legal system cannot easily circumvent the need for a solid scientific basis for deciding some important legal issues, like the risk-risk trade-offs that arise in pharmaceutical litigation and regulation. It therefore behooves us to search for ways to redirect the considerable energy and resources that are currently being devoted to bending science toward more productive uses that will increase the flow of policy-relevant research that is free of outcome-oriented biases.

Unfortunately, existing legal rules and procedures provide powerful incentives for sponsors to bend science while at the same time generating few countervailing incentives for sponsors to commission independent research and for other important actors to expose and discount bent science. This state of affairs did not come about wholly by accident. The current legal regime reflects advocates' demands for "sound science" without acknowledging the corresponding need to curb opportunities for bending science. The perverse incentives to bend science will not change until the legal system changes, and the legal system will not change until an attentive public, catalyzed by an agitated scientific community, demands it.

Fortunately, we are seeing signs in the global warming debates and the public ferment over prescription drug regulation that this may in fact be happening. The key to success will be a shift in the public understanding of the role science plays in regulation and litigation from the prevailing separatist view to a more nuanced view that recognizes the role of advocates in bending science to predetermined ends. In this regard, both the scientific journals and the popular news media have a responsibility to probe the provenance of policy-relevant science and to expose to public view the extent to which advocates' efforts have contaminated the process of generating policy-relevant science. As the public begins to recognize that the only "sound science" is independent science and that some of the science that currently drives health and environmental litigation and regulation is bent science, we may optimistically predict that the public will demand changes to the legal system aimed at reducing existing incentives for advocates to pollute the fragile stream of policy-relevant science that is critical to objective regulatory and judicial decision-making.

Notes

1. Introduction

1. See *Daubert v. Merrell Dow Pharmaceuticals, Inc.,* 509 U.S. 579, 593-594 (1993); Office of Management and Budget, "Final Information Quality Bulletin for Peer Review," 70 *Fed. Reg.* 2664, 2671 (2005). For an insightful article criticizing the courts for this undiscrimating deference to scientific peer review, see Susan Haack, "Peer Review and Publication: Lessons for Lawyers," *Stetson Law Review* 36 (2007): 789.
2. *Daubert v. Merrell Dow,* 593.
3. Susan Haack, "Scientific Secrecy and 'Spin': The Sad, Sleazy Saga of the Trials of Remune," *Law & Contemporary Problems* 69 (2006): 47, 48–49.
4. P. Terrence Gaffney, letter to Jane Brooks, April 29, 2003, website of American Chemical Society, pubs.acs.org/subscribe/journals/esthag-w/2006/feb/business/pt_weinberg.html (describing the Weinberg Group, which at the time had for twenty-three years "helped numerous companies manage issues allegedly related to environmental exposures").
5. Hugh G. Gauch, Jr., *Scientific Method in Practice* (New York: Cambridge University Press, 2003), 406–409.
6. Helen Longino, *Science as Social Knowledge* (Princeton, N.J.: Princeton University Press, 1990), 216 ; Karl R. Popper, *Conjectures and Refutations: The Growth of Scientific Knowledge* (New York: Routledge & Kegan Paul, 1992), 56.
7. Robert K. Merton, "The Ethos of Science," in Piotr Sztompka, ed., *On the Social Structure and Science* (Chicago: University of Chicago Press, 1996), 274–276.
8. See D. Melnick, et al., "Participation of Biologists in the Formulation of National Science Policy," *Federation Proceedings* 35 (1996): 1957.

9. International Committee of Medical Journal Editors, Uniform Requirements for Manuscripts Submitted to Biomedical Journals (October 2007), www.icmje.org/.
10. See Donald Kennedy, "Responding to Fraud," *Science* 314 (2006): 1353 (reporting on an independent investigation commissioned by *Science* into fraud in an article by Woo Suk Hwang that the journal published).
11. Office of Research Integrity, Scientific Misconduct Regulations, 40 C.F.R. pt. 50, subpt. A. See Nicholas Wade, "Cancer Study Was Made Up," *NYT,* January 19, 2006, A16 (fabricated data in cancer study); Nicholas Wade, "Crumpled Papers," *NYT,* January 11, 2006, D14 (fraudulent paper reporting cloning of human cells).
12. Haack, "Scientific Secrecy," 49.
13. Sheila Jasanoff, "Transparency in Public Science: Purposes, Reasons, Limits," *Law & Contemporary Problems* 69 (2006): 21, 34 (observing that scientific detachment "is especially hard to achieve in the context of highly consequential public science, in which the intellectual and social biases of experts are most likely to come into play"); International Committee of Medical Journal Editors, Uniform Requirements (recognizing that conflicts of interest can occur for reasons such as "personal relationships, academic competition, and intellectual passion").
14. This is a much narrower set of research-related categories than that covered by standard definitions of "conflict of interest," which is generally defined as a situation "in which professional judgment regarding a primary interest (such as a patient's welfare or the validity of research) tends to be unduly influenced by a secondary interest, such as financial gain." Dennis F. Thompson, "Understanding Financial Conflicts of Interest: Sounding Board," *NEJM* 329 (1993): 573.
15. Jasanoff, "Transparency," 36 n. 88.
16. See Catherine D. DeAngelis, "The Influence of Money on Medical Science," *JAMA* 296 (2006): 996.
17. *Daubert v. Merrell Dow,* 593–594.
18. See also Jasanoff, "Transparency," 24 (employing the similar term "public science" to include "policy-relevant knowledge in the broadest sense: science that underwrites specific regulatory decisions, science offered as legal evidence, science that clarifies the causes and impacts of phenomena that are salient to society, and science that self-consciously advances broad social goals").
19. Sheila Jasanoff, *The Fifth Branch: Science Advisers as Policymakers* (Cambridge, Mass.: Harvard University Press, 1990), 79–83 (referring to policy-relevant science as "regulatory science").
20. Martin Carrier, "Knowledge and Control: On the Bearing of Epistemic Values in Applied Science," in Peter Machamer & Gereon Wolters, eds., *Science, Values, and Objectivity* (Pittsburgh: University of Pittsburgh Press, 2004), 279.

21. Richard B. Stewart, "The Reformation of American Administrative Law," *Harvard Law Review* 88 (1975): 1669.
22. Mary L. Lyndon, "Information Economics and Chemical Toxicity: Designing Laws to Produce and Use Data," *Michigan Law Review* 87 (1989): 1795, 1799.
23. See David Michaels & Wendy Wagner, "Disclosure in Regulatory Science," *Science* 302 (2003): 2073.
24. See Elihu Richter et al., "Efforts to Stop Repression Bias by Protecting Whistleblowers," *International Journal of Occupational and Environmental Health,* January–March 2001, 70; Justin E. Bekelman, Yan Li, & Cary P. Gross, "Scope and Impact of Financial Conflicts of Interest in Biomedical Research," *JAMA* 289 (2003): 454.
25. Peter Huber, *Galileo's Revenge: Junk Science in the Courtroom* (New York: Basic Books, 1991); Chris Mooney, *The Republican War on Science* (New York: Basic Books, 2005); Sheldon Rampton & John Stauber, *Trust Us, We're Experts* (New York: Putnam, 2001); Bjorn Lomberg, *The Skeptical Environmentalist: Measuring the Real State of the World* (Cambridge: Cambridge University Press, 2001); Steven J. Milloy, *Junk Science Judo: Self-Defense against Health Scares and Scams* (Washington, D.C.: Cato Institute, 2001).
26. Jerry L. Mashaw & David L. Harfst, *The Struggle for Auto Safety* (Cambridge, Mass.: Harvard University Press, 1990), ch. 1 (exploring legal culture).
27. See Ian Ayres & Robert Gertner, "Filling Gaps in Incomplete Contracts: An Economic Theory of Default Rules," *Yale Law Journal* 99 (1989): 87, 91.

2. Why Bend Science?

1. Henrik Ibsen, "An Enemy of the People," in Rolf Fjelde, trans., *Ibsen: The Complete Major Prose Plays* (New York: Plume, 1978), 281, 298.
2. Ibid., 301.
3. Ibid., 386.
4. Joe Barton, letter to Michael Mann, June 23, 2005, www.energycommerce.house.gov.
5. Kansas Corn Growers Association, Triazine Network, & Center for Regulatory Effectiveness, Request for Correction of Information Contained in the Atrazine Environmental Risk Assessment, Docket no. OPP-34237A (November 25, 2002), 8, website of the Center for Regulatory Effectiveness, www.thecre.com/pdf/petition-atrazineZB.pdf, 8.
6. Tom Humber, memorandum to Ellen Merlo, re: ETS, undated [c. January 1993], Bates no. 2024713141.
7. Christopher J. Bosso, *Pesticides and Politics* (Pittsburgh: University of Pittsburgh Press, 1987), 118; Frank Graham, *Since Silent Spring* (New York: Random House, 1970), 48–59; Richard Kluger, *Ashes to Ashes* (New York: Knopf, 1996).

8. Elihu Richter et al., "Efforts to Stop Repression Bias by Protecting Whistleblowers," *International Journal of Occupational and Environmental Health,* January–March 2001, 70.
9. Justin E. Bekelman, Yan Li, & Cary P. Gross, "Scope and Impact of Financial Conflicts of Interest in Biomedical Research," *JAMA* 289 (2003): 454.
10. See, e.g., Nathin Vardi, "Poison Pills," *Forbes Global,* April 19, 2004, 7; Ford Fessenden, "Judge Orders Ephedra Maker to Pay Back 12.5 Million," *NYT,* May 31, 2003, A12; Geoffrey Cowley et al., "Sweet Dreams or Nightmare," *Newsweek,* August 19, 1991, 44; "File Shows Merck Sought to Change Vioxx," *LAT,* June 23, 2005, C3; Leila Abboud, "Lilly Denies Hiding Data Tying Prozac to Suicide," *WSJ,* January 6, 2005, D10.
11. Steven Shavell, "Liability and the Incentive to Obtain Information about Risk," *Journal of Legal Studies* 21 (1992): 259, 263.
12. Paul Brodeur, *Outrageous Misconduct: The Asbestos Industry on Trial* (New York: Knopf, 1985), 118–119, 145, 116–117; Stanton Glantz et al., *The Cigarette Papers* (Berkeley: University of California Press, 1996), 15, 58–107; Philip J. Hilts, *Smokescreen: The Truth behind the Tobacco Industry Cover-Up* (Boston: Addison-Wesley, 1996), 38–40.
13. National Research Council, National Academies of Sciences, *Toxicity Testing: Strategies to Determine Needs and Priorities* (Washington, D.C.: National Academies Press, 1984).
14. Margaret A. Berger, "Eliminating General Causation: Notes towards a New Theory of Justice and Toxic Torts," *Columbia Law Review* 97 (1997): 2135–2140; Wendy E. Wagner, "Choosing Ignorance in the Manufacture of Toxic Products," *Cornell Law Review* 82 (1997): 796.
15. Philip Hilts, *Protecting America's Health* (New York: Knopf, 2003), 53–54.
16. Ibid., 93.
17. Bosso, *Pesticides,* 55, 77.
18. Hilts, *Protecting,* 165.
19. 15 U.S.C. §2603.
20. See, e.g., 33 U.S.C. §1321(b)(5) (Clean Water Act); 42 U.S.C. §9603(a) (Comprehensive Environmental Response, Compensation and Liability Act).
21. Peter W. Huber, *Liability: The Legal Revolution and Its Consequences* (New York: Basic Books, 1988); David E. Bernstein, "The Breast Implant Fiasco," *California Law Review* 87 (1999): 459.
22. Thomas Bodenheimer, "Uneasy Alliance: Clinical Investigators and the Pharmaceutical Industry," *NEJM* 342 (2000): 1539.
23. P. Terrence Gaffney, letter to Jane Brooks, April 29, 2003, website of American Chemical Society, pubs.acs.org/subscribe/journals/esthag-w/2006/feb/business/pt_weinberg.html (letter from the Weinberg Group to the DuPont Corporation offering help in reducing the company's potential regulatory and legal liabilities with respect to the chemical PFOA). See also David Michaels, "Doubt Is Their Product," *Scientific American,* June 2005, 99.
24. Peter Grossi & Sarah Duncan, "Litigation-Driven 'Medical' Screenings: Diagnoses for Dollars," *BNA Product Safety and Liability Reporter* 33 (2005): 1027.

25. *Daubert v. Merrell Dow Pharmaceuticals, Inc.,* 509 U.S. 579, 590 (1993).
26. Thomas O. McGarity, "On the Prospect of 'Daubertizing' Judicial Review of Risk Management," *Law & Contemporary Problems* 66 (2003): 158; David Michaels and Celeste Monforton, "Manufacturing Uncertainty: Contested Science and the Protection of the Public's Health and Environment," *American Journal of Public Health* 95 (2005): S39.
27. The Data Access Act, in the Omnibus Appropriations Act for Fiscal Year 1999, Pub. L. No. 105–277, 112 Stat. 2681–2495 (1998); the Information Quality Act, sec. 515 of the Treasury and General Government Appropriations Act for Fiscal Year 2001, Pub. L. No. 106–554, 114 Stat. 2763A-153–155 (2001); Office of Management and Budget, Peer Review and Information Quality, Proposed Bulletin, August 2003, www.whitehouse.gov/omb/inforeg/peer_review_and_info_quality.pdf.; Office of Management and Budget, Proposed Risk Assessment Bulletin, January 2006, www.whitehouse.gov/omb/inforeg/infopoltech.html#iq.
28. Sidney A. Shapiro & Thomas O. McGarity, "Not So Paradoxical: The Rationale for Technology-Based Regulation," *Duke Law Journal* 1991 (1991): 729, 731–739.
29. Kluger, *Ashes,* 359–360 (quoting Anthony Colucci).
30. Michaels, "Doubt," 100.
31. Tyrone B. Hayes et al., "Hermaphroditic, Demasculinized Frogs after Exposure to the Herbicide Atrazine at Low Ecologically Relevant Doses," *Proceedings of the National Academies of Sciences* 99 (2002): 5476.
32. See Cass R. Sunstein, *Risk and Reason* (New York: Cambridge University Press, 2002), ch. 2.
33. Robert F. Blomquist, "Emerging Themes and Dilemmas in American Toxic Tort Law, 1988–91: A Legal-Historical and Philosophical Exegesis," *Southern Illinois University Law Journal* 18 (1993): 1, 43; Frank J. Macchiarola, "The Manville Personal Injury Settlement Trust: Lessons for the Future," *Cardozo Law Review* 17 (1996): 583, 583–584; Francine Schwadel, "Robins and Plaintiffs Face Uncertain Future," *WSJ,* Aug. 23, 1985, A4.
34. Joseph Sanders, "Jury Deliberation in a Complex Case: *Havner v. Merrell Dow Pharmaceuticals,*" *Justice System Journal* 16 (1993): 45, 55–56 (citing unpublished district court opinion).
35. *Corrosion Proof Fittings v. EPA,* 947 F.2d 1201, 1214 (5th Cir. 1991) (invalidating the EPA's ban of asbestos under the Toxic Substances Control Act because the agency had the burden of proving its risks outweighed its benefits and because the EPA did not undertake a thorough enough analysis).
36. Office of Information and Regulatory Affairs, Office of Management and Budget, *2006 Report to Congress on the Costs and Benefits of Regulation* (Washington, D.C.: 2006), iii.
37. Don Mayer, "The Precautionary Principle and International Efforts to Ban DDT," *South Carolina Environmental Law Journal* 9 (2000–2002): 135, 170–179; the EPA's asbestos home page providing a history of asbestos regulation under the EPA's asbestos program: www.epa.gov/abestos/pubs/ban.html. The website of the EPA also details the ban and

phase-out of the manufacture of PCBs: www.epa.gov/history/topics/pcbs/01.htm.

38. David Michaels, "Forward: Sarbanes-Oxley for Science," *Law & Contemporary Problems* 69 (2006): 1.
39. William L. Anderson, Barry M. Parsons, & Drummond Rennie, "*Daubert*'s Backwash: Litigation-Generated Science," *University of Michigan Journal of Law Reform* 34 (2001): 619, 620–621.
40. In re *Silica Prods. Liab. Litig.*, 398 F. Supp. 2d 563, 571–572 (S.D. Tex. 2005). See also Roger Parloff, "Diagnosing for Dollars," *Fortune,* June 13, 2005, 96.
41. Parloff, "Diagnosing," 101–102 (discussing plaintiffs' attorneys' demand of $1 billion from defendants to settle mass silicosis claims).
42. In re *Silica,* 581, 589, 601, 615–618, 635 (quot.); Parloff, "Diagnosing," 108–110 (cut-and-paste jobs).
43. Ibid., 102 (citing plaintiffs' attorney Mike Martin).
44. James L. Nash, "Why Are So Many Still Exposed to Silica?" *Occupational Hazards,* 66 (2004): 30.
45. *See In re Diet Drugs,* 236 F. Supp. 2d 445 (E.D. Pa. 2002) (fen-phen); *Raymark Industries, Inc. v. Stemple,* 1990 U.S. Dist. Lexis 6710 (D. Kan. 1990) (asbestos); Lester Brickman, "On the Theory Class's Theories of Asbestos Litigation: The Disconnect between Scholarship and Reality," *Pepperdine Law Review* 31 (2003): 33 (asbestos); Joseph N. Gitlin et al., "Comparison of 'B' Readers' Interpretation of Chest Radiographs for Asbestos Related Changes," *Academic Radiology* 11 (2004): 843 (asbestos); Alison Frankel, "Still Ticking: Mistaken Assumptions, Greedy Lawyers, and Suggestions of Fraud Have Made Fen-Phen a Disaster of a Mass Tort," *American Lawyer* 27 (2005): 92 (fen-phen).
46. *Daubert v. Merrell Dow Pharmaceuticals,* 43 F.3d 1311, 1317 (9th Cir. 1995).
47. Anderson et al., "*Daubert's* Backwash," 660–662.
48. *See In re Silica,* 635.
49. See Daniel J. Givelber & Anthony Robbins, "Public Health versus Court-Sponsored Secrecy," *Law & Contemporary Problems* 69 (2006): 131.
50. See Sunstein, *Risk and Reason,* ch. 4 (analyzing the "risk-of-the-month syndrome"); Dan Vergano & Cathy Lynn Grossman, "The Whole World, from Whose Hands?" *USA Today,* October 11, 2005, D6 (53 percent of Americans still believe in creationism as opposed to evolution).
51. 21 U.S.C. §§348(c)(3)(A); 379e(b)(5)(B) (2000).
52. Richard D. Lyons, "Saccharin Ban Causes Storm of Complaints," *NYT,* March 11, 1977, A1.
53. Richard A. Merrill, "FDA's Implementation of the Delaney Clause: Repudiation of Congressional Choice or Reasoned Adaptation to Scientific Progress?" *Yale Journal on Regulation* 5 (1988): 1, 29.
54. Lyons, "Saccharin," A28.

55. World Health Organization, International Agency for Research on Cancer, *Monographs on the Evaluation of Carcinogenic Risks to Humans* 73 (1999): 5.5 (concluding that the mechanism for toxicity of saccharine in rodents is rodent specific).
56. See Stephen Breyer, *Breaking the Vicious Circle: Toward Effective Risk Regulation* (Cambridge, Mass.: Harvard University Press, 1993); Sunstein, *Risk and Reason.*
57. See Michael Gough, "Science, Risks and Politics," in M. Gough, ed., *Politicizing Science: The Alchemists of Policymaking* (Stanford, Calif.: Hoover Institution, 2003), 1, 9–10 (arguing that environmental organizations' publicity efforts "can keep assertions alive despite mounting scientific evidence that they are wrong"); Steven F. Hayward, "Environmental Science and Public Policy," *Social Research* 73 (2006): 891, 903–912 (criticizing environmental groups for using science to frame overly alarmist messages).
58. Gary J. Wingenbach & Tracy A. Rutherford, "Trust, Bias, and Fairness of Information Sources for Biotechnology Issues," *Agricultural Biotechnology Forum* 8 (2005): 213 (journalists perceive environmental activist groups as generally untrustworthy, biased, and unfair in communications related to agricultural biotechnology issues). The website www.junkscience.com contains specific allegations, many of which are quite incomplete or poorly supported.
59. See Thomas P. Stossel, "Regulating Academic-Industrial Research Relationships—Solving Problems or Stifling Progress?" *NEJM* 353 (2005): 1060, 1061–1062.
60. 42 U.S.C. §4332(2)(C) (environmental impact statement requirement); 16 U.S.C. 1538, 1538 (endangered species protections).
61. Most of the facts for this account, with the exception of individual statements cited separately, are taken from Sam Roe, "Decades of Risk: U.S. Knowingly Allowed Workers to Be Overexposed to Toxic Dust," *Pittsburgh Post-Gazette,* March 30, 1999, A1.
62. Ibid.
63. Michaels, "Doubt," 98; Roe, "Decades of Risk," A1.
64. Thomas O. McGarity, "Federal Regulation of Mad Cow Disease Risks," *Administrative Law Review* 57 (2005): 289, 389 (discussing the USDA's narrow institutional mission of promoting agribusiness).
65. Union of Concerned Scientists, *Scientific Integrity in Policymaking* (Cambridge, Mass.: Union of Concerned Scientists, February 2004), 12.
66. Tarek Maassarani, *Redacting the Science of Climate Change* (Washington, D.C.: Government Accountability Project, March 2007), 8–32 (detailed description of experiences of National Oceanic and Atmospheric Administration scientists); Robert Cohen, "Muzzle on Scientists Outrages Senators," *Newark Star-Ledger,* February 8, 2007, A3 (National Oceanic and Atmospheric Administration scientists); Peter N. Spotts, "Has the White House Interfered on Global Warming Reports?" *Christian Science Monitor,* January 31, 2007, A1 (120 scientists).

67. David Willman, "Risk Was Known as FDA OK'd Fatal Drug," *LAT,* March 11, 2001, A1; David Willman, "'Fast-Track' Drug to Treat Diabetes Tied to 33 Deaths," *LAT,* December 6, 1998, A1.
68. Willman, "Risk Was Known."
69. David Willman, "Diabetes Drug Rezulin Pulled Off the Market," *LAT,* March 22, 2000, A1.
70. Office of Inspector General, Department of the Interior, *Report of Investigation: Julie MacDonald, Deputy Assistant Secretary, Fish, Wildlife and Parks* (Washington, D.C.: March 23, 2007); Felicity Barringer, "Report Says Interior Official Overrode Work of Scientists," *NYT,* March 29, 2007, A13; Juliet Eilperin, "Report Faults Interior Appointee," *WPOST,* March 30, 2007, A5; Juliet Eilperin, "Bush Appointee Said to Reject Advice on Endangered Species," *WPOST,* October 30, 2006, A3.
71. *Oversight and Government Reform Committee Majority Staff, Memorandum to Democratic Members of the Committee on Oversight and Government Reform re: Full Committee Hearing on Political Interference with Science: Global Warming,* pt. 2, March 19, 2007 (quoting White House documents and deposition of Philip Cooney) available at http://oversight.house.gov/documents/20070319101643-36472.pdf; Andrew Revkin, "Bush Aide Edited Climate Reports," *NYT,* June 8, 2005, A1.
72. Paul Giannelli, "The *Daubert* Trilogy and the Law of Expert Testimony," in Richard Lempert, ed., *Evidence Stories* (New York: Foundation, 2006), 181, 200.
73. Donald Kennedy, "Forensic Science: Oxymoron?" *Science* 302 (2003), 1625.
74. Stephen J. Carroll et al., *Asbestos Litigation Costs and Compensation: An Interim Report* (Santa Monica, Calif.: Rand Institute for Civil Justice, September 25, 2002), 16.
75. Ibid., vi, 13.
76. Brodeur, *Outrageous Misconduct,* 118–119, 116–117 (quot.), 145.
77. Michael Bowker, *Fatal Deception* (New York: Simon & Schuster, 2003), 99.
78. Ibid., 99–100.
79. *Borel v. Fibreboard Paper Prods. Corp.,* 493 F.2d 1076, 1085 (5th Cir. 1973); Bowker, *Fatal Deception,* vii ($200 billion); Dan Zegart, "Commentary: Tort Reform Advocates Play Fast and Loose with the Facts," *Montana Lawyer,* February 2005, 30.
80. See Steven J. Milloy, *Junk Science Judo: Self-Defense against Health Scares and Scams* (Washington, D.C.: Cato Institute, 2001) (offering numerous but unevenly supported accounts of corporate losses resulting from successful attempts to bend science).
81. See Claude Lenfant, "The Calcium Channel Blocker Scare: Lessons for the Future," *Circulation* 91 (1995): 2855.
82. Donald Kennedy, prologue to Rena Steinzor & Wendy Wagner, eds., *Rescuing Science from Politics* (New York: Cambridge University Press, 2006), vi.
83. Catherine D. DeAngelis, "The Influence of Money on Medical Science," *JAMA* 296 (2006): 996.

3. Where Are the Scientists?

1. Sheila Jasanoff, *The Fifth Branch: Science Advisers as Policymakers* (Cambridge, Mass.: Harvard University Press, 1990), 80.
2. Saad Z. Nagi & Ronald G. Corwin, "The Research Enterprise: An Overview," in Saad Z. Nagi & Ronald G. Corwin, eds., *The Social Contexts of Research* (Hoboken, N.J.: Wiley, 1972), 24; Adil E. Shamoo & David B. Resnik, *Responsible Conduct of Research* (New York: Oxford University Press, 2003), 280.
3. See Bernard D. Goldstein & Stuart L. Shalat, "The Causal Relation between Benzene Exposure and Multiple Myeloma," letter, *Blood* 95 (2000): 1512.
4. Robert K. Merton, *The Sociology of Science,* Norman Storer, ed. (Chicago: University of Chicago Press, 1973); Helen E. Longino, *Science as Social Knowledge* (Princeton, N.J.: Princeton University Press, 1990); David L. Hull, *Science as a Process* (Chicago: University of Chicago Press, 1988); Susan Haack, *Defending Science—Within Reason: Between Scientism and Cynicism* (Amherst, N.Y.: Prometheus, 2003), 69–71.
5. Daryl E. Chubin & Edward J. Hackett, *Peerless Science: Peer Review and U.S. Science Policy* (Albany: State University of New York Press, 1990), 216; Norman W. Storer, "Relations among Scientific Disciplines," in Nagi & Corwin, *Social Contexts,* 229, 236.
6. See Helen E. Longino, *The Fate of Knowledge* (Princeton, N.J.: Princeton University Press, 2002), 250 (research quality); Philip Kitcher, "Patterns of Scientific Controversies," in Peter Machamer, Marcello Pera, & Aristides Baltas, eds., *Scientific Controversies: Philosophical and Historical Perspectives* (New York: Oxford University Press, 2000), 31–35 (methods and theories); American Association for the Advancement of Science (AAAS), *Science for All Americans: A Project 2061 Report on Literacy Goals in Science, Mathematics, and Technology* (Waldorf, Maryland: AAAS Books, 1989), 28.
7. AAAS, *Science for All,* 28; Haack, *Defending Science,* 106–109; Peter Weingart, "Between Science and Values," in Peter Machamer & Gereon Wolters, *Science, Values, and Objectivity* (Pittsburgh: University of Pittsburgh Press, 2004), 114; Helen E. Longino, "How Values Can Be Good for Science," in Machamer & Walters, *Science, Values, and Objectivity,* 140; Steve Fuller, *Science* (Minneapolis: University of Minnesota Press, 1997); Evelyn Fox Keller, *Reflections on Gender and Science* (New Haven, Conn.: Yale University Press, 1996), 9, 12.
8. Haack, *Defending Science,* 106–9; Longino, *Science as Social Knowledge,* 74.
9. Haack, *Defending Science,* 23; Chubin & Hackett, *Peerless Science,* 134–135.
10. See, e.g., American Chemical Society, Chemist's Code of Ethics (1994), http://temp.onlineethics.org/codes/ACScode.html.
11. Chubin & Hackett, *Peerless Science,* 4; Sheila Jasanoff, "Transparency in Public Science: Purposes, Reasons, Limits," *Law & Contemporary Problems* 69 (2006): 21, 36; Elizabeth Knoll, "The Communities of Scientists and Journal Peer Review," *JAMA* 263 (1990): 1330.

12. Sheldon Krimksy, *Science in the Private Interest* (Oxford: Rowman & Littlefield, 2003), 125–140.
13. C. D. DeAngelis, P. B. Fontanarosa, & A. Flanagin, "Reporting Financial Conflicts of Interest and Relationships between Investigators and Research Sponsors," *JAMA* 286 (2001): 89. See also S. A. Rosenberg, "Secrecy in Medical Research," *NEJM* 334 (1996): 392; Drummond Rennie, "Thyroid Storm," *JAMA* 277 (1997): 1238–1243; D. Blumenthal et al., "Withholding Research Results in Academic Life Science: Evidence from a National Survey of Faculty," *JAMA* 277 (1997): 1224.
14. Longino, *Science as Social Knowledge,* 80.
15. Robert K. Merton, "The Normative Structure of Science," in Jerry Gaston, ed., *Sociology of Science* (Hoboken, N.J.: Wiley, 1973), 267, 275; David B. Resnik, *The Ethics of Science: An Introduction* (New York: Routledge, 1998), 53–72.
16. See, e.g., 40 C.F.R. §158.340.
17. Dan Fagin & Marianne Lavelle, *Toxic Deception* (Secaucus, N.J.: Birch Lane Press, 1996), 33–35.
18. William L. Anderson, Barry M. Parsons, & Drummond Rennie, "*Daubert*'s Backwash: Litigation-Generated Science," *University of Michigan Journal of Law Reform* 34 (2001): 619, 635. See also Chubin & Hackett, *Peerless Science,* 99 (noting that one of the two leading criteria used by journals to make publication decisions is "contribution to the field").
19. See generally Environmental Protection Agency, Confidentiality of Business Information, 40 C.F.R. §§2.201–2.310 (2003).
20. Martin Carrier, "Knowledge and Control: On the Bearing of Epistemic Values in Applied Science," in Machamer & Wolters, *Science, Values, and Objectivity,* 279.
21. Environmental Protection Agency, "Multi-substance Rule for the Testing of Neurotoxicity," 58 *Fed. Reg.* 40262, 40263 (1993).
22. See, e.g., Office of Pesticide Programs, Data Requirements for Registration, 40 C.F.R. §§158.290, 158.490, 158.590.
23. See Goldie Blumenstyke, "The Price of Research: A Berkeley Scientists Says a Corporate Sponsor Tried to Bury His Unwelcome Findings and Then Buy His Silence," *Chronicle,* October 31, 2003, chronicle.com/free/v50/i10/10a02601.htm.
24. Tyrone Hayes et al., "Feminization of Male Frogs in the Wild," *Nature* 419 (2002): 895.
25. James A. Carr et al., "Response of Larval *Xenopus Laevis* to Atrazine: Assessment of Growth, Metamorphosis, and Gonadal and Laryngeal Morphology," *Environmental Toxicology and Chemistry* 22 (2003): 396, 404 (acknowledging that the authors are funded by Atrazine's manufacturer, Syngenta).
26. Rebecca Renner, "Conflict Brewing over Herbicide's Link to Frog Deformities," *Science* 298 (2002): 938.
27. Environmental Protection Agency, White Paper on Potential Development

Effects of Atrazine on Amphibians, June 2003, www.pestlaw.com/x/guide/2003/EPA-20030529A.pdf.

28. Environmental Protection Agency, Report of FIFRA Scientific Advisory Panel Meeting on the Potential Developmental Effects of Atrazine on Amphibians, June 17–20, 2003, www.epa.gov/scipoly/sap/ meetings/2003/june/junemeetingreport.pdf, 17–19.
29. See ibid., 18 (referencing several studies sponsored by Syngenta as particularly flawed).
30. Luz Tavera-Mendoza et al., "Response of the Amphibian Tadpole *Xenopus Laevis* to Atrazine during Sexual Differentiation of the Ovary," *Environmental Toxicology and Chemistry* 21 (2001): 1264; Wanda L. Goleman and James A. Carr, *Response of Larval Xenopus Laevis to Atrazine Exposure: Assessment of Metamorphosis and Gonadal and Laryngeal Morphology,* laboratory study ID ECORISK no. TTU-01 (Lubbock: Institute of Environmental and Human Health, Texas Tech University, Texas Tech University Health Sciences Center, 2003).
31. Anderson et al., "*Daubert*'s Backwash," 639.
32. J. Scott Armstrong, "Peer Review of Journals: Evidence of Quality Control, Fairness, and Innovation," *Science and Engineering Ethics* 3 (1997): 63–84; Nelson Yuan-sheng Kiang, "How Are Scientific Corrections Made?" *Science and Engineering Ethics* 1 (1995): 347.
33. Arnold Relman, "Lessons from the Darsee Affair," *NEJM* 308 (1983): 1415.
34. John Abramson, "Drug Profits Infect Medical Studies," *LAT,* January 7, 2006, B17 (editorial by clinical investigator at Harvard Medical School).
35. Brian C. Martinson, Melissa S. Anderson, & Raymond de Vries, "Scientists Behaving Badly," *Nature* 435 (2005): 737; Nicholas Wade, "Korean Scientist Said to Admit Fabrication in a Cloning Study," *NYT,* December 16, 2006, A1.
36. Chubin & Hackett, *Peerless Science,* 128.
37. Anderson et al., "*Daubert*'s Backwash," 630, 634.
38. Chubin & Hackett, *Peerless Science,* 88–96 (discussing, among other problems, the "caprice" that can attend journal peer review).
39. Renner, "Conflict Brewing," 938.
40. EPA, Report of FIFRA Panel Meeting, 18, 20.
41. Jocelyn Kaiser, "Synergy Paper Misconduct," *Science* 294 (2001):763; Jocelyn Kaiser, "Synergy Paper Questioned at Toxicology Meeting," *Science* 275 (1997): 1879.
42. Sheldon Krimsky, *Hormonal Chaos: The Scientific and Social Origins of the Environmental Endocrine Hypothesis* (Baltimore: Johns Hopkins University Press, 2000), 159–163.
43. *Frye v. United States,* 293 F. 1013, 1014 (D.C. Cir. 1923).
44. Paul Giannelli, "The Admissibility of Novel Scientific Evidence: *Frye v. United States,* a Half Century Later," *Columbia Law Review* 80 (1980): 1197.
45. *Daubert v. Merrell Dow Pharmaceuticals, Inc.,* 509 U.S. 579, 590 (1993).
46. Carl Cranor, *Toxic Torts: Science, Law, and the Possibility of Justice* (New York: Cambridge University Press, 2006).

47. Jasanoff, *The Fifth Branch,* 206.
48. Center for Science in the Public Interest, *Ensuring Independence and Objectivity at the National Academies* (Washington, D.C.: 2006).
49. Anonymous senior staff, National Academies of Sciences, interview with Wagner, December 3, 2004.
50. Weingart, "Between Science and Values," 121.
51. Carrier, "Knowledge and Control," 280.
52. See National Research Council, National Academies of Sciences, *Risk Assessment in the Federal Government: Managing the Process* (Washington, D.C.: National Academies Press, 1983), 29–33.
53. Weingart, "Between Science and Values," 121.
54. Catherine D. DeAngelis, "The Influence of Money," 996, 997.
55. John D. Dingell, "Shattuck Lecture: Misconduct in Medical Research," *NEJM* 328 (1993): 1610. See also William J. Broad & Nicholas Wade, *Betrayers of the Truth* (New York: Simon & Schuster, 1982); Horace F. Judson, *The Great Betrayal: Fraud in Science* (San Diego, Calif.: Harcourt, 2004).
56. Deena Weinstein, "Commentary: Bureaucratized Science and Resistance to Whistleblowing," in Judith P. Swazey & Stephen R. Scher, eds., *Whistleblowing in Biomedical Research: Policies and Procedures for Responding to Reports of Misconduct* (Washington, D.C.: President's Commission for the Study of Ethical Problems in Medicine and Biomedical and Behavioral Research, 1981), 194–195.
57. DeAngelis, "The Influence of Money," 996.

4. Shaping Science

1. Except as otherwise noted, this narrative relies on David Willman, "Drug Trials with a Dose of Doubt," *LAT,* July 16, 2006, A1.
2. Thomas J. Walsh et al., "Liposomal Amphotericin B for Empirical Therapy in Patients with Persistent Fever and Neutropenia," *NEJM* 340 (1999): 764, 764.
3. Thomas Fisher, Gudula Heussel, & Christoph Huber, letter, *NEJM* 341 (1999): 1152.
4. Quoted in Willman, "Drug Trials."
5. Thomas J. Walsh et al., letter, *NEJM* 341 (1999): 1155.
6. Thomas J. Walsh et al., "Caspofungin versus Liposomal Amphotericin B for Empirical Antifungal Therapy in Patients with Persistent Fever and Neutropenia," *NEJM* 351 (2004): 1391.
7. Francisco M. Marty & Collene M. Lowry, letter, *NEJM* 352 (2005): 410.
8. Dimitrios Kontoyiannis & Russell E. Lewis, letter, *NEJM* 352 (2005): 410. For the authors' response, see Thomas J. Walsh, Gerald R. Donowitz, & Ben E. DePauw, letter, *NEJM* 352 (2005): 413 (arguing that "regulatory and clinical considerations" mandated a 3 mg/kg dosage: authors did not disclose that Walsh reported using levels of 7.5 to 15 in 1999).
9. Quoted in Willman, "Drug Trials."
10. David Willman, "Income from Two Sources," *LAT,* July 16, 2006, A29.

11. David Willman, "NIH Audit Criticizes Scientist's Dealings," *LAT,* September 10, 2006, A1.
12. *Subcommittee on Oversight and Investigations of the House Committee on Energy and Commerce, Hearing on Continuing Ethics and Management Concerns at NIH and the Public Health Service Commissioned Corps,* 109th Cong., 2d Sess. 355 (2006) (testimony of Dr. John Niederhuber, director, National Cancer Institute).
13. See Hugh G. Gauch, Jr., *Scientific Method in Practice* (New York: Cambridge University Press, 2003), 406–409.
14. See Adil E. Shamoo & David B. Resnik, *Responsible Conduct of Research* (New York: Oxford University Press, 2003), 139–142 (discussing and defining conflict of interest and the importance of disclosure to regain trust of the reader).
15. Peter Weingart, "Between Science and Values," in Peter Machamer & Gereon Wolters, eds., *Science, Values, and Objectivity* (Pittsburgh: University of Pittsburgh Press, 2004), 121; Martin Carrier, "Knowledge and Control: On the Bearing of Epistemic Values," in the same volume, 276.
16. Carrier, "Knowledge and Control," 276. See also John D. Graham et al., *In Search of Safety: Chemicals and Cancer Risk* (Cambridge, Mass.: Harvard University Press, 1988), 187–189 (observing distinct disciplinary or ideological groupings of scientists); Ted Greenwood, *Knowledge and Discretion in Government Regulation* (Westport, Conn.: Greenwood, 1984), 192; Nicholas A. Ashford, "Advisory Committees in OSHA and EPA: Their Use in Regulatory Decisionmaking," *Science, Technology, & Human Values* 9 (1984): 72, 77 (discussing effects of disciplinary bias on policy judgments).
17. Robert K. Merton, "The Normative Structure of Science," in Jerry Gaston, ed., *Sociology of Science* (Hoboken, N.J.: Wiley, 1973), 267, 275.
18. "One in Three Scientists Confesses to Having Sinned," *Nature* 435 (2005): 718; Rick Weiss, "Many Scientists Admit to Misconduct," *WPOST,* June 9, 2005, A3.
19. Marcia Angell, *The Truth about the Drug Companies* (New York: Random House, 2005), 100; Jennifer Washburn, *University, Inc.* (New York: Basic Books, 2005), xvi.
20. Merton, "Normative Structure," 275.
21. Ibid., 140.
22. Ibid., 140–141.
23. Sheldon Krimsky, *Science in the Private Interest* (Oxford: Rowman & Littlefield, 2003), 125–140; Shamoo & Resnik, *Responsible Conduct,* 139–142.
24. 21 U.S.C. §§321(s), 348. See also Thomas O. McGarity, "Seeds of Distrust: Federal Regulation of Genetically Modified Foods," *University of Michigan Journal of Law Reform* 35 (2002): 403. Except where otherwise indicated, the olestra description is taken from Marion Nestle, *Food Politics* (Berkeley: University of California Press, 2002), 339–343.

25. Jane Levine, Joan Dye Gussow, Diane Hastings, & Amy Eccher, "Authors' Financial Relationships with the Food and Beverage Industry and Their Published Positions on the Fat Substitute Olestra," *American Journal of Public Health* 93 (2003): 664–669.
26. See Valerio Gennaro & Lorenzo Tomatis, "Business Bias: How Epidemiologic Studies May Underestimate or Fail to Detect Increased Risks of Cancer and Other Diseases," *International Journal of Occupational and Environmental Health* 11 (2005): 356; David S. Egilman & Marion Billings, "Abuse of Epidemiology: Automobile Manufacturers Manufacture a Defense to Asbestos Liability," *International Journal of Occupational and Environmental Health* 11 (2005): 360.
27. Gennaro & Tomatis, "Business Bias," 356–57; Egilman & Billings, "Abuse of Epidemiology," 363–368.
28. Gennaro & Tomatis, "Business Bias," 357 (citing Otto Wong deposition); see also 356–57; Egilman & Billings, "Abuse of Epidemiology," 363–368.
29. *Black v. Rhone-Poulenc, Inc.,* 19 F. Supp. 2d 592, 601 (S.D. W. Va. 1998).
30. Ibid., 599.
31. Ibid., 600–601.
32. National Institute of Medicine, National Academies of Sciences, *The Future of Drug Safety: Promoting and Protecting the Health of the Public* (Washington, D.C.: National Academies Press, 2006), 2–4; Washburn, *University, Inc.,* 110.
33. Angell, *Truth,* 107.
34. Ibid., 108–9. See also Thomas Bodenheimer, "Uneasy Alliance: Clinical Investigators and the Pharmaceutical Industry," *NEJM* 342 (2000): 1539.
35. Angell, *Truth,* 108; Bodenheimer, "Uneasy Alliance."
36. Penni Crabtree, "Panel Urges FDA to Reject Maxim Skin-Cancer Drug," *San Diego Union-Tribune,* December 14, 2000, C1 (FDA advisory panel recommends against approval of skin cancer treatment Maxamine because clinical trial was biased to give it to healthier patients).
37. Angell, *Truth,* 108.
38. Ibid., 29; Jerome P. Kassirer, *On the Take* (New York: Oxford University Press, 2005), 162; comments of Dr. Bernard Goldstein on an early draft of this book, November 2006.
39. Angell, *Truth,* 29.
40. Ibid., 101.
41. Ibid., 30–31; Kassirer, *On the Take,* 9.
42. Angell, *Truth,* 162; Carl Elliott, "Pharma Buys a Conscience," *American Prospect,* September 24, 2001, 16.
43. Barry Meier, "Contracts Keep Drug Research out of Reach," *NYT,* November 29, 2004, A1.
44. Ibid.
45. Angell, *Truth,* xviii.
46. Robert Steinbrook, "Gag Clauses in Clinical-Trial Agreements," *NEJM* 352 (2005): 2160; Bodenheimer, "Uneasy Alliance."

47. Bodenheimer, "Uneasy Alliance."
48. Jennifer Washburn, "Hired Education," *American Prospect,* February, 2005, 29; Angell, *Truth,* 101.
49. Arlene Weintraub & Amy Barrett, "Medicine in Conflict," *Business Week,* October 23, 2006, 76; Paul Jacobs, "How Profits, Research Mix at Stanford Med School," *San Jose Mercury News,* July 10, 2006, A1.
50. Warren E. Leary, "Business and Scholarship: A New Ethical Quandary," *NYT,* June 12, 1889, A1; Peter G. Gosselin, "Flawed Study Helps Doctors Profit on Drug," *Boston Globe,* October 19, 1988, 1.
51. Gosselin, "Flawed Study."
52. Dan Fagin & Marianne Lavelle, *Toxic Deception* (Secaucus, N.J.: Birch Lane Press, 1996), 44.
53. Ibid., 45.
54. See Frederick Grinnell, *The Scientific Attitude* (Boulder, Colo.: Westview, 1987), 5–20.
55. *Subcomm. on Health of the Senate Labor & Public Welfare Comm., Hearings on Preclinical and Clinical Testing by the Pharmaceutical Industry,* Pt. 3, 94th Cong., 2d Sess. 25 (1976). See also Sidney A. Shapiro, "Divorcing Profit Motivation from New Drug Research: A Consideration of Proposals to Provide the FDA with Reliable Test Data," *Duke Law Journal* 1978 (1978): 155, 166–167.
56. Michael Green, *Bendectin and Birth Defects* (Philadelphia, Pa.: University of Pennsylvania Press, 1998), 85.
57. Ibid., 86.
58. Fagin & Lavelle, *Toxic Deception,* 33–35.
59. Kurt Eichenwald & Gina Kolata, "A Doctor's Drug Trials Turn into Fraud," *NYT,* May 17, 1999, A1.
60. Ibid.
61. David Healy, *Let Them Eat Prozac* (New York: New York University Press, 2004), 126.
62. Jerald L. Schnoor, "Fraud in Science: How Much Scientific Fraud Is Out There?" *Environmental Science and Technology* 40 (2006): 1375 (fraud very rare in environmental science); Rick Weiss, "Deception by Researchers Relatively Rare," *WPOST,* January 15, 2006, A19 (interview with director of federal Office of Research Integrity). But see Nicholas Wade, "Journal Faulted in Publishing Korean's Claims," *NYT,* November 29, 2006, A14 (controversy over publication of fraudulent data on stem cell research by Hwang Woo Suk in *Science*); Nicholas Wade, "Cancer Study Was Made Up," *NYT,* January 19, 2006, A16 (fabricated data in cancer study).
63. Peter Breggin, *Toxic Psychiatry* (New York: St. Martin's Press, 1994), 252; David H. Jacobs, "Psychiatric Drugging," *Journal of Mind and Behavior* 4 (1995): 421; "High Anxiety," *Consumer Reports,* January 1993, 19.
64. Deborah Nelson, "Drug Giant's Spin May Obscure Risk," *WPOST,* March 18, 2001, A14.
65. The Actonel account is derived from Jo Revill, "How the Drugs Giant and a Lone Academic Went to War," *Guardian,* December 4, 2005, A10.

66. Revill, "Drugs Giant," A10 (quoting Procter & Gamble email).
67. Ibid; Jo Revill, "Doctor Accuses Drugs Giant of 'Unethical' Secrecy," *The Observer,* December 4, 2005.
68. Revill, "Drugs Giant," A10 (quoting Procter & Gamble press release).
69. International Committee of Medical Journal Editors, Uniform Requirements for Manuscripts Submitted to Biomedical Journals, updated 2006, www.icmje.org.
70. Kevin A. Schulman et al., "A National Survey of Provisions in Clinical-Trial Agreements between Medical Schools and Industry Sponsors," *NEJM* 347 (2002): 1335. See also Robert Steinbrook, "Gag Clauses in Clinical-Trial Agreements," *NEJM* 352 (2005): 2160 (no change in previous three years).
71. Ford Fessenden, "Studies of Dietary Supplements Come under Growing Scrutiny," *NYT,* June 23, 2003, A1.
72. *Subcommitee on Oversight and Investigations and the Subcommittee on Commerce, Trade, and Consumer Protection of the House Committee on Energy and Commerce, Hearings on Issues Releating to Ephedra-Containing Dietary Supplements,* 108th Cong., 1st Sess. (2003), 148.
73. Penni Crabtree, "Suit Involving Ephedra Pill Alleges Fraud," *San Diego Union-Tribune,* March 6, 2003, C1.
74. Egilman & Billings, "Abuse of Epidemiology."
75. David Michaels, Celeste Monforton, & Peter Lurie, "Selected Science: An Industry Campaign to Undermine an OSHA Hexavalent Chromium Standard," *Environmental Health* 5 (2006): 5; Rick Weiss, "Chromium Evidence Buried," *WPOST,* February 24, 2006, A3.
76. David E. Lilienfeld, "The Silence: The Asbestos Industry and Early Occupational Cancer Research—A Case Study," *American Journal of Public Health* 81 (1991): 791, 792–793.
77. The Vioxx account is taken from David Michaels, "Doubt Is Their Product," *Scientific American,* June 2005, 100.
78. An-Wen Chan et al., "Empirical Evidence for Selective Reporting of Outcomes in Randomized Trials," *JAMA* 291 (2004): 2457, 2463; William L. Anderson, Barry M. Parsons, & Drummond Rennie, "*Daubert's* Backwash: Litigation-Generated Science," *University of Michigan Journal of Law Reform* 34 (2001): 619, 639.
79. An-Wen Chan & Douglas G. Altman, "Identifying Outcome Reporting Bias in Randomised Trials on PubMed: Review of Publications and Survey of Authors," *British Medical Journal* 330 (2005): 753; An-Wen Chan et al., "Outcome Reporting Bias in Randomized Trials Funded by the Canadian Institutes of Health Research," *Canadian Medical Association Journal* 171 (2004): 735; An-Wen Chan et al., "Empirical Evidence," 2463.
80. Shamoo & Resnik, *Responsible Conduct,* 52.
81. Anderson et al., "*Daubert*'s Backwash," 660–662.
82. Sheldon Krimksy, *Science in the Private Interest* (Oxford: Rowman & Littlefield, 2003), 171; Anderson et al., "*Daubert*'s Backwash," 646–647; S. V. McCrary et al., "A National Survey of Policies on Disclosure of Conflicts of Interest in Biomedical Research," *NEJM* 343 (2000): 1621.

83. Krimksy, *Science in the Private Interest,* 169–70; Merrill Goozner, *Unrevealed: Non-Disclosure of Conflicts of Interest in Four Leading Medical and Scientific Journals* (Center for Science in the Public Interest, 2004), 8; Julie Bell, "Fraud under a Microscope," *Baltimore Sun,* January 22, 2006, C6 (prominent neuroscience journal did not ask for conflict of interest information in 2002).
84. Bodenheimer, "Uneasy Alliance"; Drummond Rennie, Veronica Yank, & Linda Emanuel, "When Authorship Fails: A Proposal to Make Contributors Accountable," *JAMA* 278 (1997): 579; Anna Wilde Mathews, "At Medical Journals, Paid Writers Play Big Role," *WSJ,* December 12, 2005, A1.
85. Kassirer, *On the Take,* 112.
86. Healy, *Prozac,* 113.
87. Ibid., 117.
88. Ibid., 116.
89. Ibid., 116.
90. Ibid., 112.
91. Shankar Vedantam, "Comparison of Schizophrenia Drugs Often Favors Firm Funding Study," *WPOST,* April 12, 2006, A1 (quoting Dr. John Davis).
92. Green, *Bendectin,* 78.
93. Healy, *Prozac,* 117–118.
94. Kassirer, *On the Take,* 31.
95. Krimksy, *Science in the Private Interest,* 116.
96. William M. Tierney & Martha S. Gerrity, "Scientific Discourse, Corporate Ghostwriting, Journal Policy, and Public Trust," *Journal of General Internal Medicine* 20 (2005): 550 (journal refuses to publish ghostwritten article).
97. Bodenheimer, "Uneasy Alliance."
98. Healy, *Prozac,* 112.
99. Ibid., 113–114.
100. Kassirer, *On the Take,* 31.
101. Sheldon Rampton & John Stauber, *Trust Us, We're Experts* (New York: Putnam, 2001), 200.
102. Healy, *Prozac,* 114.
103. Jonathan Gabe & Michael Bury, "Anxious Times: The Benzodiazepine Controversy and the Fracturing of Expert Authority," in Peter Davis, ed., *Contested Ground* (New York: Oxford University Press, 1996), 42, 47.
104. Drummond Rennie, "Fair Conduct and Fair Reporting of Clinical Trials," *JAMA* 282 (1999): 1766; Rennie et al., "When Authorship Fails," 579; Patricia Huston & David Moher, "Redundancy, Disaggregation, and the Integrity of Medical Research," *Lancet* 347 (1996): 1024, 1025.
105. Richard Smith, "Medical Journals Are an Extension of the Marketing Arm of Pharmaceutical Companies," *PloS Medicine* 2 (May 2005): 138; Huston & Moher, "Redundancy."
106. Jerome P. Kassirer & Maria Angell, "Redundant Publication: A Reminder," *NEJM* 333 (1995): 449–450.
107. David Blumenthal, "Academic-Industrial Relationships in the Life Sciences," *NEJM* 349 (2003): 2452; Deborah E. Barnes & Lisa A. Bero, "Industry-

Funded Research and Conflict of Interest: An Analysis of Research Sponsored by the Tobacco Industry through the Center for Indoor Air Research," *Journal of Health Politics, Policy and Law* 21 (1996): 515, 516.

108. David Kessler, *A Question of Intent* (New York: Public Affairs, 2001), 198–200; Richard Kluger, *Ashes to Ashes* (New York: Knopf, 1996), 165; Alix M. Freedman and Laurie P. Cohen, "How Tobacco Firms Keep Health Questions 'Open' Year after Year," *Sacramento Bee,* February 28, 1993, F1.
109. E. Peeples, memorandum to unnamed addressees, re: Response to Cigarette/Health Controversy, February 4, 1976, Bates no. 170042567, 2.
110. Barnes & Bero, "Industry-Funded Research," 533.
111. Jacqui Drope & Simon Chapman, "Tobacco Industry Efforts at Discrediting Scientific Knowledge of Environmental Tobacco Smoke: A Review of Internal Industry Documents," *Journal of Epidemiology & Community Health* 55 (2001): 588, 592.
112. Lisa Bero, "Implications of the Tobacco Industry Documents for Public Health and Policy," *Annual Review of Public Health* 23 (2003): 267; Monique E. Muggli et al., "The Smoke You Don't See: Uncovering Tobacco Industry Scientific Strategies Aimed Against Environmental Tobacco Smoke Policies," *American Journal of Public Health* 91 (2001): 1419, 1420–1421; Drope & Chapman, "Tobacco Industry Efforts," 590.
113. Muggli et al., "Smoke You Don't See," 1421.
114. John Rupp, letter to Paul Sadler, quoted in Drope & Chapman, "Tobacco Industry Efforts," 591.
115. The cell phone account is drawn from George Carlo & Martin Schram, *Cell Phones: Invisible Hazards in the Wireless Age* (New York: Carroll & Graf, 2001), 6–9, 25–26, 97, 107, 115–118.
116. Ibid., 141 (quoting John Welsh).
117. About the Center, website of the Mickey Leland National Air Toxics Research Center, www.sph.uth.tmc.edu/mleland/Pages/about.htm.
118. Ibid.
119. HEI Board of Directors, website of the Health Effects Institute, www.healtheffects.org/board.htm; Devra Davis, *When Smoke Ran Like Water* (New York: Basic Books, 2002), 154.
120. About the Center, website of the Mickey Leland National Air Toxics Research Center.
121. Ibid.
122. Bob Burtman, "The Silent Treatment," *OnEarth,* Spring 2002, 14, 18.
123. Leigh Hopper, "The Burden of Proof," *Houston Chronicle,* January 18, 2005, A1.
124. Bruce M. Owen & Ronald Braeutigam, *The Regulation Game* (New York: HarperCollins, 1978), 7.
125. Angell, *Truth,* 103–104.
126. Elizabeth A. Boyd & Lisa A. Bero, "Assessing Faculty Financial Relationships with Industry: A Case Study," *JAMA* 284 (2000): 2209, 2211–2212; E. G. Campbell et al., "Looking a Gift Horse in the Mouth: Corporate Gifts Supporting Life Sciences Research," *JAMA* 279 (1998): 995.

127. Joseph N. Gitlin et al., "Comparison of 'B' Readers' Interpretation of Chest Radiographs for Asbestos Related Changes," *Academic Radiology* 11 (2004): 843.
128. See Kessler, *Question of Intent,* 203–4 (noting that "Special Projects had no official address, no incorporation papers, no board of directors, no by-laws, and no accountability").
129. Ibid.
130. Dan Zegart, *Civil Warriors* (New York: Delacorte, 2000), 290.
131. See Ross Gelbspan, *Boiling Point* (Cambridge, Mass.: Perseus, 1998), 40–44. See also website of Greenpeace's Exxon Secrets website, www.exxonsecrets.org/html/personfactsheet.php?id=4 (detailing Dr. Michaels's connections to the energy industry).
132. Markowitz & Rosner, *Deceit,* 47.
133. Ibid.
134. Lennart Hardell et al., "Secret Ties to Industry and Conflicting Interests in Cancer Research," *American Journal of Industrial Medicine* 50 (2007): 227; Sarah Boseley, "Renowned Cancer Scientist Was Paid by Chemical Firm for 20 Years," *Guardian,* December 8, 2006, A1.
135. Richard Wakeford, letter, *American Journal of Industrial Medicine* 50 (2006): 239; Boseley, "Renowned Cancer Scientist" (quoting Sir Richard Peto).
136. Krimsky, *Science in the Private Interest,* 19; David Willman, "Drug Maker Hired NIH Researcher," *LAT,* December 7, 1998, A1.
137. Krimsky, *Science in the Private Interest,* 19–20; David Willman, "Scientists Who Judged Pill Safety Received Fees," *LAT,* October 29, 1999, A22.
138. Willman, "Drug Maker Hired."
139. David Willman, "NIH Director Calls for Review of Scientist's Ties to Firm," *LAT,* December 16, 1998, A28; Willman, "Drug Maker Hired."
140. Krimsky, *Science in the Private Interest,* 21–22; Willman, "Drug Maker Hired."
141. David Willman, "Second NIH Researcher to Become Part of Conflict Probe," *LAT,* September 4, 1999, A12.
142. Willman, "Scientists Who Judged"; David Willman, "Second NIH Researcher."
143. Willman, "Drug Maker Hired."
144. Willman, "Second NIH Researcher."
145. Willman, "Drug Maker Hired."
146. Krimsky, *Science in the Private Interest,* 20.
147. Gardiner Harris, "Health Agency Tightens Rules Governing Federal Scientists," *NYT,* August 26, 2005, A12; Jocelyn Kaiser, "Forty-Four Researchers Broke NIH Consulting Rules," *Science* 309 (2005): 546; David Willman, "NIH Inquiry Shows Widespread Ethical Lapses, Lawmaker Says," *LAT,* July 14, 2005, A1; Willman, "Scientists Add Clout"; David Willman, "NIH Seeks 'Higher Standard,'" *LAT,* February 2, 2005, A12.
148. Harris, "Health Agency Tightens Rules"; Willman, "NIH Seeks 'Higher Standard.'"

149. Jocelyn Kaiser, "NIH Rules Make Some Pack, Others Plead," *Science* 307 (2005): 1703; Michael S. Rosenwald & Rick Weiss, "New Ethics Rules Cost NIH Another Top Researcher," *WPOST,* April 2, 2005, A1.
150. Rick Weiss, "NIH Workers Angered by New Ethics Rules," *WPOST,* February 3, 2005, A25.
151. Jocelyn Kaiser, "Final NIH Rules Ease Stock Limits," *Science* 309 (2005): 1469; Harris, "Health Agency Tightens Rules."
152. David Willman, "Scientists Add Clout."
153. Jocelyn Kaiser, "NIH Rules Rile Scientists, Survey Finds," *Science* 314 (2006): 740.
154. Roger Parloff, "Diagnosing for Dollars," *Fortune,* June 13, 2005, 50.
155. Rampton & Stauber, *Trust Us,* 236; Zegart, *Civil Warriors,* 287–303, 306.
156. See Shayne C. Gad, Sixth Triennial Toxicologist Salary Survey, Society of Toxicology, www.toxicology.org/AI/PUB/Sp05/sp05_salary.asp (table 1); Constance Holden, "Long Hours Aside, Respondents Say Jobs Offer 'as Much Fun as You can Have,'" *Science* 304 (2004): 5678.
157. Most of the Synthroid account is taken from Krimsky, *Science in the Private Interest,* 15–18. See also Washburn, *University, Inc.,* 19–20; Douglas M. Birch and Gary Cohn, "Standing Up to Industry," *Baltimore Sun,* June 26, 2001, A1.
158. Quoted in Krimsky, *Science in the Private Interest,* 15.
159. Birch & Cohn, "Standing Up."
160. Quoted in Krimsky, *Science in the Private Interest,* 16.
161. David Shenk, "Money + Science = Ethical Problems on Campus," *Nation,* March 22, 1999, 11.
162. Birch & Cohn, "Standing Up"; Shenk, "Money + Science."
163. Justin E. Bekelman, Yan Li, & Cary P. Gross, "Scope and Impact of Financial Conflicts of Interest in Biomedical Research," *JAMA* 289 (2003): 454.
164. David Blumenthal et al., "Relationships between Academic Institutions and Industry in the Life Sciences—An Industry Survey," *NEJM* 334 (1996): 368; David Blumenthal et al., "Participation of Life-Science Faculty in Research Relationships with Industry," *NEJM* 335 (1996): 1734.
165. Bernard J. Goldstein, comments on an early draft of this book, November 2006.
166. Linda Greer & Rena Steinzor, "Bad Science," *Environmental Forum,* January–February 2002, 13.
167. Bayh-Dole University and Small Business Patent Procedures Act of December 12, 1980, Pub. L. No. 96-517, 94 Stat. 3015 (codified as amended at 35 U.S.C. §§200–211), 301–307.
168. Washburn, *University, Inc.,* 60–72 (revealing history of Bayh-Dole Act). See also Angell, *Truth,* 7; Krimsky, *Science in the Private Interest,* 30.
169. Reed Abelson & Stephanie Saul, "Ties to Industry Cloud a Clinic's Mission," *NYT,* December 17, 2005, A1 (Cleveland Clinic); Julie Bell, "Industry Ties Testing Schools," *Baltimore Sun,* May 8, 2005, A1.
170. Kassirer, *On the Take,* 14, 180–181; Washburn, *University, Inc.,* 140; Re-

becca Buckman, "More Universities Increasing Support for Campus Start-Ups," *WSJ*, November 27, 2006, B1.

171. Julie Bell, "Industry Ties."
172. The EPA's Research and Development (R & D) budget in 1976 was $743 million (in FY 2004 dollars) and declined to $591 million in 2005 (in FY 2004 dollars). American Association for the Advancement of Science, Guide to R & D Funding Historical Data, www.aaas.org/spp/rd/guihist.htm. See Marc Kaufman, "Cutbacks Impede Climate Studies," *WPOST*, January 16, 2007, A1 (relating NAS report detailing 30 percent drop in funding from the National Aeronautics and Space Administration for earth sciences).
173. National Research Council, National Academies of Sciences, *Trends in Federal Support of Research and Graduate Education* (Washington, D.C.: National Academies Press, 2001), 35.
174. M. Granger Morgan, letter to Stephen L. Johnson, March 13, 2007, re: Comments on EPA's Strategic Research Directions and Research Budget for FY 2008, An Advisory Report for the U.S. Environmental Protection Agency Science Advisory Board, http://www.epa.gov/sab/pdf/sab-07-004.pdf.
175. Amy Dockser Marcus, "Funding Concerns Hit Some Cancer Trials," *WSJ*, February 7, 2007, D3.
176. Jeffrey Peppercorn et al., "Association between Pharmaceutical Involvement and Outcomes in Breast Cancer Clinical Trial," *Cancer* 109 (2006): 1243.
177. Eyal Press & Jennifer Washburn, "The Kept University," *Atlantic Monthly*, March 1, 2000, 39.
178. Krimsky, *Private Interest*, 7.
179. Ibid., 35.
180. See S. Gaylen Bradley, "Managing Conflicting Interests," in Francis L. Macrina, ed., *Scientific Integrity: An Introductory Text with Cases* (Washington, D.C.: ASM Press, 2000), 131–133; Shamoo & Resnik, *Responsible Conduct*, 7, ch. 8; National Academies of Sciences, *On Being a Scientist*, Preface (Washington, D.C.: National Academies Press, 1995) ("trust [in science] will endure only if the scientific community devotes itself to exemplifying and transmitting the values associated with ethical scientific conduct").
181. National Institute of Medicine, National Academies of Sciences, *Integrity in Scientific Research: Creating an Environment That Promotes Responsible Conduct* (Washington, D.C.: National Academies Press, 2002), ch. 5.
182. The Poynter Center may be accessed at www.poynter.indiana.edu. The Online Ethics Center for Engineering and Science may be accessed at www.onlineethics.org.
183. The Duke resource center may be accessed at www.lib.duke.edu/chem/ethics/long.html. The Columbia resource center may be accessed at ccnmtl.columbia.edu/projects/rcr.
184. Task Force on Financial Conflicts of Interest in Clinical Research, Association of American Medical Colleges, *Protecting Subjects, Preserving Trust,*

Promoting Progress II (Washington, D.C.: AAMC, 2002); Task Force on Research Accountability, Association of American Universities, *Report on Individual and Institutional Financial Conflict of Interest* (Washington, D.C.: AAU, 2001); American Society for Microbiology, Promoting Research Integrity at the American Association of Microbiology, April 10, 2000, www.asm.org/Policy/index.asp?bid=3437.

185. Susan Ehringhaus & David Korn, *U.S. Medical School Policies on Individual Financial Conflicts of Interest* (Washington, D.C.: Association of American Medical Colleges, 2004).
186. Mildred K. Cho et al., "Policies on Faculty Conflicts of Interest at US Universities," *JAMA* 284 (2000): 2203, 2208.
187. Ibid., 2205.
188. Ibid., 2205, 2207.
189. Michelle Mello, Brian R. Clarridge, & David M. Studdert, "Academic Medical Centers' Standards for Clinical-Trial Agreements with Industry," *NEJM* 352 (2005): 2202.
190. Patricia Baird, "Commentary: Getting It Right: Industry Sponsorship and Medical Research," *Canadian Medical Association Journal* 168 (2003): 1267.
191. Andrea Jones, "Psychiatrist to Leave Medical Journal Post," *Atlanta Journal-Constitution,* August 29, 2006, B1; Benedict Carey, "Correcting the Errors of Disclosure," *NYT,* July 25, 2006, F5.
192. Catherine D. DeAngelis, "The Influence of Money on Medical Science," *JAMA* 296 (2006): 1001.
193. Donald G. McNeil, Jr., "Tough-Talking Journal Editor Faces Accusations of Lenience," *NYT,* August 1, 2006, F1 (quoting Catherine D. DeAngelis).
194. Washburn, *University, Inc.,* 101 (quoting Dr. Kenneth Rothman); McNeil, "Tough-Talking" (quoting Dr. Thomas P. Stossel).
195. U.S. House Committee on Government Operations, *Are Scientific Misconduct and Conflicts of Interest Hazardous to Our Health? Nineteenth Report by the House Committee on Government Operations,* 101st Cong., 2d Sess. 16–17 (1990); Eliott Marshall, "The Florida Case: Appearances Matter," *Science* 248 (1990): 153.
196. House Committee on Government Operations, *Are Scientific Misconduct,* 17.
197. Marshall, "The Florida Case."
198. Joel Rutchick, "Clinic Executive Out," *Cleveland Plain Dealer,* August 18, 2006, A1.
199. Ibid.
200. Joel Rutchick, "Surgeon Kept Quiet about Stake in Company," *Cleveland Plain Dealer,* August 18, 2006, A1.
201. Ibid.
202. "Symptoms of Conflict," editorial, *Cleveland Plain Dealer,* December 12, 2006, B8.
203. Sarah Treffinger, "Clinic Weighs Conflicts of Interest," *Cleveland Plain Dealer,* September 21, 2006, B2.

204. See, e.g., Jeneen Interlandi, "An Unwelcome Discovery," *NYT,* October 22, 2006, G4 (University of Vermont action against Eric Poehlman); Gareth Cook, "Technology Seen Abetting Manipulation of Research," *Boston Globe,* January 11, 2006, A1 (Massachusetts Institute of Technology action against Luk Van Parijs); "FDA warns Firm on Anthrax Claim," *Dallas Morning News,* November 25, 2005, B1 (University of Texas Medical Branch action against John Heggers).
205. See Environmental Protection Agency, Data Requirements for Registration, 40 C.F.R. pt. 158 (2003) (setting forth a basic core set of over one hundred studies that would assist in determining the effects of pesticides); Environmental Protection Agency, Toxicology Data Requirements, 40 C.F.R. §158.340 (providing a table for all testing requirements and guidelines for pesticides).
206. See, e.g., "Multi-Substance Rule for the Testing of Neurotoxicity," 58 *Fed. Reg.* 40262, 40262–40263 (July 27, 1993) (to be codified at 40 C.F.R. pt. 799) (discussing ten-year history of EPA's efforts to force neurological testing on several toxic substances).
207. Smith, "Medical Journals"; Krimsky, *Science in the Private Interest,* 143, 156–59.
208. Financial Disclosure by Clinical Investigators, 21 C.F.R. pt. 54.
209. *Daubert v. Merrell Dow Pharmaceuticals, Inc.,* 509 U.S. 579, 590 (1993).
210. Angell, *Truth,* 95.
211. Smith, "Medical Journals."
212. Bekelman et al., "Scope and Impact," 454. See also Joel Lexchin et al., "Pharmaceutical Industry Sponsorship and Research Outcome and Quality: Systematic Review," *British Medical Journal* 326 (2003): 1167; Paul M. Ridker & Jose Torres, "Reported Outcomes in Major Cardiovascular Clinical Trials Funded by For-Profit and Not-for-Profit Organizations: 2000–2005," *JAMA* 295 (2006): 2270 (cardiovascular trials funded by for-profit organizations more likely to report positive findings than trials funded by not-for-profit organizations); Mohit Bhandari et. al., "Association between Industry Funding and Statistically Significant Pro-industry Findings in Medical and Surgical Randomized Trials," *Canadian Medical Association Journal* 170 (2004): 477–480. See also John Yaphe et al., "The Association between Funding by Commercial Interests and Study Outcome in Randomized Controlled Drug Trials," *Family Practice* 18 (2001): 565; Syed Ahmer et al., "Conflict of Interest in Psychiatry," *Psychiatric Bulletin* 29 (2005): 302 (manufacturer-supported clinical trials for psychiatric drugs significantly more likely to reach positive results than non-manufacturer-supported trials); Peppercorn, "Association," 1239.
213. Deborah E. Barnes & Lisa A. Bero, "Why Review Articles on the Health Effects of Passive Smoking Reach Different Conclusions," *JAMA* 279 (1998): 1566.
214. Frederick S. vom Saal & Claude Hughes, "An Extensive New Literature Concerning Low-Dose Effects of Bisphenol A Shows the Need for a New Risk Assessment," *Environmental Health Perspectives* 113 (2005): 926.

215. Leonard I. Lesser et al., "Relationship between Funding Source and Conclusion among Nutrition-Related Scientific Articles," PLoS *Medicine* 4 (2007): 5. See also Nestle, *Food Politics,* 114.
216. Gennaro & Tomatis, "Business Bias," 358.
217. See, e.g., Daniel Thau Teitelbaum et al., letter, *Blood* 95 (2000): 2995–2997.
218. See *American Journal of Public Health* 95S (2005) (special symposium supplement); *International Journal of Occupational and Environmental Health,* October–December 2005 (symposium).

5. Hiding Science

1. Sheldon Rampton & John Stauber, *Trust Us, We're Experts* (New York: Putnam, 2001), 214.
2. Bruce M. Psaty & Drummond Rennie, "Stopping Medical Research to Save Money: A Broken Pact with Researchers and Patients," *JAMA* 289 (2003): 2128; Eyal Press & Jennifer Washburn, "The Kept University," *Atlantic Monthly,* March 1, 2000, 39.
3. Sheila Jasanoff, "Transparency in Public Science: Purposes, Reasons, Limits," *Law & Contemporary Problems* 69 (2006): 21.
4. The primary source for the Olivieri story is Mariam Shuchman, *The Drug Trial* (Toronto, Ont.: Random House Canada, 2005). Other less comprehensive sources include Sheldon Krimsky, *Science in the Private Interest* (Oxford: Rowman & Littlefield, 2003), 45; Joseph Weber, "The Doctor vs. the Drugmaker," *Business Week,* November 30, 1998, 87.
5. Shuchman, *Drug Trial,* 141–142.
6. Ibid., 145.
7. Ibid., 149.
8. Ibid., 155.
9. Ibid., 171.
10. Ibid., 261.
11. Weber, "The Doctor." See also David G. Nathan & David J. Weatherall, "Academic Freedom in Clinical Research," *NEJM* 347 (2002): 1368.
12. Shuchman, *Drug Trial,* 349.
13. Ibid., 350.
14. Ibid., 367.
15. Ibid., 363.
16. National Research Council, National Academies of Sciences, *Sharing Publication-Related Data and Materials: Responsibilities of Authorship in the Life Sciences* (Washington, D.C.: National Academies Press, 2003), 4.
17. Gerard E. Markowitz & David Rosner, *Deceit and Denial* (Berkeley: University of California Press 2003), 173, 175 (quoting a March 1974 internal B. F. Goodrich memorandum).
18. W. Page Keeton et al., *Prosser and Keeton on the Law of Torts* §41, at 269–270 (St. Paul, Minn.: West, 5th ed., 1984).

19. Joseph Sanders, "From Science to Evidence: The Testimony on Causation in the Bendectin Cases," *Stanford Law Review* 46 (1993): 1, 53–54.
20. Rebecca Dresser, Wendy Wagner, & Paul Giannelli, "Breast Implants Revisited: Beyond Science on Trial," *Wisconsin Law Review* 1997 (1997): 705; Thomas Koenig & Michael Rustad, "His and Her Tort Reform: Gender Injustice in Disguise," *Washington Law Review* 70 (1995): 1, 39–46.
21. Koenig & Rustad, "His and Her Tort Reform."
22. Scott Lassman, "Transparency and Innuendo: An Alternative to Reactive Over-Disclosure," *Law & Contemporary Problems* 69 (2006): 69, 73–77.
23. Wendy E. Wagner, "Choosing Ignorance in the Manufacture of Toxic Products," *Cornell Law Review* 82 (1997): 773.
24. Melody Petersen, "Bayer Official Offers Defense in Texas Trial of Drug Suit," *NYT,* March 1, 2003, C5.
25. See, e.g., National Research Council, National Academies of Sciences, *Grand Challenges in Environmental Science* (Washington, D.C.: National Academies Press, 2000); National Research Council, National Academies of Sciences, *Identifying Future Drinking Water Contaminants* (Washington, D.C.: National Academies Press, 1999); National Research Council, National Academies of Sciences, *Building a Foundation for Sound Environmental Decisions* (Washington, D.C.: National Academies Press, 1997); National Research Council, National Academies of Sciences, *Research to Protect, Restore and Manage the Environment* (Washington, D.C.: National Academies Press, 1993); National Research Council, National Academies of Sciences, *Toxicity Testing: Strategies to Determine Needs and Priorities* (Washington, D.C.: National Academies Press, 1984).
26. National Research Council, *Toxicity Testing,* 118, fig. 2.
27. Environmental Defense Fund, *Toxic Ignorance* (New York City: Environmental Defense Fund, 1997); Environmental Protection Agency, Office of Pollution Prevention and Toxics, "What Do We Really Know about the Safety of High Production Volume Chemicals?" *BNA Chemical Regulation Reporter* 22 (1998): 261.
28. See 7 U.S.C. §136a (Federal Insecticide, Fungicide and Rodenticide Act); 21 U.S.C. §355 (Food, Drug and Cosmetics Act).
29. See 7 U.S.C. §136a-1 (Federal Insecticide, Fungicide and Rodenticide Act); 15 U.S.C. §§2603 and 2604(e) (2000) (Toxic Substances Control Act); 21 U.S.C. §346a(b)(2) (2000) (Food Quality Protection Act); 21 U.S.C. §§355, 360e(b)(1) (Food, Drug and Cosmetics Act).
30. Wendy E. Wagner, "Commons Ignorance: The Failure of Environmental Law to Produce Needed Information on Health and the Environment," *Duke Law Journal* 53 (2004): 1619, 1671–1673.
31. See 15 U.S.C. §2604(e) (2000) (Toxic Substances Control Act) (permitting the EPA to require additional safety testing if it has reason to suspect that the new or existing chemical "may present" a risk or hazard).
32. National Institute of Medicine, National Academies of Sciences, *The Future of Drug Safety* (Washington, D.C.: National Academies Press, 2006), 2–7.

33. Food and Drug Administration, "Report on the Performance of Drug and Biologics Firms in Conducting Postmarketing Commitment Studies," 71 *Fed. Reg.* 10978 (2006).
34. David Healy, *Let Them Eat Prozac* (New York: New York University Press, 2004), 11–12, 14, 36, 43, 61; Food and Drug Administration, *FDA Public Health Advisory: Suicidality in Children and Adolescents Being Treated with Antidepressant Medications* (Washington, D.C.: Food and Drug Administration, October 15, 2004); Barry Meier, "Two Studies, Two Results, and a Debate over a Drug," *NYT,* June 3, 2004, A1.
35. *Hilliard v. A. H. Robins Co.,* 196 Cal. Rptr. 117, 132 n. 21 (Cal. Ct. App. 1983); Morton Mintz, *At Any Cost: Corporate Greed, Women, and the Dalkon Shield* (New York: Knopf, 1985), 123, 133.
36. 16 U.S.C. §2605 (testing authority for toxic substances); 7 U.S.C. §136a(c)(2)(B) (testing authority for pesticides).
37. Melissa Lee Phillips, "Obstructing Authority," *Environmental Health Perspectives* 114 (2006): A706 (quoting former EPA assistant administrator Lynn Goldman) ("[w]hen you don't have the ability to regulate in your armamentarium, you are in a very weak negotiating position"); U.S. Government Accountability Office, *Chemical Regulation: Options Exist to Improve EPA's Ability to Assess Health Risks and Manage Its Chemical Review Program* (Washington, D.C.: GAO, 2005), 20–22.
38. S. A. Ridlon, letter to TSCA Public Information Office (December 12, 1984), exhibit 5 to deposition of Randy N. Roth, *Communities for a Better Environment v. Unocal Corp.* (No. 997011), 1; ARCO Chemical Company, "Methyl-tertiary Butyl Ether: Critique of the CRS, Inc./Dynamac Corporation Information Review," July 21, 1986, exhibit 25, 5.
39. Environmental Protection Agency, Achieving Clean Air *and* Clean Water: The Report of the Blue Ribbon Panel on Oxygenates in Gasoline, rep. no. 420-R-99-021 (1999), www.epa.gov/oms/consumer/fuels/oxypanel/ r99021 .pdf. See also Thomas O. McGarity, "MTBE: A Precautionary Tale," *Harvard Environmental Law Review* 28 (2004): 281.
40. Wagner, "Choosing Ignorance."
41. *West v. Johnson & Johnson Prods., Inc.,* 220 Cal. Rptr. 437, 445 (Cal. Ct. App. 1985).
42. See, e.g., *Tetuan v. A.H. Robins Co.,* 738 P.2d 1210, 1240 (Kan. 1987); Mintz, *At Any Cost,* 122–123, 134–115.
43. *Hopkins v. Dow Corning Corp.,* 33 F.3d 1116, 1127-1128 (9th Cir. 1994); Rebecca Weisman, "Reforms in Medical Device Regulation: An Examination of the Silicone Gel Breast Implant Debacle," *Golden Gate University Law Review* 23 (1993): 973, 987–988.
44. Paul Brodeur, *Outrageous Misconduct: The Asbestos Industry on Trial* (New York: Knopf, 1985), 118–119, 145, 116–117.
45. Markowitz & Rosner, *Deceit,* ch. 6.
46. Devra Davis, *When Smoke Ran Like Water* (New York: Basic Books, 2002), 196.

47. Stanton Glantz et al., *The Cigarette Papers* (Berkeley: University of California Press, 1996), 15, 58–107; Philip J. Hilts, *Smokescreen: The Truth behind the Tobacco Industry Cover-Up* (Boston: Addison-Wesley, 1996), 38–40.
48. Thomas O. McGarity, "Beyond Buckman: Wrongful Manipulation of the Regulatory Process in the Law of Torts," *Washburn Law Journal* 41 (2002): 549, 559–560.
49. Daniel K. Benjamin, Jr., et al., "Peer-Reviewed Publication of Clinical Trials Completed for Pediatric Exclusivity," *JAMA* 296 (2006): 1266.
50. *State of New York v. GlaxoSmithKline, PLC,* complaint, June 2, 2004, 6–7; Brooke A. Masters, "N.Y. Sues Paxil Maker over Studies on Children," *WPOST,* June 3, 2003, E1.
51. Susan Okie, "Missing Data on Celebrex," *WPOST,* August 5, 2001, A11.
52. Peter Lurie & Allison Zieve, "Sometimes the Silence Can Be Like the Thunder: Access to Pharmaceutical Data at the FDA," *Law & Contemporary Problems* 69 (2006): 85, 86.
53. Anna Wilde Mathews, "Bayer Delayed Report to FDA on Study Results," *WSJ,* September 30, 2006, A7; Gardiner Harris, "F.D.A. Says Bayer Failed to Reveal Drug Risk Study," *NYT,* September 30, 2006, A1.
54. Justin Blum & Eva von Schaper, "Harvard Researcher Forced Bayer to Give Drug Data," *Bloomberg News,* October 6, 2006; Gardiner Harris, "F.D.A. Says Bayer Failed to Reveal Drug Risk Study," *NYT,* September 30, 2006, A1.
55. Blum & von Schaper, "Harvard Researcher."
56. Michael Bowker, *Fatal Deception* (Berkeley, Calif.: University of California Press, 2003), 138.
57. Ibid., 137–144.
58. Carrie Johnson, "W. R. Grace Wants Trial Moved out of Montana," *WPOST,* September 3, 2005, at D2.
59. Brodeur, *Outrageous Misconduct,* 145.
60. Bob Burtman, "The Silent Treatment," *OnEarth,* spring 2002, 14, 18.
61. Frank Davidoff, "New Disease, Old Story," *Annals of Internal Medicine* 129 (1998): 327; Wade Roush et al., "Publishing Sensitive Data: Who Calls the Shots," *Science* 276 (1997): 523. See also Jennifer Washburn, *University, Inc.* (New York: Basic Books, 2005), 76–80.
62. Krimsky, *Science in the Private Interest,* 44–45; Robert R. Kuehn, "Suppression of Environmental Science," *American Journal of Law & Medicine* 30 (2004): 333, 335–336.
63. Aaron S. Kesselhelm & Michelle M. Mello, "Confidentiality Laws and Secrecy in Medical Research: Improving Public Access to Data on Drug Safety," *Health Affairs* 26 (2007): 483.
64. Jasanoff, "Transparency," 22.
65. Thomas O. McGarity & Sidney A. Shapiro, "The Trade Secret Status of Health and Safety Testing Information: Reforming Agency Disclosure Policies," *Harvard Law Review* 93 (1980): 837.
66. Kesselhelm & Mello, "Confidentiality"; McGarity & Shapiro, "Trade Secret."

67. David A. Kessler, *A Question of Intent* (New York: Public Affairs, 2001), 259.
68. Ibid., xiii.
69. Ibid., 133–137.
70. Richard Kluger, *Ashes to Ashes* (New York: Knopf, 1996), 575.
71. Ibid., 575; Kessler, *Question of Intent,* 138.
72. Sophronia S. Gregory, "Is that Smoke, or Do I Smell a Rat?," *Time,* May 9, 1994, 58.
73. Quoted in Kluger, *Ashes to Ashes,* 576.
74. Ibid. See also Kessler, *Question of Intent,* 137–138.
75. Kluger, *Ashes to Ashes,* 576. See also Kessler, *Question of Intent,* 139.
76. Kluger, *Ashes to Ashes,* 576.
77. Kessler, *Question of Intent,* 139.
78. Ibid., 137.
79. Kluger, *Ashes to Ashes,* 362.
80. Alan E. Garfield, "Promises of Silence: Contract Law and Freedom of Speech," *Cornell Law Review* 83 (1998): 261, 269–274; Terry Morehead Dworkin & Elletta Sangrey Callahan, "Buying Silence," *American Business Law Journal* 36 (1998): 151, 155.
81. D. Blumenthal et al., "Withholding Research Results in Academic Life Science: Evidence from a National Survey of Faculty," *JAMA* 277 (1997): 1224.
82. Katherine S. Squibb, "Basic Science at Risk: Why Independent Research Protections Are Critical to Creating Effective Regulations," in Wendy Wagner & Rena Steinzor, eds., *Rescuing Science from Politics: Regulation and the Distortion of Scientific Research* (New York: Cambridge University Press, 2005), 46.
83. David Blumenthal et al., "Relationships between Academic Institutions and Industry in the Life Sciences—An Industry Survey," *NEJM* 334 (1996): 368. See also Eric G. Campbell et al., "Data-Withholding in Academic Genetics: Evidence from a National Survey," *JAMA* 287 (2002): 473.
84. Elina Hemminki et al., "The Courts—A Challenge to Health Technology Assessment," *Science* 285 (1999): 203.
85. Kevin A. Schulman et al., "A National Survey of Provisions in Clinical-Trial Agreements between Medical Schools and Industry Sponsors," *NEJM* 347 (2002): 1335 (some schools unable to negotiate).
86. Robert Steinbrook, "Gag Clauses in Clinical-Trial Agreements," *NEJM* 352 (2005): 2160 (prestigious schools able to negotiate); Schulman, "National Survey" (some schools unable to negotiate).
87. Dworkin & Callahan, "Buying Silence," 156–157, 162; Garfield, "Promises," 303–306, 326–337.
88. Dworkin & Callahan, "Buying Silence," 179–180, 184; Garfield, "Promises," 275.
89. Dworkin & Callahan, "Buying Silence," 153, 186–187; Garfield, "Promises," 275, 295.
90. Miriam Shuchman, "Secrecy in Science: The Flock Worker's Lung Investigation," *Annals of Medicine* 129 (1998): 341, 344.

91. David E. Lilienfeld, "The Silence: The Asbestos Industry and Early Occupational Cancer Research—A Case Study," *American Journal of Public Health* 81 (1991): 791, 793–794.
92. Bowker, *Fatal Deception,* 53, 93; Lilienfeld, "The Silence," 794.
93. *United States v. W. R. Grace, et al.,* indictment, dated February 7, 2005, paragraphs 85–88.
94. Press & Washburn, "Kept University"; Drummond Rennie, "Thyroid Storm," *JAMA* 277 (1997): 1238; Steven A. Rosenberg, "Secrecy in Medical Research," *NEJM* 334 (1996): 392.
95. Kevin Sack & Alicia Mundy, "Over-the-Counter Peril," *LAT,* March 28, 2004, A1.
96. Ibid.
97. Garfield, "Promises," 269–274.
98. Lawrence K. Grossman, "CBS, *60 Minutes,* and the Unseen Interview," *Columbia Journalism Review,* January–February 1996, 39.
99. See 18 U.S.C. §1905 (2000) (Trade Secret Act). See also 15 U.S.C. §2613 (2000) (Toxic Substances Control Act); 42 U.S.C. §6927(b)(2) (2000) (Resource Conservation and Recovery Act); 42 U.S.C. §11045(d)(2) (2000) (Emergency Planning and Community Right to Know Act); 42 U.S.C. §7414(c) (2000) (Clean Air Act).
100. Anna Wilde Mathews & Thomas M. Burton, "After Medtronic Lobbying Push, the FDA Had Change of Heart," *WSJ,* July 9, 2004, A1.
101. Dan Ferber, "Authors Turn up Heat over Disputed Paper," *Science* 304 (2004): 1891.
102. William M. Bulkeley, "Study of Cancers at IBM Is Released," *WSJ,* October 19, 2006, A1.
103. Environmental Protection Agency, "Public Information and Confidentiality: Advance Notice of Proposed Rulemaking; Withdrawal of 1994 Proposed Rule," 65 *Fed. Reg.* 80394, 80395 (December 21, 2000).
104. See generally 40 C.F.R. §§2.201–2.310. See also Christopher J. Lewis, "Comment, When Is a Trade Secret Not So Secret?: The Deficiencies of 40 C.F.R. Part 2, Subpart B," 30 *Environmental Lawyer* (2000):143, 171–172.
105. Environmental Protection Agency, "Public Information and Confidentiality Regulations," 59 *Fed. Reg.* 60446, 60446–60447 (November 23, 1994); U.S. General Accounting Office, *Toxic Substances Control Act: Legislative Changes Could Make the Act More Effective* (Washington, D.C.: GAO, 1994), 56– 58; Hampshire Research Associates, Inc., *Influence of CBI Requirements on TSCA Implementation* (March 1992), 41 (report obtained from EPA through FOIA and on file with authors); see also Environmental Protection Agency, Confidential Business Information (CBI) Review, Pesticides: Freedom of Information Act, www.epa.gov/pesticides/foia/cbi.htm (listing environment-related information that is commonly claimed as confidential).
106. See, e.g., Warren E. Stickle, president, Chemical Producers and Distributors Association, & Bill Balek, president, International Sanitary Supply Association, letter to EPA, www.cpda.com/TeamPublish/uploads/ CPDAISSA Comments.pdf.

107. See, e.g., , 40 C.F.R. pt. 159 (Federal Insecticide Fungicide and Rodenticide Act regulations); Environmental Protection Agency, "Reporting Requirements for Risk/Benefit Information," 62 *Fed. Reg.* 49370 (1997); 21 C.F.R. §§310.305(c), 314.80(c) (Food and Drug Administration regulations); Arnold W. Reitze, Jr., & Lee D. Hoffman, "Self-Reporting and Self-Monitoring Requirements under Environmental Laws," *Environmental Lawyer* 1 (1995): 681, 739–741.
108. 15 U.S.C. §2607(c), (e).
109. 21 C.F.R. §314.80 (drugs); 21 C.F.R. pt. 803 (devices).
110. 29 C.F.R. §1910.1200(g)(5) (OSHA regulations).
111. Rampton & Stauber, *Trust Us,* 215.
112. Sara Shipley, "Study Showed Chemical Was Toxic," *St. Louis Post-Dispatch,* February 29, 2004, C1.
113. Ibid.
114. David Michaels & Celeste Monforten, "Scientific Evidence in the Regulatory System: Manufacturing Uncertainty and the Demise of the Formal Regulatory System," *Journal of Law & Policy* 8 (2005): 17, 20–21; Shipley, "Study Showed Chemical."
115. T. Christian Miller, "Limits Sought on Worker Exposure to Flavor Agent," *LAT,* July 26, 2006, A13; "Popcorn Factory Worker Wins Lawsuit," *Food Processing,* April 1, 2004, 11; Shipley, "Study Showed Chemical" (quoting Dr. David Egliman of Brown University).
116. Linda Greer & Rena Steinzor, "Bad Science," *Environmental Forum,* January–February, 2002, 28, 35.
117. 15 U.S.C. §2607(e).
118. Greer & Steinzor, "Bad Science," 35; Dorothy J. Clarke, "Court Secrecy and the Food and Drug Administration: A Regulatory Alternative to Restricting Secrecy Orders in Product Liability Litigation Involving FDA-Regulated Products," *Food & Drug Law Journal* 49 (1994): 109, 131–135.
119. 42 U.S.C. §9603(a) (Comprehensive Environmental Response, Compensation, and Liability Act); 33 U.S.C. §1321(b)(5) (Clean Water Act).
120. See, e.g., 40 C.F.R. §302.4 (hazardous substance release reporting rules).
121. See Environmental Protection Agency, Reporting Requirements (codified at 40 C.F.R. pt. 159), 49372.
122. 40 C.F.R. §159.158(a) (2003).
123. Ibid., §159.167.
124. Ibid., §159.184.
125. See Steven Shavell, "The Optimal Structure of Law Enforcement," in R. Baldwin, ed., *A Reader on Regulation* (New York: Oxford University Press, 1998), 307, 308–310.
126. Arnold W. Reitze, Jr., *Air Pollution Control Law: Compliance and Enforcement* (Washington, D.C.: Environmental Law Institute, 2001), 491. See also "Testing: Screening Studies for Evaluating Chemicals on TSCA Inventory Suggested at OTA Workshop," *BNA Chemical Regulation Reporter* 19 (1997): 105.

127. Shavell, "Optimal Structure."
128. Alicia Mundy, *Dispensing with the Truth* (New York: St. Martin's Press, 2001), 133–134.
129. Kurt Eichenwald, "Maker Admits It Hid Problems in Artery Device," *NYT,* June 13, 2003, A2.
130. Tom Abate & Todd Wallack, "Prescription for Trouble," *San Francisco Chronicle,* December 22, 2002, A1.
131. "Agency Watch, EPA's Voluntary Data," *National Law Journal,* November 4, 1996, A10.
132. From 1977 to 2006, the EPA had received 15,500 adverse event reports under the Toxic Substances Control Act. See, e.g., James W. Conrad, "Open Secrets: The Widespread Availability of Information About the Health and Environmental Effects of Chemicals," *Law & Contemporary Problems* 69 (2006): 141, 145.
133. Wendy E. Wagner & David Michaels, "Equal Treatment for Regulatory Science," *American Journal of Law & Medicine* 30 (2004): 119, 145–146.
134. DuPont, "Memorandum re C-8 Blood Sampling Results," undated draft [c. August 1981], Bates no. EID79375.
135. Robert A. Bilott, "Request for Immediate Governmental Action/Regulation Relating to DuPont's C-8 Releases in Wood County, West Virginia and Notice of Intent to Sue under the Federal Clean Water Act, Toxic Substances Control Act, and Resources Conservation and Recovery Act," March 6, 2001, EPA docket no. E.006a, Bates no. 000601, 7.
136. Environmental Working Group, DuPont Hid Teflon Pollution for Decades, December 13, 2002, www.ewg.org/issues/PFCs/20021113/20021213.php.
137. Kenneth A. Cook, letter to Michael Leavitt, November 17, 2004, www.ewg.org/issues/pfcs/20041117/index.php.
138. Michael Janofsky, "DuPont to Pay $16.5 Million for Unreported Risks," *NYT,* Dec. 15, 2005.
139. Ibid.
140. *United States v. W. R. Grace, et al.,* indictment, February 7, 2005, paragraphs 103–104.
141. Ibid.
142. Benjamin Weiser & Elsa Walsh, "Drug Firm's Strategy: Avoid Trial, Ask Secrecy," *WPOST*, Oct. 25, 1988, A1.
143. Ibid.
144. Ibid.
145. See Davan Maharaj, "Tire Recall Fuels Drive to Bar Secret Settlements," *LAT,* September 10, 2000 (discussing sealed settlements with the Dalkon Shield); Gina Kolata, "Manufacturer of Faulty Heart Valve Barred Data on Dangers, FDA Says," *NYT,* March 21, 1992, 50 (documenting sealed settlements with Bjork-Shiley heart valves). See also Daniel J. Givelber & Anthony Robbins, "Public Health versus Court-Sponsored Secrecy," *Law and Contemporary Problems* 69 (2006): 131; Alison Lothes, "Quality, Not Quantity: An Analysis of Confidential Settlements and Litigants' Economic

Incentives," *Pennsylvania Law Review* 154 (2005): 433; Laurie K. Dore, "Secrecy by Consent: The Use and Limits of Confidentiality in the Pursuit of Settlement," *Notre Dame Law Review* 74 (1999): 283.

146. Givelber & Robbins, "Public Health."
147. Andrew D. Miller, "Comment: Federal Antisecrecy Legislation: A Model Act to Safeguard the Public from Court-Sanctioned Hidden Hazards," *Boston College Environmental Affairs Law Review* 20 (1993): 371, 372.
148. Dore, "Secrecy," 301–303.
149. Elizabeth E. Spainhour, "Unsealing Settlements: Recent Efforts to Expose Settlement Agreements That Conceal Public Hazards," *North Carolina Law Review* 82 (2004): 2155, 2167. But see Arthur R. Miller, "Confidentiality, Protective Orders, and Public Access to the Courts," *Harvard Law Review* 105 (1992): 427, 480–81 (arguing that cases involving sealed settlements that were available to the author in 1992 were all resolved in ways that did not threaten the public health).
150. Richard L. Marcus, "A Modest Proposal: Recognizing (at Last) That the Federal Rules Do Not Declare That Discovery Is Presumptively Public," *Chicago-Kent Law Review* 81 (2006): 331, 339–340.
151. Ibid., 341.
152. Andrew D. Goldstein, "Sealing and Revealing: Rethinking the Rules Governing Public Access to Information Generated through Litigation, *Chicago-Kent Law Review* 81 (2006): 375, 388–389, 406.
153. Ibid., 383–84. See *Chicago Tribune Co. v. Bridgestone/Firestone, Inc.*, 263 F.3d 1304, 1314–1315 (11th Cir. 2001).
154. Dore, "Secrecy," 396–402; Lothes, "Quality," 442–445.
155. Lothes, "Quality," 435.
156. Dore, "Secrecy," 332–344.
157. Office of Inspector General, Environmental Protection Agency, *EPA's Response to the World Trade Center Collapse: Challenges, Successes, and Areas for Improvement,* report no. 2003-P-00012 (Washington, D.C.: EPA, August 21, 2003), 2.
158. Bowker, *Fatal Deception,* 278.
159. Robert S. Devine, *Bush vs. The Environment* (New York: Anchor, 2004), 182 (quoting EPA Administrator Christine Todd Whitman).
160. Ibid., 180 (quoting EPA statement).
161. Office of Inspector General, *EPA's Response,* 7.
162. Bowker, *Fatal Deception,* 282.
163. Ibid., 282–283.
164. Ibid.
165. Andrew Schneider & David McCumber, *An Air That Kills* (New York: Putnam, 2004), 336–337, 346.
166. Gisela I. Nanauch et al., "Pulmonary Function after Exposure to the World Trade Center Collapse in the New York City Fire Department," *American Journal of Respiratory & Critical Care Medicine* 174 (2006): 312.
167. John R. Balmes, "The World Trade Center Collapse," *American Journal of Respiratory & Critical Care Medicine* 174 (2006): 235.

168. Robin Herbert et al., "The World Trade Center Disaster and the Health of Workers: Five-Year Assessment of a Unique Medical Screening Program," *Environmental Health Perspectives* 114 (2006): 1853. See also John Herzfeld, "Persistent Respiratory Symptoms Found in Mount Sinai World Trade Center Study," *BNA Occupational Safety & Health Reptorter* 36 (2006): 793; Anthony DePalma, "Illness Persisting in 9/11 Workers, Big Study Finds," *NYT,* September 6, 2006, A1.
169. DiPalma, "Illness" (quoting Robin Herbert).
170. Tarek Maassarani, *Redacting the Science of Climate Change* (Government Accountability Project, March 2007), 8–32; Andrew Revkin, "Climate Expert Says NASA Tried to Silence Him," *NYT,* January 29, 2006, A1 (quoting NASA scientist James E. Hansen). See also Donald Kennedy, "The New Gag Rules," *Science* 311 (2006): 917.
171. Tom Hamburger, "Administration Tries to Rein in Scientists," *LAT,* June 26, 2004, A1. See also Susan Okie, "Tensions between CDC, White House," *WPOST,* July 1, 2002, A15 (upper-level HHS officials require scientists at the CDC to speak with "one voice" to the press through departmental officials).
172. Department of Commerce, *Department Administrative Order 219-1* (March 29, 2007), 4, available at http://www.commerce.gov/NewsRoom/PressReleases_FactSheets/PROD01_002841.
173. Timothy Egan, "Shift on Salmon Reignites Fight on Species Law," *NYT,* May 9, 2004, A1. See also Seth Shulman, *Undermining Science* (Berkeley: University of California Press, 2006), 91–92.
174. Kenneth R. Weiss, "Action to Protect Salmon Urged," *LAT,* March 26, 2004, A1. See Ransom A. Myers et al., "Hatcheries and Endangered Salmon," *Science* 303 (2004): 1980.
175. Gardiner Harris, "F.D.A. Links Drugs to Being Suicidal," *NYT,* September 14, 2004, A1; Gardiner Harris, "Antidepressants Restudied for Relation to Child Suicide," *NYT,* June 20, 2004, A20; Elizabeth Shogren, "FDA Sat on Report Linking Suicide, Drugs," *LAT,* April 6, 2004, A13.
176. McGarity & Shapiro, "Trade Secret."

6. Attacking Science

1. David Michaels & Celeste Monforton, "Manufacturing Uncertainty: Contested Science and the Protection of the Public's Health and Environment," *American Journal of Public Health* 95 (2005): S39.
2. Ibid., S40 (citing Brown & Williamson doc. no. 332506, website of Tobacco Documents online, tobaccodocuments.org/ti).
3. Ibid., S39; Linda Rosenstock & Lore J. Lee, "Attacks on Science: The Risks to Evidence-Based Policy," *American Journal of Public Health* 92 (2002): 14.
4. See, e.g., Frank Luntz, Straight Talk, The Environment: A Cleaner, Safer, Healthier America, www.luntzspeak.com/graphics/LuntzResearch.Memo.pdf, 137.

5. Takeshi Hirayama, "Non-smoking Wives of Heavy Smokers Have a Higher Risk of Lung Cancer: A Study from Japan," *British Medical Journal* 282 (1981): 183. See Thomas O. McGarity, "On the Prospect of 'Daubertizing' Judicial Review of Risk Assessment," *Law & Contemporary Problems* 66 (2003): 155, 179–199.
6. Hirayama, "Non-smoking Wives," 183.
7. Ibid.
8. Memorandum, "Non-smoking Wives of Heavy Smokers Have a Higher Risk of Lung Cancer: A Study from Japan: A Critique," January 27, 1981, Bates no. 1002647454; Scientific Affairs Division, memorandum to Horace R. Kornegay, "Recent Paper," January 26, 1981, Bates no. 2024955077.
9. H. J. Eysenck, "More Smoke," *New Scientist* 89 (1981), 494; Martin Rutsch, "Non-smoking Wives of Heavy Smokers Have a Higher Risk of Lung Cancer," *British Medical Journal* 282 (1981): 985; Theodor Sterling, "Communication," *British Medical Journal* 282 (1981): 1156.
10. Takeshi Hirayama, letter, *British Medical Journal* 282 (1981): 1393.
11. P. N. Lee, "Paper by Dr. T. in the *British Medical Journal,* 17th January 1981 'Nonsmoking wives of heavy smokers have a higher risk of lung cancer: A study from Japan': First Comments," January 20, 1981, Bates no. 521024694.
12. Marvin Kastenbaum, memorandum to Frank Colby, February 20, 1981, Bates no. 504886141.
13. P. N. Lee, "Hirayama passive smoking study: Some comments on the critique by Colby," February 19, 1981, Bates no. 504886335.
14. Marvin A. Kastenbaum, memorandum to Horace Kornegay, "Meeting with Prof. C. Tsokos Concerning the Paper," June 1, 1981, Bates no. TI0427-2891; Marvin A. Kastenbaum, memorandum to Horace Kornegay, "Suspected Error in the Paper," June 1, 1981, Bates no. TIBR0016526.
15. Kastenbaum, memorandum to Kornegay, 4.
16. Nathan Mantel, letter to Marvin A. Kastenbaum, June 5, 1981, Bates no. TINY0020382.
17. Ibid., 2.
18. Tobacco Institute, memorandum to Executive Committee and Stan Temko, undated, Bates no. 500651593.
19. Horace Kornegay, telegram to Dr. T. Sugimura, June 10, 1981, Bates no. TINY0020599.
20. Ibid.
21. Tobacco Institute, "News Release from the Tobacco Institute for Immediate use Accompanied by Text of Cablegram to Japan and Memorandum from Dr. Mantel," draft, June 10, 1981, Bates no. 500651585.
22. Samuel D. Chilcote, Jr., letter to Ernest Pepples, June 11, 1981, Bates no. TI11903139.
23. Tobacco Institute, "Promotion of Hirayama and Garfinkel Stories," June, 1981, Bates no. 03739559, tab A.
24. Peter N. Lee, letter to D. G. I. Felton, June 17, 1981, Bates no. 03739559.
25. A. I. B., memorandum to H. B. W., June 18, 1981, Bates no. 521024587.

26. J. K. Wells III, memorandum to E. Pepples, "Smoking and Health—Tim Finnegan," July 24, 1981, Bates no. 521028146.
27. Ibid.
28. "Miscalculation Reported in Study on Cancer in Wives of Smokers," *NYT,* June 15, 1981, B7.
29. "Math Mistake Claimed in Smoking-Peril Study," *Chicago Tribune,* June 15, 1981, A16.
30. "Tobacco Unit Raps Smoking Study," *Boston Globe,* June 16, 1981, A8.
31. Radio TV Reports, Inc., "Broadcast Excerpt, WINS," June 14, 1981.
32. "Tobacco Wars; Is Passive Smoking Harmful?" *Time,* July 6, 1981, 43.
33. See, e.g., Scientific Affairs Division, Tobacco Institute, "Public Smoking in Perspective," September 28, 1981, Bates no. TNWL0039291, 2.
34. See, e.g., Helen E. Longino, "How Values Can Be Good for Science," in Peter Machamer & Gereon Wolters, eds., *Science, Values, and Objectivity* (Pittsburgh: University of Pittsburgh Press, 2004), 127, 140; Steve Fuller, *Science* (Minneapolis: University of Minnesota Press, 1997).
35. Rosenstock & Lee, "Attacks on Science."
36. Devra Davis, *When Smoke Ran Like Water* (New York: Basic Books, 2002), 155–156.
37. See, e.g., Peter Weingart, "Between Science and Values," in Machamer Wolters, *Science, Values, and Objectivity,* 112, 116.
38. Sheila Jasanoff, "Research Subpoenas and the Sociology of Knowledge," *Law & Contemporary Problems* 59 (1996): 95, 98–100.
39. Sheila Jasanoff, *The Fifth Branch: Science Advisers as Policymakers* (Cambridge, Mass: Harvard University Press, 1990), 99.
40. Helen E. Longino, *Science as Social Knowledge* (Princeton, N.J.: Princeton University Press, 1990), 224–225.
41. Philip Kitcher, "Patterns of Scientific Controversies," in Peter Machamer, Marcello Pera, & Aristides Baltas, eds., *Scientific Controversies: Philosophical and Historical Perspectives* (New York: Oxford University Press, 2000), 23. See also Longino, *Science as Social Knowledge,* 79–80, 223–224.
42. U.S. Surgeon General, *Smoking and Health: Report of the Advisory Committee to the Surgeon General* (1964), quoted in Richard Kluger, *Ashes to Ashes* (New York: Knopf, 1996), 255, 258.
43. Alix M. Freedman & Laurie P. Cohen, "How Tobacco Firms Keep Health Questions 'Open' Year after Year," *Sacramento Bee,* February 28, 1993, F1.
44. Council for Tobacco Research—U.S.A., "Special Projects Fund Summary as of May 1, 1968," undated [c. May 1968], Bates no. 1005109708, 2.
45. Council on Tobacco Research, "CTR Special Projects," April 28, 1989, Bates no. 2015007194.
46. David Kessler, *Question of Intent* (New York: Public Affairs, 2001), 204 (quoting anonymous source nicknamed "Veritas").
47. Dan Fagin & Marianne Lavelle, *Toxic Deception* (Secaucus, N.J.: Birch Lane Press, 1996), 32.
48. Marion Nestle, *Food Politics* (Berkeley: University of California Press, 2002), 284.

49. Davis, *Smoke,* 151 (2002); Richard Dahl, "Spheres of Influence," *Environmental Health Perspectives* 109 (2001): A30.
50. David Michaels, Celeste Monforton, & Peter Lurie, "Selected Science: An Industry Campaign to Undermine an OSHA Hexavalent Chromium Standard," *Environmental Health* 5 (2006): 5.
51. Michaels & Monforton, "Manufacturing Uncertainty," S39.
52. Ibid.
53. American Association for the Advancement of Science, *Science for All Americans: A Project 2061 Report on Literacy Goals in Science, Mathematics, and Technology* (New York City: Oxford University Press, 1989), 28; Longino, *Fate of Knowledge,* 206; Longino, *Science as Social Knowledge,* 76–77, 80.
54. Longino, "How Values Can Be Good," 140.
55. C. Thompson, "Memorandum to William Kloepfer, Jr. and the Tobacco Institute, Inc., from Hill and Knowlton, Inc.," October 18, 1968, Bates no. TIMN0071488.
56. Kitcher, "Patterns," 31; Peter Machamer, Marcello Pera, & Aristides Baltas, "Scientific Controversies: An Introduction," in Machamer et al., *Scientific Controversies,* 15–16.
57. Kessler, *Question of Intent,* 169 (quoting memo).
58. Tom Humber, "Memorandum to Ellen Merlo re: ETS," undated [c. January 1993], Bates no. 2024713141.
59. Ibid., 2.
60. Ibid., 3.
61. Ibid., 4–5.
62. Ibid., 7.
63. Ibid.
64. Ibid., 8–9.
65. Ibid., 9.
66. See Gordon C. Rausser et al., "Information Asymmetries, Uncertainties, and Cleanup Delays at Superfund Sites," *Journal of Environmental Economics and Management* 35 (1998): 48, 49; Sidney A. Shapiro & Thomas O. McGarity, "Not So Paradoxical: The Rationale for Technology-Based Regulation," *Duke Law Journal* 1991 (1991): 729, 737–739.
67. The fenoterol account is drawn from Neil Pearce, "Adverse Reactions, Social Responses: A Tale of Two Asthma Mortality Epidemics," in Peter Davis, ed., *Contested Ground* (New York: Oxford University Press, 1996), 57, 62–67.
68. Ibid., 66 (quoting minutes of the panel meeting).
69. See Fagin & Lavelle, *Toxic Deception,* 61.
70. See *Daubert v. Merrell Dow Pharmaceuticals, Inc.,* 509 U.S. 579, 593–595 (1993).
71. R. A. Deyo, "Bruce Psaty and the Risks of Calcium Channel Blokers," *Quality & Safety in Health Care* 11 (2002): 294.

72. Bruce Psaty et. al., "The Risk of Myocardial Infarction Associated with Antihypertensive Drug Therapies," *JAMA* 274 (1995): 620.
73. Ibid.
74. Kluger, *Ashes to Ashes,* 349–358.
75. Ibid., 357 (quoting press release).
76. Ibid., 356–357.
77. Jennifer Washburn, *University, Inc.* (New York: Basic Books, 2005), 14–17; Fred Pearce, "The Great Mexican Maize Scandal," *New Scientist,* June 15, 2002, 14.
78. Washburn, *University, Inc.,* 15.
79. Kara Platoni, "Kernels of Truth," *East Bay* (*Cal.*) *Express,* May 29, 2002.
80. Ibid.
81. Will Lepkowski, "Maize, Genes, and Peer Review," Science and Policy Perspective no. 14 (Phoenix: Center for Science, Policy and Outcomes, Arizona State University), October 31, 2002.
82. "Transgenic Corn Found Growing in Mexico by Dr. Chapela," *Nature* 413 (2001): 337; Lepkowski, "Maize"; Platoni, "Kernels of Truth."
83. Lepkowski, "Maize."
84. Ignacio Chapela & David Quist, "Transgenic DNA Introgressed into Traditional Maize Landraces in Oaxaca, Mexico," *Nature* 413 (2001): 541.
85. Lepkowski, "Maize"; Platoni, "Kernels of Truth."
86. Matthew Metz, "Suspect Evidence of Transgenic Contamination," *Nature* 416 (2002): 600; Nick Kaplinsky et al., "Maize Transgene Results in Mexico Are Artifacts," *Nature* 416 (2002): 600.
87. Emma Marris, "Four Years On, No Transgenes Found in Mexican Maize," *Nature* 436 (2005): 760; John Ross, "Tales of Corn Wars," *National Catholic Reporter,* April 16, 2004, 1; Charles C. Mann, "Has GM Corn 'Invaded' Mexico?" *Science* 295 (2002): 1617.
88. Metz, "Suspect Evidence"; Kaplinsky et al., "Maize Transgene Results," 600.
89. Kaplinsky et al., "Maize Transgene Results."
90. Lepkowski, "Maize."
91. Charles C. Mann, "Transgene Data Deemed Unconvincing," *Science* 296 (2002): 236.
92. Pearce, "Great Mexican Maize Scandal"; Platoni, "Kernels of Truth;" George Monbiot, "The Fake Persuaders," *Guardian* (*London*), May 14, 2002, 15.
93. Mann, "Transgene Data," 236; Lepkowski, "Maize"; Platoni, "Kernels of Truth."
94. Platoni, "Kernels of Truth."
95. Ibid.
96. James Meek, "Science Journal Accused over GM Article," *Guardian,* June 8, 2002, 2.
97. David Quist & Ignacio H. Chapela, "Quist and Chapela Reply," *Nature* 416 (2002): 602.

98. Editorial note, *Nature* 416 (2002): 600. See also Lepkowski, "Maize" (unprecedented step).
99. Platoni, "Kernels of Truth."
100. Jocelyn Kaiser, "Calming Fears, No Foreign Genes Found in Mexico's Maize," *Science* 309 (2005): 1000; Marris, "Four Years On," 760.
101. Davis, *Smoke,* 261.
102. Ibid.
103. Fagin & Lavelle, *Toxic Deception,* 61.
104. Phillip Morris, Inc., EPA Review Draft: Health Effects of Passive Smoking: Assessment of Lung Cancer in Adults and Respiratory Disorders in Children, Comments of Philip Morris, Inc., September 28, 1990, Bates no. 503245278, 46–50.
105. Sheldon Rampton & John Stauber, *Trust Us, We're Experts* (New York: Putnam, 2001), 199; Tobacco Institute, "Public Affairs Management Plan Progress Report, Public Smoking Issue," January 1991, Bates no. TIMN 344674, 2; Tobacco Institute, "Public Affairs Management Plan Progress Report, Public Smoking Issue," April 1990, Bates no. TIMN297148, 2.
106. Henry T. Stelfox et al., "Conflict of Interest in the Debate over Calcium Channel Antagonists," *NEJM* 338 (1998): 101.
107. Jane Levine et al., "Authors' Financial Relationships with the Food and Beverage Industry and Their Published Positions on the Fat Substitute Olestra," *American Journal of Public Health* 93 (2003): 664.
108. Christopher C. Horner, Competitive Enterprise Institute, letter to Information Officer, EPA, Request for Response to/Renewal of Federal Data Quality Act Petition against Further Dissemination of "Climate Action Report 2002," February 10, 2003, www.epa.gov/quality/informationguidelines/documents/7428.pdf.
109. Chris Mooney, "The Cold Earth Society," *Mother Jones,* May–June 2005, 36, 40; Margaret Kriz, "Cold Cash and Global-Warming Research," *National Journal* 37 (2005): 985; Jeff Nesmith, "Nonprofits Push Controversial Climate Study," *Atlanta Journal-Constitution,* June 1, 2003, A1.
110. Chris Mooney, "Some Like It Hot," *Mother Jones,* May–June 2005, 36.
111. Ibid. (quoting Michael Crichton).
112. See Michael Crichton's State of Confusion, December 13, 2005, website of RealClimate, a website about "Climate Science from Climate Scientists," www.realclimate.org; Michael Crichton's State of Confusion II: Return of the Science, December 15, 2005, www.realclimate.org.
113. Lydia Saad, "Global Warming on Public's Back Burner," Gallup Poll Tuesday Briefing, April 20, 2004; Naomi Orsekes, "Beyond the Ivory Tower: The Scientific Consensus on Climate Change," *Science* 306 (2004): 1686.
114. David Michaels, "Doubt Is Their Product," *Scientific American,* June 2005, 95, 99.
115. Ibid., 96.
116. Theodor D. Sterling, "Analysis of the Report on *Cigarette Smoking and*

Health Characteristics, National Center for Health Statistics Series 10, Number 34," undated [c. September 26, 1967], Bates no. 1005109872, 7, 10, 14.

117. E. P. Finch, "Letter to Mr. R. P. Dobson," February 18, 1969, Bates no. 680205529.
118. Ibid.
119. *House Committee on Interstate and Foreign Commerce, Hearings on Cigarette Labeling and Advertising—1969,* 91st Cong., 1st Sess. 930, 931 (1969) (testimony of Theodor Sterling).
120. Ibid., 952; Joseph F. Cullman III, letter to Rep. Richardson Preyer, May 15, 1969, Bates no. 0002608588.
121. Brown & Williamson, "How Eminent Men of Medicine and Science Challenged the Smoking-and-Health Theory during Recent Hearings in the U.S. Congress," July 1, 1969, Bates no. 2015037828.
122. Fagin & Lavelle, *Toxic Deception,* 42–43.
123. Ibid., 63–64.
124. "Science Academy to Focus on Uncertainty in Review of EPA Dioxin Study," *Inside EPA Weekly Report,* March 5, 2004, 9 (NAS study commissioned); Jocelyn Kaiser, "Just How Bad Is Dioxin?" *Science* 288 (2000): 1941 (industry consultant critique); Richard Stone, "Panel Slams EPA's Dioxin Analysis," *Science* 268 (1995): 1124 (EPA Science Advisory Board proceedings); John D. Graham, "Chlorine: Asking the Right Questions," *Chemical Week,* March 15, 1995, 42 (Harvard Center for Risk Analysis critique); Jonathan Adler, "Dioxin Joins List of Costly False Alarms," *LAT,* August 19,1991, B5 (Competitive Enterprise Institute critique); Malcolm Gladwell, "Scientists Temper Views on Cancer-Causing Potential of Dioxin," *WPOST,* May 31, 1990, A3 (industry consultant critique).
125. Gerard E. Markowitz & David Rosner, *Deceit and Denial* (Berkeley: University of California Press, 2003), 257.
126. Ibid., 258 (quoting Louisiana Chemical Association press release of July 7, 1989).
127. "Former OSHA Official Says API-Funded Study Has Potential for Bias in Findings," *BNA Occupational Safety & Health Reporter* 36 (2006): 742; Dina Cappiello, "Oil Industry Funding Study to Contradict Cancer Claims," *Houston Chronicle,* April 29, 2005, A1.
128. Annamaria Baba et al., "Legislating 'Sound Science': The Role of the Tobacco Industry, *American Journal of Public Health* 95 (1995): S20. See Chris Mooney, *The Republican War on Science* (New York: Basic Books, 2005), ch. 5 (an entertaining profile of Tozzi and history of the Information Quality Act).
129. "Group Asks Bush to Issue Executive Order; Harmonizing of Guidelines, Practices Sought," *BNA Environment Reporter* 22 (1991): 605; Jim Tozzi, letter to Marc Firestone, May 7, 1996, Bates no. 2061693747.
130. Philip Morris, "Environmental Tobacco Smoke Update," February 3, 1992, Bates no. 2023004316.
131. "Project Agreement between Multinational Business Services, Inc. and Philip Morris Management Corp.," January 1, 1998, Bates no. 206477900, 2; Philip

Morris, "Projects," undated draft [c. May 1998], Bates no. 2065259668, 2. See also Myron Levin, "Stealth Lobbying Kills Secondary Smoke Proposal," *LAT,* August 17, 1995, A1.

132. Levin, "Stealth Lobbying."
133. Omnibus Appropriations Act for Fiscal Year 1999, Pub. L. No. 105–277, 112 Stat. 2681–495 (1998).
134. Jim Tozzi, "Memorandum to Matthew Winokur re: Language on Data Access in Treasury Appropriations Bill for FY '99 (H.R. 4104)," October 12, 1998, Bates no. 2065231124, 2.
135. Consolidated Appropriations Act of 2001, Pub. L. No. 106–554, §515, 114 Stat. 2763, 2763A-153–154 (2000). See Jim Stimson, "Budget Bill Provision on Data Accuracy Could Open Rules to Industry Challenges," *BNA Chemical Regulation Reporter* 25 (2001): 145.
136. Chris Mooney, "Thanks to a Little-Known Piece of Legislation, Scientists at the EPA and Other Agencies Find Their Work Questioned Not Only by Industry, but by Their Own Government," *Boston Globe,* August 28, 2005, C1.
137. Treasury and General Government Appropriations Act for Fiscal Year 2001, Pub. L. No. 105–554, §515.
138. Office of Management and Budget, "Guidelines for Ensuring and Maximizing the Quality, Objectivity, Utility, and Integrity of Information Disseminated by Federal Agencies; Republication," 67 *Fed. Reg.* 8452 (2002).
139. Michaels, "Doubt," 101.
140. Ibid.; Eric Bailey, "Activists Enlist Unlikely Ally in Bid to Legalize Pot," *LAT,* July 18, 2005, B1.
141. Competitive Enterprise Institute, Petition to Cease Dissemination of the National Assessment on Climate Change (February 20, 2003), www.cei.org/pdf/3360.pdf; Kansas Corn Growers Association, the Triazine Network, & the Center for Regulatory Effectiveness, Request for Correction of Information Contained in the Atrazine Environmental Risk Assessment, docket no. OPP-34237A (Nov. 25, 2002), website of the Center for Regulatory Effectiveness, www.thecre.com/pdf/petition-atrazine2B.pdf, 2; Environmental Protection Agency, Response to Morgan, Lewis, & Bockius Petition, Nov. 24, 2003, www.epa.gov/oeiinter/qualityguidelines/afreqcorrectionsub/12467response-morgan-lewis.pdf; Center for Regulatory Effectiveness, Request to EPA for Correction of "Technical Review of Diisononyl Phthalate (October 16, 2003), www.epa.gov/oei/qualityguidelines/afreqcorrectionsub/13166rfc.pdf.
142. J. P. Kassierer & Joe S. Cecil, "Inconsistency in Evidentiary Standards for Medical Testimony: Disorder in the Courts," *JAMA* 288 (2002): 1382.
143. *Daubert v. Merrell Dow.* See also Joseph Sanders, "Scientific Validity, Admissibility, and Mass Torts after Daubert," *Minnesota Law Review* 78 (1994): 1387, 1389.
144. *Daubert v. Merrell Dow,* 590 & n. 9.
145. Office of Management and Budget, Guidelines.

146. Carl F. Cranor, *Toxic Torts: Science, Law, and the Possibility of Justice* (New York: Cambridge University Press, 2006).
147. See, e.g., Carl F. Cranor, "The Dual Legacy of *Daubert v. Merrell-Dow Pharmaceutical*: Trading Junk Science for Insidious Science," in Wendy Wagner & Rena Steinzor, eds., *Rescuing Science from Politics: Regulation and the Distortion of Scientific Research* (New York: Cambridge University Press 2006), 120, 127–131.
148. Ibid.
149. Sidney A. Shapiro, "The Information Quality Act and Environmental Protection: The Perils of Reform by Appropriations Rider," *William & Mary Environmental Law & Policy Review* 28 (2004): 339, 358.
150. Davis, *Smoke,* 155.
151. Office of Management and Budget, Guidelines, 8460 §V.3.b.ii.B.i. (2002).
152. Ibid., §V.8.
153. Sheila Jasanoff, "Transparency in Public Science: Purposes, Reasons, Limits," *Law & Contemporary Problems* 69 (2006): 21, 29.
154. Sheldon Krimsky, "The Weight of Scientific Evidence in Policy and Law," *American Journal of Public Health* 95 (2005): S129; Cranor, "Dual Legacy," 134–140.
155. Sanders, "Scientific Validity," 1390.
156. Thomas O. McGarity, "On the Prospect of 'Daubertizing' Judicial Review of Risk Assessment," *Law & Contemporary Problems* 66 (2003): 155, 172–178.
157. See Janet Raloff, "Benched Science," *Science News* 168 (2005): 232 (reporting the view of Lloyd Dixon of the Rand Institute for Civil Justice that the impact of *Daubert* and its progeny has been "profound").
158. Lucinda Finley, "Guarding the Gate to the Courthouse: How Trial Judges Are Using Their Evidentiary Screening Role to Remake Tort Causation Rules," *DePaul Law Review* 49 (1999): 335, 336; Michael H. Gottesman, "From Barefoot to Daubert to Joiner: Triple Play or Double Error?" *Arizona Law Review* 40 (1998): 753, 769–770; Raloff, "Benched Science" (reporting the view of Professor Margaret Berger that plaintiffs must "present the most compelling witnesses available . . . or risk having their case dismissed by a judge").
159. *Siharath v. Sandoz Pharmaceuticals Corp.,* 131 F. Supp. 2d 1347 (N.D. Ga. 2001), aff'd sub nom. *Rider v. Sandoz Pharmaceuticals Corp.,* 295 F.3d 1194 (11th Cir. 2002).
160. *Siharath,* 1370.
161. See Davis, *Smoke,* xvii; Thomas O. McGarity, "A Proposal for Linking Culpability and Causation to Ensure Corporate Accountability for Toxic Risks," *William & Mary Environmental Law & Policy Review* 26 (2001): 1, 19.
162. Davis, *Smoke,* 136–138.
163. Rampton & Stauber, *Trust Us,* 302.

164. Ibid., 305–306.
165. Davis, *Smoke,* 138.

7. Harassing Scientists

1. See Robert R. Kuehn, "Scientific Speech: Protecting the Right of Environmental Scientists to Express Professional Opinions," *Environmental Law Reporter* 35 (2005): 10857; Brian Martin, "The Scientific Straightjacket: The Power Structure of Science and the Suppression of Environmental Scholarship," *Ecologist,* January–February 1981, 33, 35.
2. Michael Bowker, *Fatal Deception* (Berkeley: University of California Press, 2003), 99.
3. Ibid., 99–100.
4. Elihu Richter et al., "Efforts to Stop Repression Bias by Protecting Whistleblowers," *International Journal of Occupational and Environmental Health* 7 (2001): 70.
5. Union of Concerned Scientists & Public Employees for Environmental Responsibility, *Survey of NOAA Fisheries Services of Employees* (Washington, D.C., 2005); Public Employees for Environmental Responsibility, *2003 PEER Survey of EPA Region 8 Employees* (Washington, D.C., 2003).
6. Neal Lane, "Politics and Science: A Series of Lessons," *Social Research* 73 (2006): 861, 868.
7. Kuehn, "Scientific Speech," 10858.
8. George Akerlof, "The Market for 'Lemons': Qualitative Uncertainty and the Market Mechanism," *Quarterly Journal of Economics* 84 (1970): 488.
9. Robert K. Merton, "The Matthew Effect in Science," *Science* 159 (1968): 56.
10. Research Triangle Institute, *Survey of Accused but Exonerated Individuals in Research Misconduct Cases* (Rockville, MD: DHHS, ORI, 1996), 17, 20.
11. Robert R. Kuehn, "Suppression of Environmental Science," *American Journal of Law & Medicine* 30 (2004): 333, 334.
12. Mike Rossner & Kenneth M. Yamada, "What's in a Picture? The Temptation of Image Manipulation," *Journal of Cell Biology* 166 (2004): 11 (observing that "[b]eing accused of misconduct initiates a painful process that can disrupt one's research and career").
13. Sheldon Krimsky, *Science in the Private Interest* (Oxford: Rowman & Littlefield, 2003), 190; Gerard E. Markowitz & David Rosner, *Deceit and Denial* (Berkeley: University of California Press, 2003), 135; Sheldon Rampton & John Stauber, *Trust Us, We're Experts* (New York: Putnam, 2001), 95.
14. Rampton & Stauber, *Trust Us,* 96.
15. Krimsky, *Science in the Private Interest,* 190.
16. Markowitz & Rosner, *Deceit,* 136; Devra Davis, *When Smoke Ran Like Water* (New York: Basic Books, 2002), 127–128; Rampton and Stauber, *Trust Us,* 96.
17. Krimsky, *Science in the Private Interest,* 190.
18. Davis, *Smoke,* 128.

19. William Edward Daniel, "Science, Integrity, and Investigators' Rights: Current Challenges," *Journal of Regulatory Toxicology & Pharmacology* 24 (1996): S152, S157–S158; Markowitz and Rosner, *Deceit,* 136–137.
20. Davis, *Smoke,* 129.
21. Richard Jackson et al., "Will Biomonitoring Change How We Regulate Toxic Chemicals?" *Journal of Law, Medicine and Ethics* 30 (2002): 177, 179.
22. See, e.g., Heinz Award, www.heinzawards.net/recipients.asp?action=detail&recipientID=29; Center for Science in the Public Interest Scientific Integrity Award 2003, www.cspinet.org/integrity/press/200211191.html. But see www.foxnews.com/story/0,2933,91600,00.html.
23. 42 U.S.C. §289b.
24. 42 C.F.R. §93.102 (2004). See Council of Science Editors, *CSE's White Paper on Promoting Integrity in Scientific Journal Publications* (Reston, Va: CSE, 2006), 45 (description and history of the Office of Research Integrity).
25. Ibid.
26. See, e.g., University of Texas, Misconduct in Science and Other Activities, www.utexas.edu/policies/hoppm/11.B.01.html, sec. I.
27. Office of Research Integrity, Analysis of Institutional Policies for Responding to Allegations of Scientific Misconduct (2000), ori.dhhs.gov/documents/institutional_policies.pdf, table 3-2.
28. Kuehn, "Scientific Speech," 10866.
29. Paul M. Fischer, "Science and Subpoenas: When Do the Courts Become Instruments of Manipulation?" *Law & Contemporary Problems* 59 (1996): 159, 159; Herbert L. Needleman, "Salem Comes to the National Institute of Health: Notes from Inside the Crucible of Scientific Integrity," *Pediatrics* 90 (1992): 977.
30. Ralph A. Bradshaw, president, Federation of American Societies for Experimental Biology, letter to Donna Shalala, Secretary, DHHS, July 2, 1996, www.faseb.org/opar/hhslet2.html; Ralph A. Bradshaw, letter to William F. Raub, science advisor, Office of Science Policy, DHHS, May 13, 1996, www.faseb.org/opar/cristat.html.
31. Fischer, "Science and Subpoenas"; Needleman, "Salem." See also Richard C. Paddock, "Tobacco Funding of Research Reviewed," *LAT,* March 28, 2007, A1 (scientist funded by tobacco industry filed scientific misconduct charges against prominent tobacco industry critic).
32. Richard A. Deyo et al., "The Messenger under Attack—Intimidation of Researchers by Special-Interest Groups," *NEJM* 336 (1997): 1176, 1176–77; W. Daniell, "Science, Integrity, and Investigators' Rights: Current Challenges," *Journal of Regulatory Toxicology and Pharmacology* 24 (1996): S152.
33. Daryl E. Chubin & Edward J. Hackett, *Peerless Science: Peer Review and U.S. Science Policy* (Albany: State University of New York Press, 1990), 138–153.
34. Ross Gelbspan, *The Heat Is On* (Cambridge, Mass.: Perseus, 1997), 22, 79.
35. Ibid., 80.
36. Ibid.

37. Ibid.
38. Susan K. Avery et al., letter to Ben Santer, July 15, 1996, website of the University Corporation for Atmospheric Research, www.ucar.edu/communications/quarterly/summer96/insert.html, attachment 1.
39. Ibid.
40. Gelbspan, *Heat,* 22, 79–81.
41. William Saletan, "Brokeback Mutton," *WPOST,* February 4, 2007, B2; Andy Dworkin, "Sheep Study Touches off Backlash," *Oregonian,* February 2, 2007, A1; John Schwartz, "Of Gay Sheep, Modern Science and Bad Publicity," *NYT,* January 25, 2007, A1; Andy Dworkin, "OSU, OHSU Study of 'Male-Oriented' Rams Gains High-Profile Foe," *Oregonian,* November 4, 2006, D1.
42. David Kessler, *Question of Intent* (New York: Public Affairs, 2001), 249–50 (RJR-prepared questions for FDA head were asked by Rep. Bilirakis at congressional hearing); Chubin & Hackett, *Peerless Science,* 146.
43. Rodney W. Nichols, "R & D Outlook: Selected Issues on National Policies for Science and Technology," in *Policy Analysis on Major Issues: A Compilation of Papers Prepared for the Commission on the Operation of the Senate,* 94th Cong., 2d Sess. (Washington, D.C.: GPO Comm. Print, 1977), 72, 94.
44. See Chris Mooney, *The Republican War on Science* (New York: Basic Books, 2005), 55–64.
45. *Subcommittee on Energy and Environment of the House Committee on Science, Hearings on Scientific Integrity and Federal Policies and Mandates: Case Study 3—EPA's Dioxin Reassessment,* 104th Cong., 1st Sess. (1995); *Subcommittee on Energy and Environment of the House Committee on Science, Hearings on Scientific Integrity and Public Trust: The Science behind Federal Policies and Mandates: Case Study 2—Climate Models and Projections of Potential Impacts of Global Climate Policy,* 104th Cong., 1st Sess. (1995); *Subcommittee on Energy and Environment of the House Committee on Science, Hearings on Scientific Integrity and Public Trust: The Science Behind Federal Policies and Mandates: Case Study 1—Stratospheric Ozone: Myths and Realities,* 104th Cong., 1st Sess. (1995).
46. *Subcommittee on Energy and Environment, Case Study 1,* 28.
47. Gelbspan, *Heat,* 77.
48. George E. Brown, Jr., *Environmental Science under Siege: Fringe Science and the 104th Congress,* Report by Rep. George E. Brown, Jr., Ranking Democratic Member to the Democratic Caucus of the Committee on Science, U.S. House of Representatives (October 23, 1996).
49. Kessler, *Question of Intent,* 154.
50. Joe Barton, letter to Michael Mann, June 23, 2005, www.energycommerce.house.gov.
51. Ibid., 3.
52. Michael E. Mann, letter to Joe Barton, July 15, 2003, www.realclimate.org/Mann_response_to_Barton.pdf.
53. Ibid., 2.

54. Ralph J. Cicerone, president, National Academies of Sciences, letter to Reps. Joe Barton & Ed Whitfield, July 15, 2005, www.realclimate.org/Cicerone _to_Barton.pdf; Alan I. Leshner, chief executive officer, American Association for the Advancement of Science, letter to Rep. Joe Barton, July 13, 2005, www.aaas.org/news/releases/2005/0714letter.pdf. See also "Hunting Witches," editorial, *WPOST,* July 23, 2005, A16.
55. Andrew Schneider & David McCumber, *An Air That Kills* (New York: Putnam, 2004), 124.
56. Kara Platoni, "Kernels of Truth," *East Bay* (*Cal.*) *Express,* May 29, 2002.
57. Jennifer Washburn, *University, Inc.* (New York: Basic Books, 2005), 16.
58. Rex Dalton, "Ecologist's Tenure Hailed as Triumph for Academic Freedom," *Nature* 435 (2005): 390; Jennifer Jamall, "Chapela Granted Tenure," *Daily Californian,* May 23, 2005.
59. George Canan & Penelope Pring, "Strategic Lawsuits against Political Participation," *Social Problems* 35 (1988): 506. See also Joseph Brecher, "The Public Interest and Intimidation Suits: A New Approach," *Santa Clara Law Review* 28 (1988): 105 (SLAPPs described as "public interest intimidation suits").
60. Geoffrey Cowley et al., "Sweet Dreams or Nightmare," *Newsweek,* August 19, 1991, 44.
61. Michael L. Millenson, "Consumer Concerns Put Focus on FDA Safety Reviews," *Chicago Tribune,* January 26, 1992 (seven million prescriptions).
62. Geoffrey Cowley, "More Halcion Headaches," *Newsweek,* March 7, 1994, 50.
63. Cowley et al., "Sweet Dreams," 44.
64. Gina Kolata, "Maker of Sleeping Pill Hid Data on Side Effects, Researchers Say," *NYT,* January 20, 1992, 1.
65. Clare Dyer, "Upjohn Sues for Libel," *British Medical Journal* 304 (1992): 273; Kolata, "Maker."
66. Cowley, "More Halcion Headaches."
67. Ibid.
68. Chris Mihill & Clare Dyer, "Sleeping Drug Withdrawn after Side-Effect Fears," *Guardian,* October 3, 1991.
69. Dyer, "Upjohn Sues."
70. Clare Dyer, "A Libel Writ That May Tie Tongues," *Guardian,* January 29, 1992.
71. Clare Dyer, "Both Sides Win in Upjohn Libel Case," *British Medical Journal* 308 (1994): 1455.
72. Kuehn, "Scientific Speech," 10864.
73. Washburn, *University, Inc.,* 103–108 (relating controversy between Immune Response Corporation and James Kahn).
74. David Kessler, *Question of Intent,* 169 (quoting memo).
75. Timothy Lytton, *Catholic Clergy and Public Litigation* (Cambridge, Mass.: Harvard University Press, forthcoming 2008); Jules Lobel, "Courts as Forums for Protest," *University of California at Los Angeles Law Review* 52 (2004): 477.

76. Lytton, *Catholic Clergy.*
77. Lobel, "Courts," 481–482.
78. Frank Clifford, "Lawsuit Threat Halts Dump Site Tests Environment," *LAT,* Nov. 22, 1996, A3.
79. Ibid.
80. Kuehn, "Scientific Speech," 10864.
81. Ibid., 10862–10864.
82. Federal Rules of Civil Procedure, Rule 11.
83. Ibid., Rule 11(c)(2).
84. Stephen L. Kling, "Missouri's New Anti-SLAPP Law," *Journal of the Missouri Bar Association* 61 (2005): 124.
85. The Joe Camel description is taken from Fischer, "Science and Subpoenas."
86. See "Court-Ordered Disclosure of Academic Research: A Clash of Values of Science and Law," symposium, *Law & Contemporary Problems* 59 (1996): 1.
87. Steve Wing, "Social Responsibility and Research Ethics in Community Driven Studies of Industrial Hog Production," *Environmental Health Perspectives* 110 (2002): 437, 441.
88. Federal Rules of Civil Procedure, Rule 45 (providing authority to subpoena third-party research if it is relevant to ongoing litigation).
89. Michael Traynor, "Countering the Excessive Subpoena for Scholarly Research," *Law & Contemporary Problems* 59 (1996): 119.
90. Federal Rules of Civil Procedure, Rule 45.
91. Marcia Angell, *Science on Trial* (New York: Norton, 1996), 145–146.
92. Catherine Arnst, "Keeping Lawyers out of the Lab," *Business Week,* February 5, 2007, 54 (quoting Dr. Brad Racette).
93. Brad A. Racette, "The Impact of Litigation on Neurologic Research," *Neurology* 67 (2006): 2124; Arnst, "Keeping Lawyers Out."
94. Steven Picou, "Compelled Disclosure of Scholarly Research: Some Comments on 'High Stakes Litigation,'" *Law & Contemporary Problems* 59 (1996): 149, 155.
95. Wing, "Social Responsibility."
96. Jocelyn Kaiser, "Lead Paint Experts Face a Barrage of Subpoenas," *Science* 309 (2005): 362.
97. Markowitz & Rosner, *Deceit.*
98. Jon Weiner, "Cancer, Chemicals and History," *Nation,* February 7, 2005, 19, 21.
99. Public Law no. 105–277, 112 Stat. 2681–495 (1998).
100. National Research Council, National Academies of Sciences, *Access to Research Data in the 21st Century: An Ongoing Dialogue among Interested Parties: Report of a Workshop* (Washington, D.C.: National Academies Press, 2002), viii.
101. Office of Management and Budget, OMB Circular A-110, "Uniform Administrative Requirements for Grants and Agreements with Institutions of Higher Education, Hospitals, and Other Non-Profit Organizations," 64 *Fed. Reg.* 54926 (1999).

102. Maureen L. Condic and Samuel B. Condic, "The Appropriate Limits of Science in the Formation of Public Policy," *Notre Dame Journal of Law, Ethics & Public Policy* 17 (2003): 157, 170.
103. Kuehn, "Scientific Speech," 10866.
104. Richard Kluger, *Ashes to Ashes* (New York: Knopf, 1996), 160–165.
105. Ibid., 161–162.
106. Kessler, *Question of Intent,* 207.
107. Frank G. Colby, "Memorandum to Peter J. van Every," July 22, 1982, Bates no. 500534218 (Wynder paid $100,000 per year by German tobacco industry).
108. Dan Fagin & Marianne Lavelle, *Toxic Deception* (Secaucus, N.J.: Birch Lane Press, 1996), 62; Kuehn, "Scientific Speech," 10858–10860.
109. Geoffrey Lean, "How I Told the Truth and Was Sacked," *Independent,* March 7, 1999, 11.
110. Martin Teitel & Kimberly A. Wilson, *Genetically Engineered Food: Changing the Nature of Nature* (Rochester, Vt.: Inner Traditions, 1999), 52; Maria Margaronis, "The Politics of Food," *Nation,* December 27, 1999, 11.
111. Kessler, *Question of Intent,* 61.
112. Ibid.
113. Sylvia Pagan Westphal, "Bausch & Lomb Solution Recall Exposes Risks for Eye Infections," *WSJ,* July 26, 2006, A1.
114. Barnaby Feder, "Lens Care Solution Is Faulted," *NYT,* August 23, 2006, C1; Alison Young, "Bausch & Lomb's Solution Infected Eyes, CDC Says," *Atlanta Journal-Constitution,* August 31, 2006, A1.
115. *Senate Committee on Finance, Hearings on FDA, Merck and Vioxx: Putting Patient Safety First?,* 108th Cong., 2d Sess. (2004) (testimony of Dr. Gurkipal Singh); Carolyn Abraham, "The Painful Battle Over the 'Wonder Drug,'" *Globe & Mail,* February 19, 2005, A8; Snigdha Prakash, Documents Suggest Merck Tried to Censor Vioxx Critics, transcript, National Public Radio, June 9, 2005, www.npr.org/templates/story/story.php?storyId=4696609; Snigdha Prakash, Did Merck Try to Censor Vioxx Critics? transcript, National Public Radio, June 9, 2005, www.npr.org/templates/story/story.php?storyId=4696711.
116. Frederick R. Anderson, "Science Advocacy and Scientific Due Process," *Issues in Science and Technology* 16 (Summer 2000): 71, 74.
117. Needleman, "Salem."
118. Douglas M. Birch & Gary Cohn, "Standing up to Industry," *Baltimore Sun,* June 26, 2001, A1.

8. Packaging Science

1. 41 C.F.R. §101-6.100(b)(2)(iii). The obligation of fair balance arises under section 5(c) of the Federal Advisory Committee Act, 5 U.S.C. App. II §5(c).
2. See U.S. General Accounting Office, *EPA's Science Advisory Board Panels: Improved Policies and Procedures Needed to Ensure Independence and Balance* (Washington, D.C.: GAO, June 2001); U.S. General Accounting

Office, *Federal Advisory Committees, Additional Guidance Could Help Agencies Better Ensure Independence and Balance* (Washington, D.C.: GAO, April 2004).

3. Seth Shulman, *Undermining Science* (Berkeley: University of California Press, 2006), 35–36; Dan Ferber, "Overhaul of CDC Panel Revives Lead Safety Debate," *Science* 298 (2002): 732.
4. Shulman, *Undermining Science,* 36.
5. Edward J. Markey, "Turning Lead into Gold" (October 8, 2002), website of Mindfully.org, a group that purports to be apolitical and dedicated to providing honest and useful information, www.mindfully.org/Health/2002/Lead-Into-Gold-MARKEY8oct02.htm.
6. Government Reform Minority Office, Lead Poisoning Advisory Committee, website of House Oversight Committee, www.oversight.house.gov/features/politics_and_science/example_lead_poisoning.htm.
7. Ibid.
8. CDC Childhood Lead Poisoning Prevention Program (CDC CLPPP), Frequently Asked Questions, "Why Not Change the Blood Lead Level of Concern at This Time?" www.cdc.gov/nceh/lead/spotLightst/changeBLL.htm.
9. Dan Ferber, "Critics See a Tilt in a CDC Science Panel," *Science* 297 (2002): 1456.
10. *Chlorine Chemistry Council v. EPA,* 206 F.3d 1286 (D.C. Cir. 2000).
11. Robert Steinbrook, "Science, Politics, and Federal Advisory Committees," *NEJM* 350 (2004): 1454, 1457.
12. Mark R. Powell, *Science at EPA: Information in the Regulatory Process* (Washington, D.C.: RFF Press, 1999), 40.
13. Steinbrook, "Advisory Committees."
14. Steinbrook, "Advisory Committees," 1456.
15. Ibid.
16. Eliot Marshall, "Hit List at the EPA?" *Science* 219 (1983): 1303.
17. Shulman, *Undermining Science,* chapter 8; Steinbrook, "Advisory Committees," 1454.
18. Union of Concerned Scientists, Preeminent Scientists Protest Bush Administration's Misuse of Science, press release, (February 18, 2004), www.ucsusa.org/news/press_release/preeminent-scientists-protest-bush-administrations-misuse-of-science.html.
19. General Accounting Office, *EPA's Science Advisory Board*; General Accounting Office, *Federal Advisory Committees.*
20. American Association for the Advancement of Science, AAAS Resolution Regarding Membership on Federal Advisory Committees (2003), www.aaas.org/news/releases/2003/0305fair2.shtml; American Public Health Association, Ensuring the Scientific Credibility of Government Public Health Advisory Committees (2003), statement 2003–6, www.apha.org/legislative/policy/2003/2003-006.pdf.
21. Rick Weiss, "HHS Seeks Science Advice to Match Bush Views," *WPOST,* September 17, 2002, A1; Aaron Zitner, "Advisors Put under a Microscope,"

LAT, December 23, 2003, A1; Sarah Begley, "Now, Science Panelists Are Picked for Ideology Rather Than Expertise," *WSJ,* December 6, 2002, B1; Ferber, "Critics," 1456; Donald Kennedy, "An Epidemic of Politics," *Science* 299 (2003): 625.

22. Office of Management and Budget, "Final Information Quality Bulletin for Peer Review," 70 *Fed. Reg.* 2664, 2675 §II (2005).
23. Office of Management and Budget, "Proposed Bulletin on Peer Review and Information Quality," 68 *Fed. Reg.* 54023, 54027, §3 (2003).
24. General Accounting Office, *EPA's Science Advisory Board Panels,* 6, 16–18.
25. Steinbrook, "Advisory Committees," 1455.
26. "Religion and Medicine Don't Mix on Fed Panel," editorial, *Seattle Times,* February 25, 2003, B4; Dirk Johnson, "A Nation Bound by Faith," *Newsweek,* February 24, 2003, 18; David Michaels et al., "Advice without Dissent," editorial, *Science* 298 (2002): 703.
27. Michaels, "Advice without Dissent."
28. General Accounting Office, *Federal Advisory Committees,* 5.
29. Michaels, "Advice without Dissent."
30. General Accounting Office, *EPA Science Advisory Board,* 18–19.
31. Ibid., apps. 2 and 3.
32. See ibid., 29, 38, 41, 46.
33. Ibid., 16–19, 29–30, 43–44.
34. Food and Drug Administration, Draft Guidance on Disclosure of Conflicts of Interest for Special Government Employees Participating in FDA Product Specific Advisory Committees (January 2002), www.fda.gov/oc/guidance/advisorycommittee.html. See also Peter Lurie et al., "Financial Conflict of Interest Disclosure and Voting Patterns at Food and Drug Administration Drug Advisory Committee Meetings," *JAMA* 295 (2006): 1921, 1922.
35. Lurie et al., "Financial Conflict of Interest Disclosure."
36. Ricardo Alonso-Zaldivar, "Senator Questions FDA Panelists' Ties to Industry," *LAT,* February 26, 2005, A16; Center for Science in the Public Interest, Conflicts of Interest on COX-2 Panel, www.cspinet.org/new/200502251.html. But see "Pitt Doctor Disputes Report of FDA Conflict," *Pittsburgh Post-Gazette,* February 26, 2005, B11.
37. Food and Drug Administration, *Draft Guidance for the Public, FDA Advisory Committee Members, and FDA Staff on Procedures for Determining Conflicts of Interest and Eligibility for Participation in FDA Advisory Committees* (Washington, D.C.: DHHS, FDA, 2007); Food and Drug Administration, "Draft Guidance for the Public, FDA Advisory Committee Members, and FDA Staff on Procedures for Determining Conflicts of Interest and Eligibility for Participation in FDA Advisory Committees; Availability," 72 *Fed. Reg.* 13805 (2007).
38. General Accounting Office, *EPA's Science Advisory Board;* General Accounting Office, *Federal Advisory Committees.*
39. Michael Jacobson, Introduction to *Lifting the Veil of Secrecy: Corporate Support for Health and Environmental Professional Associations, Charities,*

and Industry Front Groups (Washington, D.C.: Center for Science in the Public Interest, 2003), ix.

40. Jo Revill & Paul Harris, "Tackling Obesity: America Stirs up a Sugar Rebellion," *Observer,* January 18, 2004, 20.
41. International Life Sciences Institute home page, www.ilsi.org/about/index.cfm?pubentityid=3.
42. The account of the FAO consultation is drawn from Sarah Boseley, "Sugar Industry's Cash Sweetener to UN Food Report," *Guardian,* October 9, 2004, 14.
43. Ibid.
44. Ibid.
45. Ibid.
46. Tim Lougheed, "WHO/ILSI Affiliation Sustained," *Environmental Health Perspectives* 144 (2006): A521; Sarah Boseley, "WHO 'Infiltrated by Food Industry,'" *Guardian,* January 9, 2003, 1; Laura Burton, "Food: A Spoonful of Propaganda," *Guardian,* April 12, 2002, 7.
47. Lougheed, "WHO/ILSI Affiliation."
48. Charles Fulwood, "Alar Report Right from the Start, but You'd Never Know It," *Public Relations Quarterly* 41 (1996): 9.
49. Gerard E. Markowitz & David Rosner, *Deceit and Denial* (Berkeley: University of California Press, 2003), 288 (quots. are from the ACSH).
50. Jim Motavalli, "Science for Sale?" *E,* March 1, 2000, 23.
51. "The ACSH," *Consumer Reports,* May 1994, 316.
52. Ibid.; Fulwood, "Alar Report"; Motavalli, "Science for Sale."
54. "The ACSH," 316.
54. Pat Phibbs, "Council on Science, Health Urges Changes in Use of Animals to Predict Risk to Humans," *BNA Product Safety and Liability Reporter* 33 (2005): 128.
55. Jacobson, introduction, ix.
56. ACOEM Overview, www.acoem.org/aboutACOEM.aspx.
57. American College of Occupational and Environmental Medicine, Adverse Human Health Effects Associated with Molds in the Indoor Environment (2002), www.acoem.org/guidelines.aspx?id=850&print=1.
58. David Armstrong, "Amid Suits over Mold, Experts Wear Two Hats," *WSJ,* January 9, 2007, A1.
59. Lisa Cosgrove et al., "Financial Ties between DSM-IV Panel Members and the Pharmaceutical Industry," *Psychotherapy & Psychosomatics* 75 (2006): 154; Judith Graham, "Top Mental Health Guide Questioned," *Chicago Tribune,* April 20, 2006, C8; Shankar Vedantam, "Experts Defining Mental Disorders Are Linked to Drug Firms," *WPOST,* April 20, 2006, A7.
60. Alicia Mundy, "Hot Flash, Cold Cash," *Washington Monthly,* January–February 2003, 35.
61. Jacqui Drope & Simon Chapman, "Tobacco Industry Efforts at Discrediting Scientific Knowledge of Environmental Tobacco Smoke: A Review of Internal Industry Documents," *Journal of Epidemiology & Community Health* 55 (2001): 588, 589.

62. Tobacco Institute, "Public Affairs Management Plan Progress Report, Public Smoking Issue," April 1990, Bates no. TIMN297141, 1.
63. Diana Avedon, "Memorandum to Regional Vice Presidents re: Resources—Witnesses," December 3, 1992, Bates no. TIDN0025717.
64. Tobacco Institute, "Agenda, ETS Coordinating Committee," June 28, 1990, Bates no. 507851044, 3–4.
65. Drope & Chapman, "Tobacco Industry Efforts."
66. Chris Mooney, "Blinded by Science," *Columbia Journalism Review,* November–December 2004, 5.
67. See, e.g., Ross Gelbspan, *The Heat Is On* (Cambridge, Mass.: Perseus, 1997), 44; Chris Mooney, "The Cold Earth Society," *Mother Jones,* May–June 2005, 41; Fact Sheet: Sallie Baliunas, website of Greenpeace, www.exxonsecrets.org; Eli Kintisch, "Global Warming Skeptic Argues U.S. Position in Suit," *Science* 308 (2005): 482; Fact Sheet: Frederick Seitz, www.exxonsecrets.org.; Fact Sheet: Patrick J. Michaels, www.exxonsecrets.org.
68. See, e.g., Tobacco Institute, "Public Affairs Management Plan Progress Report," March 1991, Bates no. TIMN344588, 2; Tobacco Institute, "Public Affairs Management Plan Progress Report, Public Smoking Issue," April 1990, Bates no. TIMN297141.
69. See, e.g., Philip Morris, "Goal: Generated More Balance Media Presentation of the ETS Issue," March 5, 1991, Bates no. 2033610038.
70. Veterinary Quality Assurance Task Force, Report, March 29, 2007, us.iams.com/global/Vet_Quality_Assurance_Report.htm.
71. Glenn Hess, "CIR Panel Finds Phthalates Safe for Cosmetics Applications," *Chemical Market Reporter,* November 25, 2002, 1.
72. Ibid.
73. Joseph Odorcich et al., "Letter to Joseph A. Califano and Ray Marshall," March 20, 1978, Bates no. DO83048.
74. Sam Roe, "Death of a Safety Plan," *Pittsburgh Post-Gazette,* March 31, 1999, A1.
75. Ian Higgens et al., letter to Joseph A. Califano and Ray Marshall, February 10, 1978, Bates no. ME007469.
76. Thomas O. McGarity, "Resisting Regulation with Blue Ribbon Panels," *Fordham Urban Law Journal* 33 (2006): 1157, 1162–1171.
77. Alicia Mundy, *Dispensing with the Truth* (New York: St. Martin's Press, 2001), 109.
78. Ibid., 109–110.
79. Kara Sissel, "Study Dismisses Bisphenol-A as Endocrine Disrupter," *Chemical Week,* September 8, 2004, 28.
80. Ibid.
81. Frederick S. vom Saal & Claude Hughes, "An Extensive New Literature Concerning Low-Dose Effects of Bisphenol A Shows the Need for a New Risk Assessment," *Environmental Health Perspectives* 113 (2005): 926.
82. Motavalli, "Science for Sale," 23.
83. Bette Hileman, "Panel Ranks Risks of Common Phthalate," *Chemical & Engineering News,* November 14, 2005, 32.

84. Environmental Protection Agency, Office of Pollution Prevention and Toxics, *Draft Risk Assessment of the Potential Human Health Effects Associated with Exposure to Perfluorooctanoic Acid and Its Salts* (Washington, D.C.: EPA, January 4, 2005).
85. Brian Ford, "ACSH Challenges Animal Tests as Cancer Indicator," *Chemical News & Intelligence,* April 3, 2005, 1; "Teflon-Production Chemical Does Not Pose Health Risk to General Population, Science Panel Finds," *Medical News Today,* March 19, 2005, www.medicalnewstoday.com. See also Henry Miller, editorial, *Chicago Sun-Times,* August 29, 2005, 49.
86. Stephanie Saul, "Panel Urges Limits on Use of a Heart Drug," *NYT,* June 15, 2005, C5.
87. Stephanie Saul, "Expert Panel Gives Advice That Surprises a Drug Maker," *NYT,* August 9, 2005, C1.
88. Jonathan Gabe & Michael Bury, "Anxious Times: The Benzodiazepine Controversy and the Fracturing of Expert Authority," in Peter Davis, ed., *Contested Ground* (New York: Oxford University Press, 1996), 47.
89. Philip Morris, "Revised 1993 Budget: European Consultant Program," November 13, 1992, Bates no. 2028398173, 4.
90. Travis Madsen & Sujatha Jahagirdar, *Perchlorate and Children's Health* (Los Angeles: Environment California Research and Policy Center, 2005), 17; Peter Waldman, "The Fight to Limit Regulation of a Military Pollutant," *WSJ,* December 29, 2005, A1; Stan Lim, "The Stakes Are High," *Riverside (CA) Press-Enterprise,* December 19, 2004, A1.
91. "Perchlorate State-of-the-Science Symposium 2003," preliminary program brochure, undated [c. September 2003] available at http://www.percholoratesymposium.com/prelim_program.pdf.
92. University of Nebraska Medical Center, "Scientists Reach Consensus on Key Scientific Issues Related to Perchlorate," press release, October 8, 2003.
93. Ibid.
94. Jim Tozzi, "Memorandum to Jim Boland, Tom Borelli, and Ted Lattanzio," December 29, 1993, Bates no. 2024207174, 2 (touting the virtues of industry funding of a conference through an institute that did not take funding from corporations); British American Tobacco Company, "Environmental Tobacco Smoke: Improving the Quality of Public Debate: Strategies," November, 1990, Bates no. 507958704, 17.
95. Susan Okie, "Scientists Question Objectivity of Top NIH Nutrition Official," *WPOST,* December 24, 1985, A6.
96. Paul Singer, "Food Fight," *National Journal* 36 (2004): 3760.
97. Barry Commoner, "Keynote Address: The Political History of Dioxin," Second Citizens Conference on Dioxin (July 30, 1994), website of Green Parties world wide, www.greens.org/s-r/078/07-03.html; Center for Health, Environment and Justice, *Behind Closed Doors* (Falls Church, Va: CHEJ, 2001), 4.
98. Commoner, "Political History"; Leslie Roberts, "Flap Erupts over Dioxin Meeting," *Science* 251 (1991): 866.

99. Roberts, "Flap Erupts Over Dioxin."
100. "A Tale of Science and Industry," *Rachel's Hazardous Waste News,* no. 248, August 28, 1991.
101. Leslie Roberts, "Dioxin Risks Revisited," *Science* 251 (1991): 624.
102. Christine Gorman, "The Double Take on Dioxin," *Time,* August 26, 1991, 52.
103. Keith Schneider, "U.S. Backing Away from Saying Dioxin Is a Deadly Peril," *NYT,* August 15, 1991, A1.
104. Jocelyn Kaiser, "Panel Backs Dioxin Reassessment," *Science* 290 (2000): 1071.
105. Martin B. Powers and Otto P. Preuss, memorandum to J. E. Gulick re: Proposed Program of Filling Need for New and Accurate Beryllium Health and Safety Information, January 23, 1987, Bates no. E0070413, 1.
106. Ibid., 1.
107. Ibid., 1.
108. Ibid., 1.
109. Ibid., 2.
110. Brush Wellman, "Five-Year Strategic Plan, Office of Environmental Affairs (1987–1991)," undated [c. 1986], Bates no. D101541.
111. Sam Roe, "Thought Control: Brush Devised Strategy to Shape Knowledge," *Pittsburgh Post-Gazette,* April 2, 1999, A1.
112. Richard C. Davis, "Letter to James Slawski," February 15, 1991, Bates no. D101328.
113. Sam Roe, "Thought Control."
114. Ibid.
115. British American Tobacco Company, "Environmental Tobacco Smoke," 20.
116. Sheldon Rampton & John Stauber, *Trust Us, We're Experts* (New York: Putnam, 2001), 200.
117. Tobacco Institute, "EPA Projects," August 23, 1991, Bates no. 2025481023.
118. Deborah E. Barnes & Lisa A. Bero, "Why Review Articles on the Health Effects of Passive Smoking Reach Different Conclusions," *JAMA* 279 (1998): 1566, 1568.
119. Melody Petersen, "Court Papers Suggest Scale of Drug's Use," *NYT,* May 30, 2003, C1.
120. David H. Jacobs, "Psychiatric Drugging: Forty Years of Pseudo-Science, Self-Interest, and Indifference to Harm," *Journal of Mind and Behavior* 16 (1995): 421.
121. Richard Stone, "Dioxin Report Faces Scientific Gauntlet," *Science* 265 (1994): 1650; "Dioxin and the EPA: The Science and Politics of Regulation," *Environmental Review Newsletter,* May 1995, www.environmentalreview.org/vol2/mattison.html.
122. Jeff Nesmith, "Global Warming Study Sets off Storm," *Austin American-Statesman,* July 7, 2003, A1; Jeff Nesmith, "Nonprofits Push Controversial Climate Study," *Atlanta Journal-Constitution,* June 1, 2003, A1; Antonio Regalado, "Global Warming Skeptics Are Facing Storm Clouds," *WSJ,* July 31,

2003, 1; William Weiss, "Smoking and Cancer: A Rebuttal," *American Journal of Public Health* 65 (1975): 954.

123. Quoted in Chris Mooney, "Some Like It Hot," *Mother Jones,* May–June 2005, 39.
124. David Michaels, *Manufacturing Uncertainty* (New York: Oxford University Press, forthcoming 2008). Names of other publicly identified sponsors for the organization are available at the website of the International Society for Regulatory Toxicology and Pharmacology, www.isrtp.org/sponsors.htm. They include two large risk management consulting groups and Frito-Lay.
125. Michaels, *Manufacturing Uncertainty.*
126. Ibid.
127. See International Society for Regulatory Toxicology and Pharmacology, Submission Instructions, www.isrtp.org/nonmembers/directions.htm
128. Paul D. Thacker, "Enviromental Journals Feel Pressure to Adopt Disclosure Rules," *Environmental Science & Technology* 40 (2006): 6873.
129. Ibid.

9. Spinning Science

1. John H. Cushman, Jr., "After *Silent Spring,* Industry Put Spin on All It Brewed," *NYT,* March 26, 2001, A14.
2. Sheldon Rampton & John Stauber, *Trust Us, We're Experts* (New York: Putnam, 2001), chs. 1–2.
3. Gerard E. Markowitz & David Rosner, *Deceit and Denial* (Berkeley: University of California Press, 2003), 8.
4. Rampton & Stauber, *Trust Us,* 22–23.
5. Ibid., 28–29.
6. Luis R. Varela, "Assessment of the Association between Passive Smoking and Lung Cancer," May 1987, Bates no. 2028463883; Thomas J. Borelli, "Untitled Response to Questions Posed by Richard Kluger on November 12, 1991," undated [c. November 1991], Bates no. 2021162632; Thomas Borelli, memorandum to Alexander Holtzman re: Varela Dissertation, March 31, 1992, Bates no. 2023037601. See Richard Kluger, *Ashes to Ashes* (New York: Knopf, 1996), 694–696.
7. Varela, "Assessment," 154.
8. "Philip Morris Starts Offensive with Report," *Richmond (VA) Times-Dispatch,* May 5, 1990.
9. Alan Leviton, "Letter to Marvin Bienstock," February 15, 1990, Bates no. 2021162804, 3.
10. Ibid.
11. Ibid.
12. Borelli, "Untitled Response," 1; Borelli, memorandum to Holtzman, 1.
13. Thomas J. Borelli, letter to William Farland, March 9, 1990, Bates no. 2023546021; Philip Morris, draft letter to William Farland, February 26, 1990, Bates no. 2023989053 (fax heading on the draft letter indicates that it was faxed from Newman to Tom Borelli).

14. William K. Reilly, letter to Joseph F. Cullman, June 15, 1990, Bates no. 2026127716; Environmental Protection Agency, "EPA's Statement on Current Activities Related to Environmental Tobacco Smoke (Passive Smoking)," May 9, 1990, Bates no. 2046667023; Rudy Abramson and Myron Levin, "EPA Will Link Second-Hand Smoke, Cancer," *LAT,* May 9, 1990, 1 (quoting Robert Axelrad).
15. Environmental Protection Agency, "EPA's Statement."
16. Newman Partnership, Ltd, memorandum to Tom Borelli re: Varela News Conference, March 19, 1990, Bates no. 2023550279.
17. Newman Partnership, Ltd, & Edelman Group, memorandum to Tom Borelli re: Varela Study Communications Strategy, draft, March 23, 1990, Bates no. 2021162582.
18. Newman Partnership, Ltd, "Memorandum to Tom Borelli," 1.
19. Ibid.
20. Thomas J. Borelli, "Form Letter to Scientists," April 30, 1990, Bates no. 2021162900. See also Thomas J. Borelli, letter to J. Donald Millar, April 30, 1990, Bates no. 2046667026.
21. Joe Cullman, letter to William K. Reilly, April 30, 1990, Bates no. 2023388013.
22. Terry Atlas, "Cig Maker Airs Chilling EPA Draft on Second Hand Smoke," *Philadelphia Daily News,* May 10, 1990.
23. John Craddock, "A New Tact? Report on Passive Smoking Marks New Aggressiveness for Philip Morris," *Saint Petersburg (FL) Times,* April 29, 1990.
24. Craddock, "A New Tact."
25. Ibid.
26. Mike Allen, "Philip Morris Starts Offensive with Report," *Richmond (VA) Times-Dispatch,* May 5, 1990.
27. Environmental Protection Agency, "EPA's Statement."
28. Abramson & Levin, "EPA Will Link." See Philip J. Hilts, "Wide Peril Is Seen in Passive Smoking," *NYT,* May 10, 1990, A25; Tobacco Institute, "Public Affairs Management Plan Progress Report, Media Relations," May 1990, Bates no. TIMN194311, 1 (noting that "[t]he story, carried on the *Los Angeles Times* wire service prompted other media to run the story").
29. Terry Atlas, "EPA Raises Estimate of Deaths," *Chicago Tribune,* May 10, 1990, 3.
30. William Kloepfer, Jr., "Public Relations in the Nation's Capital," speech was delivered in Washington, D.C., September 16, 1971, Bates no. TIMN0122370.
31. Ibid., 5–6.
32. Ibid., 8.
33. Ibid.
34. Peter Huber, *Galileo's Revenge: Junk Science in the Courtroom* (New York: Basic Books, 1991).
35. Rampton & Stauber, *Trust Us,* 252.
36. Elisa K. Ong & Stanton P. Glantz, "Constructing 'Sound Science' and

'Good Epidemiology': Tobacco, Lawyers, and the Public Relations Firms," *American Journal of Public Health* 91 (2001): 1749; Rampton & Stauber, *Trust Us,* 258.

37. Marian Burros & Donald G. McNeil, Jr., "U.S. Inspections for Mad Cow Lag Those Done Abroad," *NYT,* December 24, 2003, A19.
38. Sheryl Gay Stolberg, "Administration Backs a Food-Labeling Delay," *NYT,* January 22, 2004, A18.
39. Frank Lutz, Straight Talk, The Environment: A Cleaner, Safer, Healthier America, www.luntzspeak.com/graphics/LuntzResearch.Memo.pdf, 137–138.
40. See Michael Gough, "Science, Risks and Politics," in M. Gough, ed., *Politicizing Science: The Alchemists of Policymaking* (Stanford, Calif.: Hoover Institution, 2003), 1, 9–10 (arguing that environmental organizations' publicity efforts "can keep assertions alive despite mounting scientific evidence that they are wrong"); Steven F. Hayward, "Environmental Science and Public Policy," *Social Research* 73 (2006): 891, 903–912 (criticizing environmental groups for using science to frame overly alarmist messages).
41. Ronald E. Gots, *Toxic Risks: Science, Regulation, and Perception* (Boca Raton, Fla.: CRC Press, 1993) 248–249.
42. Joseph D. Rosen, "Much Ado about Alar," *Issues in Science & Technology* 5 (Fall 1990): 85, 87–88.
43. American Council on Science and Health, "Our Food Supply Is Safe," advertisement, *NYT,* April 5, 1989, A11 (Apple Institute waged a damage control campaign that included "full-page advertisements in more than thirty newspapers saying "An apple a day is still good advice"); Julie Kosterlitz, "The Food Lobby's Menu," *National Journal* 22 (1990): 2336.
44. Leslie Roberts, "Pesticides and Kids," *Science* 243 (1989): 1280.
45. Kosterlitz, "Food Lobby's Menu"; Leslie Roberts, "Alar: The Numbers Game," *Science* 243 (1989): 1430.
46. Gots, *Toxic Risks,* 251.
47. Rampton & Stauber, *Trust Us,* 229.
48. Gots, *Toxic Risks,* 251.
49. See Daniel Sarewitz, *Frontiers of Illusion: Science, Technology, and the Politics of Progress* 88, 93 (Philadelphia: Temple University Press, 1996); Chris Mooney, "Beware Sound Science: It's Doublespeak for Trouble," *WPOST,* February 29, 2004, B2; Celia Campbell-Mohn & John Applegate, "Federal Risk Legislation: Some Guidelines," *Harvard Environmental Law Review* 23 (1999): 93, 105–107 & app. 2.
50. David Michaels, "Manufacturing Uncertainty: Contested Science and the Protection of the Public's Health and Environment," *American Journal of Public Health* 95 (2005): S41.
51. Alvin M. Weinberg, "Science and Trans-Science," *Minerva* 10 (1972): 209, 220. See Howard Latin, "Regulatory Failure, Administrative Incentives, and the New Clean Air Act," *Environmental Law* 21 (1991): 1647, 1662; Frederica Perera & Catherine Petito, "Formaldehyde: A Question of Cancer Policy?" *Science* 216 (1982): 1285, 1290.

52. See U.S. Congress, Office of Technology Assessment, *Identifying and Regulating Carcinogens: Background Paper* (Washington, D.C.: GPO, 1987), 106 (discussing seven air toxic standards promulgated by the EPA and extraordinary delays associated with them); U.S. General Accounting Office, *Delays in EPA's Regulation of Hazardous Air Pollutants* (Washington, D.C.: GPO, 1983), i (noting that four of thirty-seven hazardous substances identified for possible regulation in 1977 had been regulated by 1983).
53. Quoted in Rampton & Stauber, *Trust Us,* 58.
54. Bob Norman, "The GOP's Brain," *New Times Broward–Palm Beach (FL),* April 22, 2004, A1.
55. Marc Pilisuk et al., "Public Perception of Technological Risk," *Sociology of Science Journal* 24 (1987): 403, 407.
56. National Science Board, *Science & Engineering Indicators—2004* (Arlington, Va.: National Science Foundation, 2004), 7–27.
57. See Jon D. Miller, *The American People and Science Policy: The Role of Public Attitudes in the Policy Process* (Burlington, Mass.: Elsevier, 1983), 92.
58. Rampton & Stauber, *Trust Us,* 268.
59. Fred Pearce, "Experts with a Price on Their Heads," *Guardian,* May 7, 1998, A2; John H. Cushman, "Industrial Group Plans to Battle Climate Treaty," *NYT,* April 26, 1998, A1.
60. Chris Mooney, "Blinded by Science," *Columbia Journalism Review,* November–December 2004, 6; Cushman, "Industrial Group."
61. Ryan Myers, "Newspaper Publisher Says Claims of Jury Tampering 'Ridiculous,'" *Beaumont (TX) Enterprise,* April 14, 2007, A1.
62. Quoted in Trudy Lieberman, *Slanting the Story* (New York: New Press, 2000), 34.
63. David Kessler, *Question of Intent* (New York: Public Affairs, 2001), xiii.
64. Howard Marder, "Letter to James E. Gulick," February 21, 1989, Bates no. RD006548.
65. Ibid., 1–2.
66. Ibid., 3.
67. Ibid.
68. Ibid.
69. Hill & Knowlton, Division of Scientific, Technical, and Environmental Affairs, memorandum, undated [c. February 1989], Bates no. WM033491, 3, 6.
70. Ibid., 17–19.
71. Ibid., 20–22.
72. Ibid., 25–26.
73. Ibid., 4–15.
74. Rampton & Stauber, *Trust Us,* 22.
75. Ross Gelbspan, *The Heat Is On* (Cambridge, Mass.: Perseus, 1997), 37.
76. P. Terrence Gaffney, letter to Jane Brooks, April 29, 2003, pubs.acs.org/subscribe/journals/esthag-w/2006/feb/business/pt_weinberg.html.
77. Ibid.

78. Louis Jacobson, "PR's Brass-Knuckled Boys," *National Journal* 34 (2002): 1960.
79. Ibid., 1960, 1961.
80. Kessler, *Question of Intent,* 208.
81. Markowitz & Rosner, *Deceit,* 7.
82. Rampton & Stauber, *Trust Us,* 272.
83. Tom Humber, memorandum to Steve Parrish re: EPA Decision-Day Plan, November 25, 1991, Bates no. 2021183574, 6.
84. Tobacco Institute, "Public Affairs Operating Plan—1991," undated [c. December 1990], Bates no. 507660330, 5.
85. Ibid.
86. Susan M. Stuntz, "Memorandum to the Members of the EPA/OSHA Task Force," April 15, 1991, Bates no. 2021172926.
87. Greenpeace, *Denial and Deception: A Chronicle of ExxonMobil's Efforts to Corrupt the Debate on Global Warming* (Washington, D.C.: Greenpeace, May 2002), 3.
88. Rampton & Stauber, *Trust Us,* 283.
89. Ibid., 283.
90. Ibid., 147.
91. Dan Fagin & Marianne Lavelle, *Toxic Deception* (Secaucus, N.J.: Birch Lane Press, 1996), 57.
92. Lieberman, *Slanting,* 4.
93. Curtis Moore, "Rethinking the Think Tanks," *Sierra,* July–August 2002, 56, 58.
94. John H. Cushman, "Industrial Group Plans to Battle Climate Treaty," *NYT,* April 26, 1998, A1; David Malakiff, "Advocacy Mailing Draws Fire," *Science* 280 (1998): 195.
95. Rampton & Stauber, *Trust Us,* 278, 281–282.
96. Ibid., 281–282.
97. Moore, "Rethinking," 56, 57; Chris Mooney, "Some Like It Hot," *Mother Jones,* May–June 2005, 36, 39.
98. Moore, "Rethinking," 56, 58.
99. Rampton & Stauber, *Trust Us,* 306.
100. Ong & Glantz, "Constructing," 1750; Jonathan M. Samet & Thomas A. Burke, "Turning Science into Junk: The Tobacco Industry and Passive Smoking," *American Journal of Public Health* 91 (2001): 1742, 1743.
101. Rampton & Stauber, *Trust Us,* 239 (quoting Philip Morris memo). See also Chris Mooney, *The Republican War on Science* (New York: Basic Books, 2005), 65–69.
102. Rampton & Stauber, *Trust Us,* 270.
103. Sam Fulwood, "Budget Ax Brings on Fencing Match," *LAT,* January 30, 1995, A19.
104. National Sleep Foundation, Mission and Goals, www.sleepfoundation.org/site/c.huIXKjM0IxF/b.2418935/k.E1AF/Mission_and_Goals.htm.
105. Dorsey Griffith & Steve Wiegand, "A Little Too Cozy? Not-for-Profits May

Have Undisclosed Ties to For-Profit Drug Companies," *Pittsburgh Post-Gazette,* July 13, 2005, D2.

106. Ibid. See also Trudy Lieberman, "Bitter Pill," *Columbia Journalism Review,* July–August, 2005, 45; Duff Wilson, "Many New Drugs Have Strong Dose of Media Hype," *Seattle Times,* June 29, 2005, A1.
107. See American Diabetes Association, www.diabetes.org (description of ADA); Cadbury Schweppes, "Cadbury Schweppes Americas Beverages Joins American Diabetes Association in the Fight Against Diabetes, Obesity," press release, April 21, 2005.
108. Marc Santora, "In Diabetes Fight, Raising Cash and Keeping Trust," *NYT,* November 25, 2006, A1; "Diabetes Association Defends Cadbury Schweppes Deal," *Corporate Crime Reporter* 20 (2005): 19 (interview with Richard Kahn, chief scientific and medical officer, American Diabetes Association).
109. Tom Humber, "Memorandum to Ellen Merlo re: ETS," undated [c. January 1993], Bates no. 2024713141, 3.
110. Ibid.
111. Ibid.
112. Ibid.
113. Thomas O. McGarity, "On the Prospect of Daubertizing Judicial Review of Risk Assessment," *Law & Contemporary Problems* 66 (2003): 155.
114. Tom Humber, memorandum to Steve Parrish re: "Decision Day Plan," November 25, 1991, Bates no. 2021183574.
115. Ibid.
116. Thomas Humber, letter to Steve Parrish, November 26, 1991, Bates no. 2021183593, 1.
117. Ibid., 1.
118. Ibid., 2.
119. Ibid., 6.
120. Peter Brimelow & Leslie Spencer, "You Can't Get There from Here," *Forbes,* July 6, 1992, 59.
121. Ibid., 60.
122. Ibid., 64.
123. Ibid.
124. Lisa Antilla, "Climate of Skepticism: U.S. Newspaper Coverage of the Science of Climate Change," *Global Environmental Change* 15 (2005): 338, 350; Maxwell T. Boykoff & Jules M. Boykoff, "Balance as Bias: Global Warming and the US Prestige Press," *Global Environmental Change* 14 (2003): 125, 126.
125. Mooney, *Republican War,* 267.
126. Gelbspan, *Heat Is On,* 33.
127. Ben Goldacre, "Don't Dumb Me Down," *The Guardian,* September 8, 2005, 4.
128. Jane Gregory, *Science in Public: Communication, Culture, and Credibility* (New York: Plenum Press, 1993), 106–107.
129. Rampton & Stauber, *Trust Us,* 22.

130. Campaign for Tobacco-Free Kids, "As Historic FDA Ruling Approaches, New Poll Shows Overwhelming Public Support for Government Regulation of Tobacco Marketing Aimed at Kids," press release, August 7, 1996, Bates no.2070909276 (providing a Fenton Communications contact).
131. Philip Morris Companies, Inc., "Consulting Fees, Burson-Marsteller," undated [c. January 1992], Bates no. 2046989196.
132. Gelbspan, *Heat Is On,* 33. See also Mooney, "Blinded," 4–5.
133. Paul D. Thacker, "The Junkman Climbs to the Top," *Environmental Science and Technology Online,* May 11, 2005, pubs.acs.org/subscribe/journals/esthag-w/2005/may/business/pt_junkscience.html.
134. Mooney, *Republican War,* ch. 8; Rampton & Stauber, *Trust Us,* 248, 251.
135. Rampton & Stauber, *Trust Us,* 248.
136. Thacker, "Junkman"; Mooney, *Republican War,* 83, 113, 126.
137. Rampton & Stauber, *Trust Us,* 23.
138. Jim Morris, "In Chemical Industry, Image Counts for a Lot," *Houston Chronicle,* October 23, 1998, 2.
139. Quoted in Rampton & Stauber, *Trust Us,* 58.
140. Roger A. Pielke, "When Scientists Politicize Science," *Regulation,* Spring 2006, 28, 31.
141. Lieberman, *Slanting,* 5.

10. Restoring Science

1. Neil K. Komesar, *Imperfect Alternatives: Choosing Institutions in Law* (Chicago: University of Chicago Press, 1995), 8, 167–168, 192.
2. Ibid., 8, 164–168.
3. William M. Sage, "Regulating through Information: Disclosure Laws and American Health Care," *Columbia Law Review* 99 (1999): 1701; Bradley Karkkainen, "Information as Environmental Regulation: TRI and Performance Benchmarking, Precursor to a New Paradigm?" *Georgetown Law Journal* 89 (2001): 257.
4. Council of Science Editors, *CSE's White Paper on Promoting Integrity in Scientific Journal Publications* (Reston, VA: 2006), 9; International Committee of Medical Journal Editors, *Uniform Requirements for Manuscripts Submitted to Biomedical Journals* (ICMJE, February 2006).
5. Adil E. Shamoo & David B. Resnik, *Responsible Conduct of Research* (New York: Oxford University Press, 2003), 139–42; but see Robert Hahn, *The False Promise of "Full Disclosure,"* AEI-Brookings Joint Center working paper 02–02 (Washington, D.C.: AEI-Brookings, November 2002).
6. Catherine D. DeAngelis et al., "Reporting Financial Conflicts of Interest and Relationships between Investigators and Research Sponsors," *JAMA* 286 (2001): 89; Jeffrey M. Drazen & Gregory D. Curfman, "Financial Associations of Authors," *NEJM* 346 (2002): 1901.
7. Jerome P. Kassirer, *On the Take* (New York: Oxford University Press,

2005), 21; Merrill Goozner, *Unrevealed: Non-disclosure of Conflicts of Interest In Four Leading Medical and Scientific Journals* (Washington, D.C.: Center for Science in the Public Interest, 2004), 10; Linda Greer & Rena Steinzor, "Bad Science," *Environmental Forum,* January–February 2002, 13.

8. Council of Science Editors, *CSE's White Paper,* 2.
9. International Committee of Medical Journal Editors, *Uniform Requirements.*
10. Drazen & Curfman, "Financial Associations," 1901.
11. Kassirer, *On the Take,* 21–22.
12. Catherine D. DeAngelis, "The Influence of Money on Medical Science," *JAMA* 296 (2006): 996–998.
13. Donald G. McNeil, "Tough-Talking Journal Editor Faces Accusations of Leniency," *NYT,* August 1, 2006, F1; Paul D. Thacker, "Journals Feel Pressure to Adopt Disclosure Rules," *Environmental Science & Technology,* 40 (2006): 6873, pubs.acs.org/subscribe/journals/esthag-w/2006/sep/policy/pt_disclosure.html.
14. David Armstrong, "Medical Reviews Face Criticism over Lapses," *WSJ,* July 19, 2006, B1.
15. DeAngelis, "The Influence of Money"; McNeil, "Tough-Talking."
16. Peter Dizikes, "Under the Microscope," *Boston Globe,* January 22, 2006, E1 (quoting editors of *NEJM* and the *Lancet*).
17. Sheldon Krimsky, *Science in the Private Interest* (Oxford: Rowman & Littlefield, 2003), 169–170.
18. Goozner, *Unrevealed,* 8.
19. Ibid., 7.
20. Jennifer Washburn, *University, Inc.* (New York: Basic Books, 2005), 101 (quoting Dr. Kenneth Rothman).
21. McNeil, "Tough-Talking" (quoting Dr. Thomas P. Stossel).
22. Armstrong, "Medical Reviews," B1.
23. Arlene Weintraub, "How the Journals Are Cracking Down," *Business Week,* October 23, 2006, 76 (quoting *JAMA* editor Catherine D. DeAngelis).
24. See Jennifer A. Henderson & John J. Smith, "Financial Conflict of Interest in Medical Research: Overview and Analysis of Federal and State Controls," *Food & Drug Law Journal* 57 (2002): 445, 455 (noticing in the area of biomedical research that "both federal and state controls provide a relative lack of prospective guidance as to what constitutes acceptable institutional conflict policy").
25. Financial Disclosure by Clinical Investigators, 21 C.F.R. §54 (2003).
26. See ibid. (making no distinction between sponsor-controlled research and research where the sponsor relinquishes control).
27. Publication Policies for Science, www.sciencemag.org/help/authors/policies.dtl.
28. International Committee of Medical Journal Editors, *Uniform Requirements.*
29. 31 U.S.C. §3729.
30. Krimsky, *Science in the Private Interest,* 168; Kevin A. Schulman et al.,

"A National Survey of Provisions in Clinical-Trial Agreements between Medical Schools and Industry Sponsors," *NEJM* 347 (2002): 1335–1341.

31. Anna Wilde Mathews, "More Data to Ensure Full Disclosure of Drug Risks, Trial Gaps," *WSJ,* May 10, 2005, A1.
32. Patricia Huston & David Moher, "Redundancy, Disaggregation, and the Integrity of Medical Research," *Lancet* 347 (1996): 1024, 1026.
33. Sheila Jasanoff, "Transparency in Public Science: Purposes, Reasons, Limits," *Law & Contemporary Problems* 69 (2006): 21, 41.
34. *Daubert v. Merrell Dow Pharmaceuticals, Inc.,* 509 U.S. 579, 590 (1993).
35. See *National Bank of Commerce v. Dow Chemical Co.,* 965 F. Supp. 1490, 1516 (E.D. Ark. 1996); *Daubert v. Merrell Dow Pharmaceuticals Corp.,* 43 F.3d 1311, 1317 (9th Cir. 1995) (*Daubert* II); William L. Anderson, Barry M. Parsons, & Drummond Rennie, "*Daubert*'s Backwash: Litigation-Generated Science," *University of Michigan Journal of Law Reform* 34 (2001): 619, 658 n. 212 (citing many cases); but see *Worthington City Schools v. ABCO Insulation,* 616 N.E.2d 550 (Ohio Ct. App. 1992) (court allows testimony based on study prepared for litigation).
36. In re *Silica Products Liability Litigation,* 398 F. Supp. 2d 563 (S.D. Tex. 2005).
37. Anderson et al., "*Daubert*'s Backwash," 654–655.
38. Ibid., 659 & n. 216 (citing cases).
39. National Research Council, National Academies of Sciences, *Sharing Publication-Related Data and Materials: Responsibilities of Authorship in the Life Sciences* (Washington, D.C.: National Academies Press, 2003), 4 (advocating a "uniform principle for sharing integral data and materials expeditiously" or UPSIDE); National Research Council, National Academies of Sciences, *Responsible Science* (Washington, D.C.: National Academies Press, 1992), 11 (noting that scientists "are generally expected to exchange research data as well as unique research materials that are essential to the replication or extension of reported findings"); Robert K. Merton, *The Sociology of Science,* ed. Norman Storer (Chicago: University of Chicago Press, 1973).
40. Trudo Lemmens, *Piercing the Veil of Corporate Secrecy about Clinical Trials,* Hastings Center Report, September 1, 2004, 14.
41. Robert Steinbrook, "Gag Clauses in Clinical-Trial Agreements," *NEJM* 352 (2005): 2160, 2161; Gregory M. Lamb, "Veil of Secrecy to Lift on Drug Tests," *Christian Science Monitor,* September 13, 2004, 11 (quoting Arthur Caplan, director of the Center for Bioethics at the University of Pennsylvania).
42. Elizabeth T. Fontham et al., "Lung Cancer in Nonsmoking Women: A Multicenter Case-Control Study," *Cancer Epidemiology Biomarkers Prevention* 1 (1991): 35–43.
43. Richard A. Carchman, letter to Elizabeth T. H. Fontham, November 12, 1996, Bates no. 2060567119.
44. Ibid.; Jim J. Tozzi, draft letter to Elizabeth Fontham, September 10, 1996, Bates no. 2063640314.

45. Study investigators, letter to Richard A. Carchman, March 11, 1997, Bates no. 2063778589.
46. Philip Morris, "Summary of Analysis of Fontham Study Results," May 10, 1995, Bates no. 2050761128.
47. Annamaria Baba, Daniel Cook, Thomas McGarity, & Lisa Bero, "Legislating 'Sound Science': Role of the Tobacco Industry," *American Journal of Public Health* 95 (2005): S20, S22.
48. *Bluitt v. R. J. Reynolds Tobacco Co.,* 1994 U.S. Dist. LEXIS 16933 (E.D. La. 1994).
49. *Wolpin v. Philip Morris Inc.,* 189 F.R.D. 418 (C.D. Cal. 1999).
50. Baba et al., "Legislating," S22.
51. Philip Morris, "Force Field Analysis," draft, undated [c. May 1997], Bates no. 2081324814. See also Baba et al., "Legislating," S22.
52. Philip Morris, "Force Field Analysis."
53. The Shelby Amendment was passed as a rider to the Omnibus Consolidated and Emergency Appropriations Act for Fiscal Year 1999, Pub. L. No. 105–277, 112 Stat. 2681 (1998) and requires the OMB to amend Circular A-110 to require "[f]ederal awarding agencies to ensure that all data produced under an award will be made available to the public through the procedures established under the Freedom of Information Act."
54. Richard Shelby, "Accountability and Transparency: Public Access to Federally Funded Research Data," *Harvard Journal on Legislation* 37 (2000): 369, 380.
55. 112 Stat. at 2681 (1998).
56. See Office of Management and Budget, Circular A-110, "Uniform Administrative Requirements for Grants and Agreements with Institutions of Higher Education, Hospitals, and Other Non-profit Organizations," 64 *Fed. Reg.* 54926, 54926 (1999) (announcing the final guidelines and discussing the comments received on the draft guidelines).
57. Ibid., 54926.
58. National Research Council, National Academies of Sciences, *Access to Research Data in the 21st Century: An Ongoing Dialogue among Interested Parties: Report of a Workshop* (Washingon, D.C.: National Academies Press, 2002), viii.
59. Ibid., 27.
60. David Healy, *Let Them Eat Prozac* (New York: New York University Press, 2004), 278. See also Lamb, "Veil of Secrecy" (quoting similar views of Arthur Caplan, director of the Center for Bioethics at the University of Pennsylvania).
61. 5 U.S.C. §552.
62. Thomas O. McGarity, "Federal Regulation of Mad Cow Disease Risks," *Administrative Law Review* 57 (2005): 292, 318–319.
63. See, e.g., FDA Modernization Act of 1997, 21 U.S.C. §301; H.R. Conf. Rep. 105–399 105th Cong. 1st Sess. (1997) (stating requirements for registry); Shankar Vedantam, "Drugmakers Prefer Silence On Test Data," *WPOST,* July 6, 2004, A1.

64. Vedantam, "Drugmakers Prefer Silence."
65. Ibid.
66. Ibid.
67. Greer & Steinzor, "Bad Science," 14.
68. David Michaels, "Foreword: Sarbanes-Oxley for Science," *Law and Contemporary Problems* 60 (2006): 1, 16–17.
69. Thomas O. McGarity & Sidney A. Shapiro, "The Trade Secret Status of Health and Safety Testing Information: Reforming Agency Disclosure Policies," *Harvard Law Review* 93 (1980): 837, 876–878.
70. Eliot Marshall, "Epidemiologists Wary of Opening up Their Data," *Science* 290 (2000): 28, 29.
71. See Mary L. Lyndon, "Secrecy and Access in an Innovation Intensive Economy: Reordering Information Privileges in Environmental, Health, and Safety Law," *Colorado Law Review* 78 (2007): 465.
72. National Research Council, *Access to Research Data* (Washington, D.C.: National Academies Press, 2002) viii, 2, 6, 14; American Association for the Advancement of Science, AAAS Policy Brief: Access to Data (August 2005), www.aaas.org/spp/cstc/briefs/accesstodata/index.shtml#final.
73. David Willman, "Drug Tied to Deaths Is Pulled," *LAT,* August 9, 2001, A1.
74. David Brown, "Cholesterol Drug Taken off Market," *WPOST,* August 9, 2001, A1.
75. Bruce M. Psaty et al., "Potential for Conflict of Interest in the Evaluation of Suspected Adverse Drug Reactions," *JAMA* 292 (2004): 2622, 2623, table 1.
76. Ibid.
77. Ibid., 2628.
78. Ibid.
79. Melody Petersen, "Bayer Official Offers Defense in Texas Trial of Drug Suit," *NYT,* March 1, 2003, C5.
80. Psaty et al., "Potential for Conflict," 2628.
81. Melody Petersen & Alex Berenson, "Papers Indicate That Bayer Knew of Dangers of Its Cholesterol Drug," *NYT,* February 22, 2003, A1.
82. See National Research Council, National Academies of Sciences, *Adverse Drug Event Reporting: The Roles of Consumers and Health-Care Professionals: Workshop Summary* (Washington, D.C.: National Academies Press, 2007), 2; Margaret M. Gary & Donna J. Harrison, "Analysis of Severe Adverse Events Related to the Use of Mifepristone as an Abortifacient," *Annals of Pharmacotherapy* 40 (2006): 191, 195.
83. See 7 U.S.C. §136d(a)(2) (2000) (Federal Insecticide, Fungicide and Rodenticide Act); 15 U.S.C. §2607(c), (e) (2000) (Toxic Substances Control Act); 21 C.F.R. §§310.305(c), 314.80(c) (Food, Drug and Cosmetics Act regulations); 29 C.F.R. §1910.1200(g)(5) (OSHA regulations). See generally Arnold W. Reitze, Jr. & Lee D. Hoffman, "Self-Reporting and Self-Monitoring Requirements under Environmental Laws," *Environmental Lawyer* 1 (1995): 681, 739–741.

84. Michael Baram, "Making Clinical Trials Safer for Human Subjects," *American Journal of Law & Medicine* 27 (2001): 253, 262.
85. See generally 40 C.F.R. pt. 159 (2003) (outlining reporting requirements for risk/benefit information); Environmental Protection Agency, "Reporting requirements for Risk/Benefit Information," 62 *Fed. Reg.* 49370 (September 19, 1997) (codified at 40 C.F.R. pt. 159).
86. The EPA's current website of Toxic Substances Control Act adverse reports could serve as a model in this regard. See EPA, TSCA8(e) and FYI Submissions, www.epa.gov/opptintr/tsca8e/doc/8esub/2003/8e0102_011503.htm.
87. Bob Roehr, "Stronger Sanctions Needed against Companies That Suppress Data," *British Medical Journal* 329 (2004): 132.
88. See Environmental Protection Agency, "Reporting Requirements for Risk/Benefit Information," 49388.
89. See Environmental Protection Agency, "Reporting Requirement for Risk/Benefit Information; Amendment and Correction," 62 *Fed. Reg.* 49388 (September 19, 1997) (codified at 40 C.F.R. pt. 159) (omitting "agents" from definition of registrants).
90. Under the Toxic Substances Control Act, "any person who has possession of a study" is among those required to report relevant health and safety studies on a toxic substance to the EPA. 15 U.S.C. §2607(d) (2000). A clearer definition of "study" could impose substantially greater demands on both researchers and sponsors.
91. See, e.g., Drummond Rennie, "Thyroid Storm," *JAMA* 277 (1997): 1238.
92. S. 470, H.R. 3195, 109th Cong., 1st Sess. (2005). See Aaron S. Kesselhelm & Michelle M. Mello, "Confidentiality Laws and Secrecy in Medical Research: Improving Public Access to Data on Drug Safety," *Health Affairs* 26 (2007): 483; Steinbrook, "Gag Clauses," 2161.
93. See e.g., Section 801 of the Food and Drug Administration Amendments Act of 2007, H. R. 3580, 110th Congress, 1st Sess (2007).
94. See, e.g., ibid., §§915 and 921.
95. See, e.g., ibid., §901.
96. Jennifer Washburn, "Universities for Sale," *LAT,* July 21, 2006, 13.
97. Chris B. Pascal, "Managing Data for Integrity: Policies and Procedures for Ensuring the Accuracy and Quality of Data in the Laboratory," *Science and Engineering Ethics* 12 (2006): 23, 30–32 (describing policies at three institutions).
98. 46 C.F.R. Subpart A (responsibility for protection of human subjects); 42 C.F.R. §93.300 (responsibility for identifying and investigating scientific misconduct).
99. U.S. General Accounting Office, *University Research: Most Federal Agencies Need to Better Protect against Financial Conflict of Interest* (Washington, D.C.: GAO, 2003), 5–6; see also Washburn, *University, Inc.,* 100.
100. General Accounting Office, *University Research,* 3.
101. Mildred K. Cho et al., "Policies on Faculty Conflicts of Interest at US Universities," *JAMA* 284 (2000): 2203, 2205.

102. General Accounting Office, *University Research,* 9–10.
103. Susan Okie, "A Stand for Scientific Independence," *WPOST,* August 5, 2001, A1 (quoting Lisa Bero).
104. Pascal, "Managing Data," 25; see also Troyen A. Brennan et al., "Health Industry Practices That Create Conflicts of Interest," *JAMA* 295 (2006): 429, 430 (suggesting that academic medical centers should "more strongly regulate, and in some cases prohibit, many common practices that constitute conflicts of interest with drug and medical device companies").
105. Washburn, *University, Inc.,* 100 (rules vary); Cho et al., "Policies on Faculty Conflicts," 2208 (rules are vague).
106. Washburn, *University, Inc.,* 98.
107. Ibid., 99–100.
108. Ibid., 96–97.
109. Ibid.
110. Eli Kintisch, "BP Bets Big on UC Berkeley for Novel Biofuels Center," *Science* 315 (2007): 747; Lynnley Browning, "BMW's Custom-Made University," *NYT,* August 29, 2006, C1.
111. Rick DelVecchio, "Cal Sees BP Deal as Landmark," *San Francisco Chronicle,* February 2, 2007, B1. See also Jennifer Washburn, "Big Oil Buys Berkeley," *LAT,* March 24, 2007, C23.
112. Washburn, *University, Inc.,* 146.
113. Ibid., 159–160.
114. Ibid., 140; Rebecca Buckman, "More Universities Increasing Support for Campus Start-Ups," *WSJ,* November 27, 2006, B1.
115. Washburn, *University, Inc.,* 236–237.
116. See Peter D. Blumberg, "Comment: From 'Publish or Perish' to 'Profit or Perish': Revenues from University Technology Transfer and the 501(c)(3) Tax Exemption," *University of Pennsylvania Law Review* 145 (1996): 89.
117. David Willman, "Drug Trials with a Dose of Doubt," *LAT,* July 16, 2006, A1; David Willman, "Income from Two Sources," *LAT,* July 16, 2006, A29; David Willman, "Drug Maker Hired NIH Researcher," *LAT,* December 7, 1998, A1; David Willman, "Scientists Who Judged Pill Safety Received Fees," *LAT,* October 29, 1999, A22; David Willman, "Second NIH Researcher to Become Part of Conflict Probe," *LAT,* September 4, 1999, A12.
118. Chris Mooney, *The Republican War on Science* (New York: Basic Books, 2005), 252 (noting that "[t]he evidence suggests that many journalists reporting on science issues fall easy prey to sophisticated public relations campaigns").
119. Ross Gelbspan, *The Heat Is On* (Cambridge, Mass: Perseus, 1998), 57–58.
120. Maxwell T. Boykoff & Jules M. Boykoff, "Balance as Bias: Global Warming and the US Prestige Press," *Global Environmental Change* 14 (2003): 125, 126.
121. Quoted in ibid., 159.
122. Ibid., 126.
123. Liisa Antilla, "Climate of Skepticism: U.S. Newspaper Coverage of the

Science of Climate Change," *Global Environmental Change* 15 (2005): 338, 350.

124. Mooney, *Republican War,* 253.
125. Sheldon Rampton & John Stauber, *Trust Us, We're Experts!* (New York: Putnam, 2001), 304.

11. Reforming Science Oversight

1. Sheila Jasanoff, "Transparency in Public Science: Purposes, Reasons, Limits," *Law and Contemporary Problems* 69 (2006): 22, 43–44.
2. Sheila Jasanoff, *The Fifth Branch: Science Advisers as Policymakers* (Cambridge, Mass.: Harvard University Press, 1990), 234–237 (although couching this role in different terms).
3. Ibid., 234–237.
4. Ibid., 61–83 (discussing peer review of agency science and highlighting the role played by science advisory boards).
5. See, e.g., Philip Boffey, *The Brain Bank of America* (New York: McGraw Hill, 1975), 245 (arguing that many NAS reports are "mediocre or flawed by bias or subservient to the funding agencies"); Stephen Hilgartner, *Science on Stage: Expert Advice as Public Drama* (Stanford, Calif.: Stanford University Press, 2000); Center for the Science in the Public Interest, *Ensuring Independence and Objectivity at the National Academies* (Washington, D.C.: CSPI, 2006).
6. Neal Lane, "Politics and Science: A Series of Lessons," *Social Research* 73 (2006): 861, 870.
7. National Research Council, National Academies of Sciences, *Access to Research Data in the 21st Century: An Ongoing Dialogue Among Interested Parties: Report of a Workshop* (Washington, D.C.: National Academies Press, 2002), 8–12.
8. Ibid., 10–11.
9. Ibid.
10. What Is the Health Effects Institute? www.healtheffects.org/about.htm.
11. HEI Board of Directors, www.healtheffects.org/board.htm.
12. HEI's Research Committee, www.healtheffects.org/committees.htm.
13. Jasanoff, *Fifth Branch,* 209–216.
14. National Research Council, *Access,* 12.
15. Ibid., 11–12.
16. National Research Council, National Academies of Sciences, Ensuring the Quality of Data Disseminated by the Federal Government: Workshop no. 2, (March 22, 2002), www7.nationalacademies.org/stl/4-22-02_Transcript.doc, 18 (presentation by Robert O'Keefe, Health Effects Institute).
17. 5 U.S.C. app. §5(b)(2).
18. U.S. General Accounting Office, *Federal Advisory Committees: Additional Guidance Could Help Agencies Better Ensure Independence and Balance* (Washington, D.C.: GAO, 2004).
19. 41 C.F.R. §102–3.105.

20. National Academies of Sciences, Background Information and Conflict of Interest Declaration (BI/COI Form A).
21. 21 U.S.C. §355(n)(4).
22. Food and Drug Administration, Draft Guidance on Disclosure of Conflicts of Interest for Special Government Employees Participating in FDA Product Specific Advisory Committees (January 2002), www.fda.gov/oc/guidance/advisorycommittee.html. See also Peter Lurie et al., "Financial Conflict of Interest Disclosure and Voting Patterns at Food and Drug Administration Drug Advisory Committee Meetings," *JAMA* 295 (2006): 1921, 1922.
23. Gardiner Harris, "F.D.A. Rule Limits Role of Advisers Tied to Industry," *NYT,* March 22, 2007, A1.
24. These recommendations are consistent with those of a group of nineteen scientists and public interest groups regarding current science advisory board processes. See, e.g., David Bellinger et. al, letter to Amy L. Comstock, director, United States Office of Government Ethics (March 10, 2003), website of Center for Science in the Public Interest, cspinet.org/new/pdf/oge_letter _final.pdf.
25. Sheldon Krimsky, *Science in the Private Interest* (Oxford: Rowman & Littlefield, 2003), 228; Linda Greer & Rena Steinzor, "Bad Science," *Environmental Forum,* January–February, 2002, 13, 15.
26. Krimsky, *Science in the Private Interest,* 204.
27. General Accounting Office, *Federal Advisory Committees: Additional Guidance,* 43.
28. Jasanoff, *Fifth Branch,* 245 (also mentioning selection of advisory committee members by nongovernmental organizations as a possible but not necessarily foolproof path to reform).
29. Bruce Bimber, *The Politics of Expertise in Congress: The Rise and Fall of the Office of Technology Assessment* (Albany: State University of New York Press, 1996), 3. See also Lawrence McCray, Doing Believable Knowledge Assessment for Policymaking: How Six Prominent Organizations Go about It, draft (February 11, 2004), web.mit.edu/cis/petp_wp.html (describing "high end" independent scientific organizations and exploring how they have managed to maintain their credibility on contested issues of regulatory science).
30. Jasanoff, *Fifth Branch,* ch. 11; Mark Powell, *Science at EPA* (Washington, D.C.: RFF Press, 1999), 139.
31. 5 U.S.C. §706(1).
32. 42 U.S.C. §7409.
33. 42 U.S.C. §7409(d)(2).
34. *Whitman v. American Trucking Ass'n, Inc.,* 531 U.S. 457 (2001); *American Trucking Ass'n, Inc. v. EPA,* 283 F.3d 355 (D.C. Cir. 2002).
35. Environmental Protection Agency, "National Ambient Air Quality Standards for Particulate Matter, Proposed Rule," 71 *Fed. Reg.* 2620 (January 17, 2006). Steven D. Cook, "EPA Tightens Part of Particulate Standard but Leaves Most Existing Limits Unchanged," *BNA Environment Reporter* 37 (2006): 1945.

36. Miguel Bustillo & Marla Cone, "EPA Issues New Plan to Limit Soot," *LAT,* December 21, 2005, B1; Michael Janofsky, "Regulations Are Proposed to Cut Back Particulates," *NYT,* December 21, 2005, A26.
37. "EPA Staff Push Agency for Tighter Particulate Pollution Controls," *Risk Policy Report,* June 6, 2006, 23; "Sweeping EPA NAAQS Review May Erode Staff Role in Setting Standards," *Risk Policy Report,* January 10, 2006, 1.
38. Bustillo & Cone, "EPA Issues New Plan"; Janofsky, "Regulations Are Proposed."
39. Dawn Reeves, "EPA Rejection of Advisers' PM Advice Tees up Tensions in Ozone Review," *Inside EPA Weekly Report,* September 29, 2006, 1, 8.
40. Jeff Nesmith, "Advisors Rap EPA over Air Pollution," *Atlanta Journal-Constitution,* January 14, 2006, A10.
41. Reeves, "EPA Rejection."
42. Janet Wilson, "EPA Panel Advises Agency Chief to Think Again," *LAT,* February 4, 2006, B1.
43. Ibid.
44. Bill Lambrecht, "Scientists See Clean Air Decisions as Latest Snub," *St. Louis Post-Dispatch,* February 27, 2006, A1.
45. Environmental Protection Agency, "Particulate Matter"; Steven D. Cook, "CASAC Plans Formal Protest of Decisions by EPA Regarding Fine, Coarse Particulates," *BNA Environment Report* 37 (2006): 1989; Jane Kay, "EPA Ignores Advice for Annual Limits on Tiny Soot," *San Francisco Chronicle,* September 22, 2006, A3.
46. Steven D. Cook, "EPA Tightens."
47. Erik Stokstad, EPA Draws Fire Over Air-Review Revisions, *Science* 314 (2006): 1672.
48. "Advisors Oppose EPA Plan to Limit Panel's Role in NAAQS Process," *Risk Policy Report,* July 4, 2006, 27.
49. See Laural L. Hooper, Joe S. Cecil, & Thomas E. Willging, *Neutral Science Panels: Two Examples of Panels of Court-Appointed Experts in the Breast Implants Product Liability Litigation* (Washington, D.C.: Federal Judicial Center, 2001), 44–50; Joe S. Cecil & Thomas E. Willging, *Court-Appointed Experts: Defining the Role of Experts Appointed under Federal Rule of Evidence 706* (Washington, D.C.: Federal Judicial Center, 1993).
50. In re *Silica Products Liability Litigation,* 398 F. Supp. 2d 563, 571–72 (S.D. Tex. 2005); In re *Diet Drugs,* 236 F. Supp. 2d 445 (E.D. Pa. 2002). See also Alison Frankel, "Still Ticking: Mistaken Assumptions, Greedy Lawyers, and Suggestions of Fraud Have Made Fen-Phen a Disaster of a Mass Tort," *American Lawyer* 27 (2005): 92.
51. William L. Anderson, Barry M. Parsons & Drummond Rennie, "*Daubert*'s Backwash: Litigation-Generated Science," *University of Michigan Journal of Law Reform* 34 (2001): 619, 630 n. 72.
52. See Hooper et al., *Neutral Science Panels,* 44–50; Cecil & Willging, *Court-Appointed Experts,* 39–44.

53. For efforts of two important groups to address these questions, see American Association for the Advancement of Science, Court-Appointed Scientific Experts (CASE) Project, www.aaas.org/spp/case/case.htm; Cecil & Willging, *Court-Appointed Experts;* Hooper et al., *Neutral Science Panels.*
54. See Environmental Protection Agency, *Peer Review Handbook* (Washington, D.C.: EPA, 2006); Memorandum from Elissa R. Karpf, Deputy Assistant Inspector General for External Audits to Assistant and Regional Administrators, EPA OIG Report no. 1999-P-217 (1999), www.epa.gov/oigearth/reports/1999/9P00217.pdf (emphasizing the need to ensure the independence of peer reviewers).
55. National Research Council, National Academies of Sciences, *Strengthening Science at the U.S. Environmental Protection Agency: Research Management and Peer Review Practices* (Washington, D.C.: National Academies Press, 2000). See also Expert Panel on the Role of Science at EPA, U.S. Environmental Protection Agency, *Safeguarding the Future: Credible Science, Credible Decisions* (Washington, D.C.: EPA, 1992), 18.
56. This stands in contrast to the more formulaic guidelines provided by the OMB on the subject. See Office of Management and Budget, "Final Information Quality Bulletin for Peer Review," 70 *Fed. Reg.* 2664 (January 14, 2005).
57. Juliet Eilperin, "USGS Scientists Object to Stricter Review Rules: Prepublication Policy Seen as Cumbersome," *WPOST,* December 14, 2006, A29.
58. Cornelia Dean, "Scientists form Group to Support Science-Friendly Candidates," *NYT,* September 28, 2006, A18.
59. See Bob Ward, British Royal Society, letter to Nick Thomas, Exxon, September 4, 2006, image.guardian.co.uk/sys-files/Guardian/documents/2006/09/19/LettertoNick.pdf.
60. Joint Science Academies, Statement: Global Response to Climate Change, www.royalsoc.ac.uk/document.asp?latest=1&id=3222.
61. Joe Barton, letter to Michael Mann, June 23, 2005, energycommerce.house.gov.
62. Alan I. Leshner, American Association for the Advancement of Science, letter to Rep. Joe Barton, July 13, 2005, www.aaas.org/news/releases/2005/0714letter.pdf, 1; Ralph J. Cicerone, president, National Academies of Sciences, letter to Joe Barton, July 15, 2005, www.realclimate.org/Cicerone_to_Barton.pdf, 1.
63. Leshner, letter to Barton, 1.
64. Barton, letter to Mann, 1.
65. Ibid.
66. Ibid., 1–2.
67. Leshner, letter to Barton, 1.
68. Michael Bender et al., letter to Chairman Barton and Whitfield, July 15, 2005, www.realclimate.org/Scientists_to_Barton.pdf, 1.
69. Leshner, letter to Barton, 1.

70. Michael Mann, "Letter to Chairman Barton," July 15, 2005, www.real climate.org/Mann_response_to_Barton.pdf, 2, 4–5.
71. Ibid., 5.
72. Robert R. Kuehn, "Suppression of Environmental Science," *American Journal of Law & Medicine* 30 (2004): 333, 352.
73. Ibid., 365.
74. Paul M. Fischer, "Science and Subpoenas: When Do the Courts Become Instruments of Manipulation?" *Law & Contemporary Problems* 59 (1996): 159, 159.
75. National Academies of Sciences, National Academy of Engineering, Institute of Medicine, *Responsible Science: Ensuring the Integrity of the Research Process* (Washington, D.C.: National Academies Press, 1992), 29–30.
76. Kuehn, "Suppression," 365.
77. Frederick R. Anderson, "Science Advocacy and Scientific Due Process," *Issues in Science and Technology* 16 (Summer 2000): 71.
78. R. A. Deyo, "Bruce Psaty and the Risks of Calcium Channel Blockers," *Quality and Safety in Health Care* 11 (2002): 294–296; American Association of University Professors, *Institutional Responsibility for Legal Demands on Faculty* (Washington, D.C.: AAUP 1998), reprinted in American Association of University Professors, *Policy Documents and Reports* (Washington, D.C.: AAUP, 9th ed. 2001), 130.
79. National Academies of Sciences, *Responsible Science* (Washington, D.C.: National Academies Press, 1992), 1; 15 (stressing that "[i]ndividuals who, in good conscience, report suspected misconduct in science deserve support and protection"). See Janeen Interlandi, "An Unwelcome Discovery," *New York Times Magazine,* October 22, 2006, 98 (detailing facts underlying guilty plea by University of Vermont scientist Eric Proehlman for fabricating data on obesity, menopause, and aging).
80. National Academies of Sciences, *Responsible Science,* 1:11 (stressing the need for balance in administering research integrity procedures).
81. Chris Mooney, *The Republican War on Science* (New York: Basic Books, 2005), ch. 8.
82. Wendy Wagner, "The 'Bad Science' Fiction: Reclaiming the Debate over the Role of Science in Public Health and Environmental Regulation," *Law & Contemporary Problems* 66 (2003): 63.
83. Margaret Clune, "Ossifying Ossification: Why the Information Quality Act Should Not Provide for Judicial Review," *Environmental Law Reporter* 36 (2006): 10430; Wagner, "'Bad Science' Fiction," 89, 101–111.
84. U.S. Government Accountability Office, *Information Quality Act: Expanded Oversight and Clearer Guidance by the Office of Management and Budget Could Improve Agencies' Implementation of the Act* (Washington, D.C.: GAO, 2006).
85. Sidney A. Shapiro, "The Information Quality Act and Environmental Protection: The Perils of Reform by Appropriation," *William & Mary Environmental Law & Policy Review* 28 (2004): 339, 349–351.

86. *Subcommittee on Regulatory Affairs of the House Committee on Government Reform, Hearings on Improving Federal Government Information Quality,* 109th Cong., 1st Sess. (July 20, 2005) (Testimony of Sidney A. Shapiro, University Distinguished Chair in Law, Wake Forest University); Thomas O. McGarity et al., Truth and Science Betrayed: The Case against the Information Quality Act, publication no. 502, Center for Progressive Regulation, website of the Center for Progressive Reform, www.progressive reform.org/articles/iqa.pdf.
87. David Michaels, "Doubt Is Their Product," *Scientific American,* June 2005, 100.
88. Ibid.
89. See Nathin Vardi, "Poison Pills," *Forbes Global,* April 19, 2004, 7; Ford Fessenden, "Judge Orders Ephedra Maker to Pay Back 12.5 Million," *NYT,* May 31, 2003, A12.
90. Although the documents from the litigation were sealed, civil litigation brought against the manufacturer of Halcion by a murderess for its adverse side effects raised regulator suspicions that the drug might be harmful. See Geoffrey Cowley et al., "Sweet Dreams or Nightmare," *Newsweek,* August 19, 1991, 44; Clare Dyer, "Upjohn Sues for Libel," *British Medical Journal* 304 (1992): 273.
91. See Sheryl Gay Stolberg, "U.S. to Prohibit Supplement Tied to Health Risks," *NYT,* December 31, 2003, A1 (ephedra); Chris Mihill & Clare Dyer, "Sleeping Drug Withdrawn after Side-Effect Fears," *Guardian,* October 3, 1991, A1 (Halcion).
92. See "File Shows Merck Sought to Change Vioxx," *LAT,* June 23, 2005, C3; Heather Won Tesoriero, "Attorneys Question Disclosure by Merck of Vioxx-Study Deaths," *WSJ,* 28 September 2005, D4; Leila Abboud, "Lilly Denies Hiding Data Tying Prozac to Suicide," *WSJ,* January 6, 2005, D10.
93. See EPA Consent Agreement and Proposed Final Order to Resolve DuPont's Alleged Failure to Submit Substantial Risk Information under the Toxic Substances Control Act (TSCA) and Failure to Submit Data Requested under the Resource Conservation and Recovery Act (RCRA), December 14, 2005, website of the News Journal, Wilmington, Delaware, www.delawareonline.com/assets/pdf, 3.
94. Michael Janofsky, "DuPont to Pay $16.5 Million for Unreported Risks," *NYT,* December 15, 2005, A28.
95. Andrew Schneider, "W. R. Grace, Officials Are Indicted in Asbestos Health Case," *St. Louis-Post-Dispatch,* February 8, 2005, A1.
96. See Sara Shipley, "Study Showed Chemical Was Toxic," *St. Louis Post-Dispatch,* February 29, 2004, C1.
97. David Michaels & Celeste Monforton, "Scientific Evidence in the Regulatory System: Manufacturing Uncertainty and the Demise of the Formal Regulatory System," *Journal of Law and Policy* 13 (2005): 17, 23–24.
98. Ibid., 18–24; Chris Bowman, "Flavoring Agent Destroys Lungs," *Sacramento Bee,* July 30, 2006, A1.

99. Wendy Wagner, "When All Else Fails: Regulating Risky Products through Tort Litigation," *Georgetown Law Review* 95 (2007): 693.
100. See Michael Ravnitzky & Jeanne Weigum, "Filtered or Unfiltered Information: Choices in How to Make the Minnesota Tobacco Document Depository Records More Accessible to the Public," *William Mitchell Law Review* 25 (1999): 715.
101. David Kessler, *Question of Intent* (New York: Public Affairs, 2001), 7.
102. See Teresa Moran Schwartz, "Prescription Products and the Proposed Restatement (Third)," *Tennessee Law Review* 61 (1994): 1357, 1386 (noting that unlike the FTC, the FDA does not have administrative subpoena power and must therefore build enforcement cases by searching the literature and consulting with experts).
103. See Neil K. Komesar, *Imperfect Alternatives: Choosing Institutions in Law, Economics, and Public Policy* (Chicago: University of Chicago Press, 1994), ch. 5.
104. See, e.g., Thomas Koenig & Michael Rustad, "His and Her Tort Reform: Gender Injustice in Disguise," *Washington Law Review* 70 (1995): 1, 39–46.
105. See Marc J. Scheineson & Shannon Thyme Klinger, "Lessons from Expanded Government Enforcement Efforts against Drug Companies," *Food and Drug Law Journal* 60 (2005): 1, 10–11.
106. 29 U.S.C. §666(e).
107. See Paul J. Quirk, *Industry Influence in Federal Regulatory Agencies* (Princeton, N.J.: Princeton University Press, 1981), 4–21; Cass R. Sunstein, "Constitutionalism after the New Deal," *Harvard Law Review* 101 (1987): 421, 448–49. But see David B. Spence, "The Shadow of the Rational Polluter: Rethinking the Role of Rational Actor Models in Environmental Law," *California Law Review* 89 (2001): 917.
108. Mich. Comp. L. Ann. §600.2946(5).
109. Food and Drug Administration, "Requirements on Content and Format of Labeling for Human Prescription Drug and Biological Products," 71 *Fed. Reg.* 3922 (2006). See also Robert S. Adler & Richard A. Mann, "Preemption and Medical Devices: The Courts Run Amok," *Missouri Law Review* 59 (1995): 895; Richard C. Ausness, "'After You, My Dear Alphonse!' Should the Courts Defer to the FDA's New Interpretation of §360k(A) of the Medical Device Amendments?" *Tulane Law Review* 80 (2006): 727; Margaret H. Clune, *Stealth Tort Reform: How the Bush Administration's Aggressive Use of the Preemption Doctrine Hurts Consumers,* white paper no. 403 (Washington, D.C.: Center for Progressive Reform, October 2004).
110. Margaret A. Berger & Aaron D. Twerski, "Uncertainty and Informed Choice: Unmasking *Daubert*," *Michigan Law Review* 104 (2005): 257.
111. See, e.g., *Northern Spotted Owl v. Hodel,* 716 F. Supp. 479 (W.D.Wash. 1988).
112. Anderson, "Science Advocacy," 76.
113. See, e.g., "Bruce Psaty," 294–296; American Association of University Professors, *Institutional Responsibility for Legal Demands on Faculty.*

12. Final Thoughts

1. Howard Latin, "Ideal versus Real Regulatory Efficiency: Implementation of Uniform Standards and 'Fine-Tuning' Regulatory Reforms," *Stanford Law Review* 37 (1985): 1267; Sidney A. Shapiro & Thomas O. McGarity, "Not So Paradoxical: The Rationale for Technology-Based Regulation," *Duke Law Journal* (1991): 729; Wendy E. Wagner, "The Triumph of Technology-Based Standards," *Illinois Law Review* 2000 (2000): 83.
2. See, e.g., National Research Council, National Academies of Sciences, *Building a Foundation for Sound Environmental Decisions* (Washington, D.C.: National Academies Press, 1997); Jane Lubchenco, "Entering the Century of the Environment: A New Social Contract for Science," *Science* 279 (1998): 491, 495.
3. Leslie Roberts, "Learning from an Acid Rain Program," *Science* 251 (1991): 1302.
4. See, e.g., *Industrial Union Dep't v. American Petroleum Inst.,* 448 U.S. 607 (1980); *Corrosion Proof Fittings v. EPA,* 947 F.2d 1201, 1215 (5th Cir. 1991); *Gulf South Insulation v. United States Consumer Prod. Safety Commission,* 701 F.2d 1137, 1146 (5th Cir. 1983).
5. Wendy Wagner, "The Science Charade in Toxic Risk Regulation," *Columbia Law Review* 95 (1995): 1613.
6. See, e.g., Sidney A. Shapiro & Robert L. Glicksman, *Risk Regulation at Risk: Restoring a Pragmatic Approach* (Stanford, Calif.: Stanford University Press, 2003), 191–192.
7. See, e.g., Clean Water Act, 33 U.S.C. §1311(b); Clean Air Act, 42 U.S.C. §7412(b).
8. Clean Air Act, Title IV, 42 U.S.C. §§7651–7651o.
9. Restatement (Second) of Torts §§821D, 821F, 822.
10. See *Renken v. Harvey Aluminum, Inc.,* 226 F. Supp. 169 (D. Ore. 1963); Robert V. Percival et al., *Environmental Regulation: Law, Science, and Policy* (New York: Aspen, 2006), 85–86 (describing Supreme Court's science-blind approach to restricting sulfur emissions from plants in resolving public nuisance case).
11. *Industrial Union Dep't.,* 448 U.S. at 653.
12. Food Quality Protection Act (FQPA), 21 U.S.C. §346a; see also Thomas O. McGarity, "Politics by Other Means: Law, Science, and Policy in EPA's Implementation of the Food Quality Protection Act," *Administrative Law Review* 53 (2001): 103 (discussing problems in implementation of FQPA standard).
13. Ian Ayres & Robert Gertner, "Filling Gaps in Incomplete Contracts: An Economic Theory of Default Rules," *Yale Law Journal* 99 (1989): 87, 91.
14. The difficulties involved in trading off primary health benefits against other types of health losses in protective regulation are explored in John D. Graham & Jonathan Weiner, eds., *Risk vs. Risk: Tradeoffs in Protecting Health and the Environment* (Cambridge, Mass: Harvard University Press, 1995).

15. See, e.g., Nicole Gaouette, "Congressional Hearing Heats up over Changes to Climate Reports," *LAT,* March 20, 2007, A10; *House Committee on Natural Resources, Oversight Hearing on "Endangered Species Act Implementation: Science or Politics?"* 111th Cong., 1st Sess. (2007).
16. Securities Exchange Act of 1934, 15 U.S.C. §§78a et seq.; Louis Lowenstein, "Financial Transparency and Corporate Governance: You Manage What You Measure," *Columbia Law Review* 96 (1996): 1335 (discussing virtues of corporate financial reporting in U.S.).
17. Emergency Community Planning and Right to Know Act, 42 U.S.C. §11023. See Bradley Karkkainen, "Information as Environmental Regulation: TRI and Performance Benchmarking, Precursor to a New Paradigm?" *Georgetown Law Journal* 89 (2001): 257.
18. The Food and Drug Administration Amendments Act of 2007, Public Law no. 110–185.
19. See Andrew Bridges, "Bush Signs Drug Safety Bill into Law," *Associated Press,* September 27, 2007, http://ap.google.com; Gardiner Harris, "Senate Takes up Bill to Change Drug Agency Operations," *NYT,* May 1, 2007, A18.
20. Sheldon Krimsky, "Publication Bias, Data Ownership and the Funding Effect in Science: Threats to the Integrity of Biomedical Research," in Wendy Wagner & Rena Steinzor, eds., *Rescuing Science from Politics* (New York: Cambridge University Press, 2006), 61. See also Jennifer Washburn, *University, Inc.* (New York: Basic Books, 2005), 237–239; Marcia Angell, *The Truth about the Drug Companies: How They Deceive Us and What to Do About It* (New York: Random House, 2004), 245; Katherine S. Squibb, "Basic Science at Risk: Why Independent Research Protections Are Critical to Creating Effective Regulations," in Wagner & Steinzor, *Rescuing Science from Politics,* 46.
21. Krimsky, "Publication Bias," 61.
22. "Sound science" reforms generally call for greater mechanisms of political oversight over the quality of science used for regulation, such as unrestricted peer review and cost-benefit analysis, and sometimes raise the agency's evidentiary burden for justifying regulation. See, e.g., H.R. 9, 104th Cong., 1st Sess. (1995) (prominent House "sound science" bill that did not pass the Senate); Data Access Act, rider in the Omnibus Appropriations Act for Fiscal Year 1999, Pub. L. No. 105–277, 112 Stat. 2681–495 (1998); Information Quality Act, appropriations rider in section 515 of the Treasury and General Government Appropriations Act for Fiscal Year 2001, Pub. L. No. 106–554, 114 Stat. 2763A-153–55 (2001); Office of Management and Budget, "Final Information Quality Bulletin for Peer Review," 70 *Fed. Reg.* 2664 (2005). See also Alan C. Raul & Julie Zampa, "Regulatory *Daubert:* A Proposal to Enhance Judicial Review of Agency Science by Incorporating *Daubert* Principles into Administrative Law," *Law & Contemporary Problems* 66 (2003): 7 (recommending regulatory form of *Daubert* for judicial review).

Index